Chemistry and Physics of
MATRIX-ISOLATED SPECIES

Chemistry and Physics of
MATRIX-ISOLATED SPECIES

Edited by:

Lester ANDREWS

Department of Chemistry
University of Virginia
McCormick Road
Charlottesville, VA 22901
U.S.A.

Martin MOSKOVITS

Department of Chemistry
University of Toronto
Toronto, Ontario M5S 1A7
Canada

1989

NORTH-HOLLAND
AMSTERDAM · OXFORD · NEW YORK · TOKYO

© Elsevier Science Publishers B.V., 1989

ISBN: 0 444 70549 x

Published by:

Physical Sciences and Engineering Division
Elsevier Science Publishers B.V.
P.O. Box 103
1000 AC Amsterdam
The Netherlands

Sole distributors for the USA and Canada:

Elsevier Science Publishing Company, Inc.
655 Avenue of the Americas
New York, NY 10010
USA

Library of Congress Cataloging-in-Publication Data

Chemistry and physics of matrix-isolated species
    edited by Lester Andrews, Martin Moskovits.
      p.   cm.
    Includes bibliographies and index.
    ISBN 0-444-70549-X
    1. Matrix isolation spectroscopy.   I. Andrews, Lester, 1942–
II. Moskovits, Martin, 1943–
QD96.M33C48   1989
543′.0858—dc19   88-32993
                                          CIP

Printed in The Netherlands

Preface

Matrix-isolation spectroscopy, as a technique for studying unstable species, has had tremendous lasting powers. Since 1954, matrix-isolation has been able to assimilate new technology rather than being replaced by it. Conferences celebrating its results have become thriving enterprises. Several books concerned with matrix-isolation appeared in the early 1970s. Those written or edited by Meyer, Hallam, and Moskovits and Ozin are perhaps the best known. There was, in addition, a book outlining parallel developments in the Soviet Union, written by Sergeev and Batyuk, that appeared in the late 1970s. Those books referred to matrix-isolation as a "new technique". The past decade has seen many developments in matrix-isolation spectroscopy, making it a well-established and productive technique.

The chapters in this book describe very different types of chemistry and physics of matrix-isolated species. They range from synthesis of new chemical species to physical properties of stable molecules. As such the nature and style of the chapters varies with the subject material covered.

The cover illustration is taken from a study that represents the vast coverage of matrix-isolation spectroscopy. The weak chemical $H_2 \cdots HF$ complex is held together by electrostatic physical interactions, and the neon-matrix infrared spectrum provided a useful guide to the subsequent high-resolution gas-phase investigation, which are discussed in chapter 2.

We would like to thank our former and current students and research associates for the many years of hard work, endless discussions and dozens of ideas that they generated. Special thanks are due to John Harvell, CTI Cryogenics, for describing the cryogenic refrigerator operation and to Frances Wintrob and Teresa Tyree for their able and patient work on the manuscript.

Lester Andrews
Martin Moskovits

Contents

CHAPTER 1

Introduction and Experimental Developments

Lester ANDREWS

Chemistry Department
University of Virginia
Charlottesville, VA
USA

and

Martin MOSKOVITS

Chemistry Department
University of Toronto
Toronto, Ontario
Canada

Chemistry and Physics of
Matrix-Isolated Species
L. Andrews and M. Moskovits, eds

Contents

1. Introduction

The matrix-isolation technique for producing and trapping new chemical species has been applied to an ever increasing range of chemical and physical problems since its inception (Whittle et al. 1954). In the thirteen years since the last monograph (Moskovits and Ozin 1976) appeared, there have been a substantial number of new developments and applications of the matrix technique. Active research workers in this field have had international meetings attracting around 100 scientists on alternate years since 1977. Much of the intense research activity of the last decade in matrix-isolation spectroscopy is described in the present monograph of 12 chapters. These chapters differ in approach to the subject owing to differences in the chemical and physical nature of the research described which ranges from synthetic chemistry to solid-state physics. Any experimental technique is closely related to the development of new apparatus, and the next section will describe innovations in instrumentation such as the closed-cycle cryogenic cooler that has revolutionized the technique. In addition, each chapter describes many of the specialized methods used to prepare, trap and study particular new subject species.

The second chapter describes spectroscopy of molecular ions and hydrogen bonded complexes. Much of the early matrix work on small cations and base···HF complexes was done by this author. There has been a virtual explosion of gas-phase studies on these subjects in the 1980s, and this chapter compares some species in the gas and matrix phases and demonstrates the complementary nature of these research techniques.

Chapter 3 focuses on the application of matrix isolation to the study of metal clusters. This has been one of the most fruitful applications of matrix technology to a field of research that is currently very hot. The sister-discipline of nozzle-jet generated metal clusters will be discussed and the complementary nature of the information produced by the two techniques will be mentioned.

Free radicals and reactive molecules are discussed in chapter 4. Techniques are described for preparing these exotic species and examples are given of specific atom–molecule reactions, including those with H, C, Si and O atoms, to produce interesting new species that are most important to chemistry.

Chapter 5 presents laser-excitation studies of small molecules and metal dimers using both matrix-isolation and gas-phase methods. Laser evaporation and vacuum ultraviolet photolysis techniques are employed in these studies.

The recent work by Frei and Pimentel in infrared mode selective photo-chemistry of matrix-isolated molecules is described in chapter 6. A brief review of vibrationally stimulated reactions is given followed by a discussion of selective vibronically induced bimolecular reactions with red and near infrared photons. Ultraviolet wavelength controlled photoisomerism, which is important for both molecular energy and information storage, is described.

Chapter 7 describes production methods for neutral and charged free radicals for study by electron spin resonance in solid neon. These species include high-temperature molecules and small molecular anions and cations. Comparisons with gas-phase measurements are included, and the matrix results may be used to predict unobserved gas-phase magnetic parameters and electronic structure.

A general account of recent matrix work on reactive organic species since 1978 is given in chapter 8. These topics include hypovalent organic species, molecules with strained bonds, organosilicon, and carbonyl compounds. This chapter clearly demonstrates the steadily increasing contribution that matrix-isolation spectroscopy has made to understanding reaction mechanisms and bonding in unstable organic species.

Chapter 9 covers inorganic and organometallic photochemistry. This work in the main was done to discover photochemical pathways and structures of primary and secondary photoproducts. This chapter describes unimolecular and bimolecular photoprocesses in matrices, and the role of intermolecular complexes in matrix photochemistry.

The next chapter on reactions of first-row transition metal atoms and small clusters in matrices by the Rice group covers important chemistry. Reactions with water, methane, ethylene and acetylene are of fundamental and practical significance.

Chapter 11 presents energy transfer and lifetime studies. A little known fact is that F. Legay developed the CO_2 laser to pursue energy transfer studies in matrices. Since then he and his group have reported classic studies in energy transfer. Recent developments in this area are described.

The final chapter describes solid-state effects in matrices. Topics include matrix shifts, defects in matrices, anharmonic phenomena of guest and host, and high-pressure effects and are presented from the solid-state physics point of view in light of their relevance to matrix-isolation spectroscopy.

The chapters in this book describe many of the contributions of matrix-isolation spectroscopy to chemistry and physics in the last decade, and as such predict the continued evolution of this technique over the next decade. Certainly, the commercial availability of closed-cycle coolers to replace liquid hydrogen and liquid helium has made the technique available to substantially more workers, and further refinements in apparatus or developments of new instruments (such as commercial lasers and FTIR instruments in the 1970s) will foster new applications and new chemical and physical studies.

It is difficult to predict future developments beyond the logical evolution path. The availability of synchrotron radiation facilitates vacuum-ultraviolet spectroscopy and recent matrix studies in this region foretell more in the future (Boyle et al. 1986, Schroeder et al. 1987, Bahrdt et al. 1987).

The matrix-isolation technique is gaining acceptance as a routine analytical method as evidenced by the commercial availability of an integrated gas chromatograph/matrix infrared system (Cryolect, Mattson Instruments, Inc.). The sharp nature of matrix infrared absorptions provides fingerprint identification for many molecules.

We are confident that the matrix-isolation method will make further contributions to chemistry as a medium for studying chemical reactions to make many types of new molecular species including organic, inorganic, metallic and clusters of these species. The recently prepared LiC_6H_6 (Manceron and Andrews 1988) and P_4O (Andrews and Withnall 1988) species provide examples. It may be possible to develop novel synthetic methods which require low-temperature matrix chemistry that will lead to new and interesting materials. The matrix method will also provide a means for future physical studies involving high-energy and high-intensity radiation interacting with small molecules.

2. Experimental developments

This section will describe experimental methods that have come into general use in the matrix-isolation field since the last monograph (Moskovits and Ozin 1976). It is intended to facilitate further work in matrix-isolation spectroscopy.

2.1. Instrumentation

The improved performance of commercial Fourier transform infrared instruments in the middle 1970s has increased the information chemical spectroscopists can get from matrix-isolation experiments. In particular, increased resolution and decreased time required to obtain high-quality spectra over a large spectral region have been of great advantage in matrix-isolation experiments. The FTIR instrument produces improved signal-to-noise spectra in the $3000-4000$ cm^{-1} region over grating instruments. The large body of work done at Virginia on HF complexes (Andrews 1984) was made possible by FTIR. Another advantage of FTIR is the possibility to focus the IR beam on a small spot of sample, this enables samples which require special preparation to be examined in the small area where a high concentration of new species are trapped. Such is the case with the matrix-electron impact apparatus (Süzer and Andrews 1988) to be described later in this chapter.

2.2. Cryogenic refrigeration

The matrix-isolation research field has been revolutionized by the closed-cycle refrigerator. It is necessary to have worked with liquid hydrogen to fully appreciate this statement. The Gifford–McMahon refrigeration cycle was developed by A.D. Little engineers to provide continuous cooling of parametric amplifiers, and "Cryodyne" refrigerators were marketed in the mid-1960s (Gifford and McMahon 1960). This refrigerator was first used for matrix-isolation experiments in 1967 (Andrews 1967, 1968).

Thermodynamically, the ideal Gifford–McMahon refrigeration cycle consists of two constant-pressure heat transfer processes linking an isothermal compression and an isentropic expansion. The main working components are a compressor, and a cylinder with integral displacer to shuttle the working fluid helium gas through a regenerative heat exchanger. The provision of valves and a surge volume allow the compressor and displacer to operate at different speeds and at separation distances of up to several thousand feet without loss of performance, thus offering significant gains in life and flexibility of operation.

In the compressor, helium is compressed to approximately 20 atm, cooled to ambient temperature and transferred to a cold head. High reliability industrial compressors, converted for helium service, provide the clean compressed helium. A key subsystem is required for oil management. Oil is injected into the helium during compression for cooling and sealing purposes, then the oil is stripped from the helium before delivery to the cold head. The compressor module will typically operate in excess of 20 000 h before requiring overhaul.

Within the cold head, fig. 1 (often referred to as the expander or refrigerator unit), a displacer piston shuttles the working fluid through the regenerative heat exchanger in the proper sequence to produce refrigeration at two stages, and feeds the expanded helium back to the compressor at approximately 5 atm. (Operated in reverse, the "cold" head becomes very hot!) Since the displacer piston is driven with a slow speed (72 RPM) motor, it is capable of running continuously for 20 000–30 000 h before routine maintenance. Also, the displacer drive mechanism is lightly loaded because the pressure is largely balanced across the displacer piston.

Perhaps the most important subsystem of a Gifford–McMahon refrigerator is the high-efficiency regenerative heat exchanger, which can be separate from or integral with the displacer piston body. The exchanger uses tightly packed bronze mesh or lead spheres to transfer, store, and release heat at the proper time to effect refrigeration. Once assembled, the life of the regenerator is virtually infinite, barring calamity.

The Gifford–McMahon refrigerators are extremely reliable. The first system used at Virginia was retired after 58 000 hours because it could not cool below 16 K. In 58 000 hours of use the bottom operating temperature gradually increased from 13 to 16 K! In the late 1960s the original design was modified to

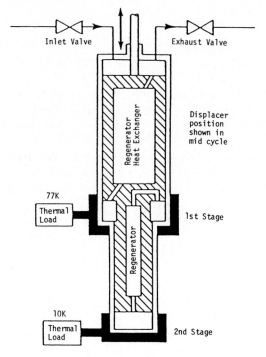

Inlet Valve

Exhaust Valve

Regenerator
Heat Exchanger

Displacer
position
shown in
mid cycle

77K

Thermal
Load

1st Stage

Regenerator

10K

Thermal
Load

2nd Stage

Fig. 1. Schematic view of the internal workings of the expander in a two-stage closed-cycle cryogenic refrigerator driven by gaseous helium. (Courtesy of CTI Cryogenics.)

provide airborne infrared detector coolers, and smaller refrigerators attained temperatures just below 10 K. Similar refrigerators are marketed in the US today by CTI-Cryogenics ("Cryodyne") and APD-Cryogenics ("Displex"), which perform extremely well for matrix-isolation experiments.

Although liquid helium lacks the chemical properties of liquid hydrogen, the use of liquid helium for 4 K experiments requires special double dewars or continuous flow cryostats *and* a supply of liquid helium. A three-stage closed-cycle refrigeration system ("Heliplex", APD-Cryogenics) capable of 1 W of refrigeration at 4.2 K is available. This device combines a two-stage closed-cycle cooler and a Joule–Thomson expansion cycle. Two gaseous helium circuits, supplied by two compressors, with the heat of compression removed by water cooling, enter the refrigeration unit shown in fig. 2. One stream expands and cools progressively colder stages in a two-stage regenerative expander and returns to the first compressor. The second stream – the Joule–Thomson circuit – is successively cooled by the two external heat stations of the expander refrigerator and by its own return stream after J–T expansion at the 4.2 K cooling station. The stream then returns to the compressor for recycling.

The Heliplex operates well for matrix-isolation experiments and is capable of

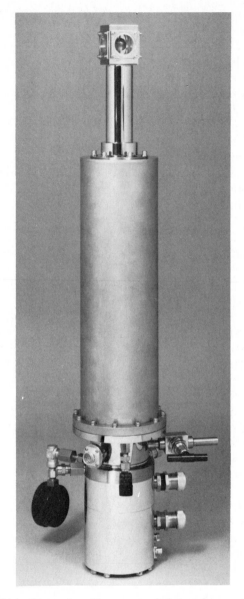

Fig. 2. "Heliplex" refrigeration system. The vacuum jacket encloses a two-stage closed-cycle refrigerator supplied by one helium compressor and a closed-cycle Joule–Thomson circuit supplied by a second helium compressor. (Courtesy of APD-Cryogenics.)

trapping H_2 for matrix reactions in solid neon (Hunt and Andrews 1987). One advantage of the Heliplex design is the J–T needle valve, which can adjust the mass flow and the cooling capacity in the J–T circuit. During cool down (which takes about 3 h) this valve is opened more than when operating at 4.2 K. The J–T valve can be adjusted to cool the system down to 3.6 K, *but* this requires subatmospheric operation of the J–T circuit. In an excellent but not perfect system, a very minute air leak will ultimately contaminate the helium fluid and result in plugging the J–T circuit. Replacing the helium gas in the entire system requires a special manifold, half of a cylinder of 99.995% helium gas ($< -80°F$ dew point), and half a day. The latter indicates the main disadvantage of the Heliplex system; the J–T circuit cooling coil has a very small cross section and minute contamination in the system (from the helium or the compressor oil) can result in a plug that requires replacing the helium and/or more extensive decontamination. The Heliplex is substantially larger and more cumbersome (four helium lines) than the two-stage expander refrigerators (two helium lines). Nevertheless, the Heliplex system once completely decontaminated and carefully maintained, provides reliable 4 K refrigeration.

2.3. Specialized techniques

Methods for preparing and trapping molecular ions are described in the chapters by Andrews and by Knight. A new technique for extracting molecular ions from a discharge for matrix trapping has been developed by Suzer and Andrews (1988). This experiment employs electron impact of the matrix sample condensing inside of a ring isolated on the cold window and facilitates the use of a focussed FTIR spectroscopic probe of this small spot of prepared sample. Figure 3 shows the thermionic source built from stainless steel and ceramic insulating parts using a thin (4–7 mil) tungsten wire filament as the electron source. Electrons boiling from the hot filament were attracted to a stainless steel ring mounted on the CsI window. Filament currents and various voltages were adjusted to obtain an electron current of 10–200 µA measured on the stainless steel ring. Typical voltages were $+20$ V on the ring, $+15$ V on the control electrode and -30 V on the filament. Electron energy was varied by the filament bias supply and was maintained in the 50–200 eV range. The geometrical arrangement was fixed so that light emitted by the filament did not fall on the CsI window, whereas electrons were bent onto the window by the applied voltage on the ring.

Electron impact on the matrix condensing inside the ring produces essentially a hollow cathode discharge similar to that widely used in gas-phase studies. Reagent molecules present at high dilution are subjected to discharge conditions, and the ion–molecule reaction products that survive are trapped in the solid matrix for spectroscopic study. The observed products are expected to parallel those observed in gas-phase studies. In the H_2O system, the OH radical,

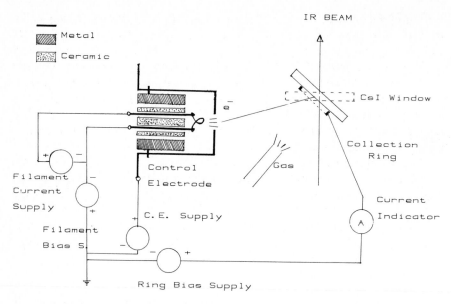

Fig. 3. Schematic diagram of electron-impact source used for matrix-isolation studies.

the OH$^-$ anion, and matrix-solvated proton Ar$_n$H$^+$ were trapped and observed by FTIR (Süzer and Andrews 1988).

2.4. Pumping

The quality of vacuum used in matrix isolation has improved steadily since 1976. The need for better vacuum, and therefore for reduced possibility of contamination has been made even more acute with the increased use of laser sources to excite emissions from the matrix. Fluorescence by even small concentrations of impurities can often dominate a spectrum, especially when low-intensity signals such as Raman emission are sought. Although the oil diffusion pump is still widely used, its territory is being invaded by the turbomolecular pump. Several manufacturers (Balzers, Edwards, Leybold-Heraeus, Varian) now offer "turbo" pumps. Originally rather unreliable and restricted to a vertical mounting position, modern turbo pumps are now considerably more dependable and most can be mounted in any orientation. Certain turbo pump failures still remain catastrophic, however and the price of turbo pumps per ℓ/s is still rather high. The turbo pump is essentially a large fan that turns at a very high rate. Molecules colliding with the fan blades are accelerated towards the rear of the pump where another pump, either a mechanical pump or a diffusion pump, maintains a low pressure. Below a certain pressure the turbo pump is able to sustain a given compression ratio between its inlet and its outlet. This ratio is usually of the order of 10^{-3} for hydrogen. Hence backing a turbo pump with a mechanical pump results in ultimate pressures of the order of 10^{-7} Torr. By using a diffusion pump as a backing device or two tandem turbo pumps one can achieve pressures in the 10^{-10} Torr range (with baking), and even lower pressures if titanium sublimation is used as well. For vacuums in that range, proper ultrahigh vacuum techniques must be used, including stainless steel chamber construction, metal gasket sealing, proper UHV welding methods and UHV compatible materials inside the chamber. For many matrix applications this order of cleanliness is unnecessary; for others it is crucial.

Other pumping methods are also available that can produce UHV conditions. Most of these are not as useful in matrix isolation where the system is subjected to sizeable gas loads. These include ion pumping and cryopumping. The former is particularly poor in handling large gas loads. The latter (a crude form of matrix isolation in its own right) is usable but may result in large quantities of frozen gases to eliminate from the cryopump surface after several experiments.

2.5. "Light" sources

Lasers have become prevalent in matrix spectroscopy. Excimer-pumped, and YAG-pumped dye lasers have become routine items. These can furnish intense, tunable, pulsed radiation from about 330 nm to approximately 760 nm, in spectral portions depending on the dye, and the laser medium used. The excimer (or tripled YAG) pump laser also provides a useful ultraviolet line whose precise frequency depends on the laser medium used. Ultraviolet can also be obtained by doubling the dye laser output using crystals whose efficiencies vary from 1–2% up to 20% for certain new and costly crystals. Here, too, a number of crystals must be used in order to cover a sizeable wavelength range. Inexpensive nitrogen-pumped dye lasers are also available if very low laser intensity is all that is required. Several new solid-state lasers (such as the alexandrite laser) have become available recently and many more are expected in the near future. Moreover, high-power and high-repetition rate lasers such as the copper vapor laser with a green fundamental line are now available. Infrared lasers or laser techniques have also come on stream. F-center lasers provide lines in the near IR, while diode lasers are now available that can cover a sizeable portion of the infrared in chunks of about $20–100 \, cm^{-1}$, depending on the center frequency. The diodes are tuned either by changing the temperature of the diode or the current through it. Tunable IR is also possible by parametric oscillation: mixing a tunable and a fixed visible frequency in a non-linear material, thereby producing an infrared difference frequency. Likewise, tunable, vacuum ultraviolet laser radiation can be generated rather easily by four-wave mixing in one of a number of metal vapors. All of these techniques are discussed in any one of several recent texts (Siegman 1986).

For lifetime measurements, one can obtain, off-the-shelf, cavity-dumped, mode-locked lasers pumped either by an Ar^+ laser or by a CW YAG laser, that delivers pulse trains with pulses several picosecond in duration. Coupled with this, there has been an enormous improvement in data collection of transients. Transient digitizers and box-car averagers with windows several picoseconds wide are now available. Those with windows of the order of 1 ns, are now routine. Multichannel-plate photomultipliers with rise times of the order of 100 ps are also available (at a price).

The other major "light" source to have become routine in the past decade or so is the synchrotron. Despite the disadvantage of having to transport ones experiment to the source, the synchrotron can provide intense, pulsed, polarized radiation spanning the portion of the spectrum from the visible to the X-ray, making hitherto impossible or impractical matrix experiments eminently possible. These include far-UV absorption studies, ultraviolet and X-ray photoelectron spectroscopy, and EXAFS.

Laser vaporization has also begun to affect the complexion of matrix isolation. Although used in the early 1970s (Von Gustorf et al. 1975), it is only rather recently, with the availability of reliable, pulsed YAG lasers that laser vaporization has become routine. With it high-temperature species generated from the most refractory of substances become available for subsequent laser vaporization.

2.6. Computers

The availability of low-cost computers has also left its mark. With it multi-surface matrix-isolation in which a large number of matrices of varying composition and condensation conditions but all belonging to a single series, can be generated and automatically positioned by computer in subsequent spectroscopic analysis. This scheme has been refined further by Mattson Instruments, Inc., allowing the deposition of matrices of varying concentration on a rotating cold-drum; the computer then keeps track of the precise drum position corresponding to a matrix of a given composition.

Computers have also had impact on matrix isolation in another way. The advent of very large computers on one hand, and very powerful small computers on the other has made the calculation of matrix effects using both quantum-chemical and molecular-dynamics computations methods possible. The nature of the interaction between the guest molecule and the host, leading to matrix frequency shifts, degeneracy removal, and splitting due to matrix sites can now be understood to a depth hitherto impossible. These calculations are still in their early stages, and much more will be forthcoming.

Perhaps the most important single application of computers in matrix-isolation spectroscopy has been in the dedicated operation and processing required for the Fourier transform infrared instrument. As discussed at the beginning of this section, FTIR instruments have made possible higher resolution and improved signal-to-noise measurements. The most important experimental development for the matrix-isolation technique, in our opinion, has been the closed-cycle cryogenic refrigerator. Perhaps the second most significant experimental advance has been Fourier transform infrared instrumentation, which has certainly been made possible by the development of low-cost computers.

References

Andrews. L., 1967, J. Phys. Chem. **71**, 2761.

Andrews, L., 1968, J. Chem. Phys. **48**, 972.

Andrews, L., 1984, J. Phys. Chem. **88**, 2940.

Andrews, L., and R. Withnall, 1988, J. Am. Chem. Soc. **110**, 5606.

Bahrdt, J., P. Gurtler and N. Schwenter, 1987, J. Chem. Phys. **86**, 6108.

Boyle, M.E., B.E. Williamson, P.N. Schatz, J.P. Marks and P.A. Snyder, 1986, Chem. Phys. Lett. **125**, 349.

Gifford, W.E., and H.O. McMahon, 1960, Adv. Cryogenic Engineering **5**, 354.

Hunt, R.D., and L. Andrews, 1987, J. Chem. Phys. **86**, 3781.

Manceron, L., and L. Andrews, 1988, J. Am. Chem. Soc. **110**, 3840.

Moskovits, M., and G.A. Ozin, eds, 1976, Cryochemistry (Wiley–Interscience, New York).

Schroeder, W., H. Wiggenhauser, W. Schrittenlacher and D.M. Kolb, 1987, J. Chem. Phys. **86**, 1147.

Siegman, A.E., 1986, Lasers (University Science Books, Mill Valley, CA).

Süzer, S., and L. Andrews, 1988, J. Chem. Phys. **88**, 916.

Von Gustorf, E.A.K., O. Jaenicke, O. Wolfbeis and C.R. Eady, 1975, Angew. Chem. Int. Ed. Engl. **14**, 278.

Whittle, E., D.A. Dows and G.C. Pimentel, 1954, J. Chem. Phys. **22**, 1943.

CHAPTER 2

Absorption Spectroscopy of Molecular Ions and Complexes in Noble-Gas Matrices

Lester ANDREWS

Chemistry Department
McCormick Road
University of Virginia
Charlottesville, VA 22901
USA

Chemistry and Physics of
Matrix-Isolated Species
L. Andrews and M. Moskovits, eds

Contents

1. Introduction

The matrix-isolation technique is particularly well-suited for the study of transient species, which react rapidly or decompose under normal chemical conditions. Molecular ions and complexes are two types of transient species studied extensively in our laboratory at Virginia. Matrix studies provide a valuable complement to gas-phase work in a number of ways. First, band positions can be located in the matrix to provide a guide for high-resolution measurements, and second, more information can often be obtained from a complete low-resolution spectrum of the reactive species than from high-resolution measurements on a system of bands.

Early matrix isolation studies of molecular ions have been reviewed by Jacox (1978) and the author (Andrews 1979, 1983a). Charged species in matrices may be described as "isolated" or "chemically bound" with respect to the counterion. The first ionic species characterized in matrices, $Li^+O_2^-$, is of the latter type with Li^+ and O_2^- electrostatically bound together in an isosceles triangular structure (Andrews 1968, 1969). The next molecular ions identified in matrices, $B_2H_6^-$ and C_2^-, were "isolated" from their counterions by a number of matrix atoms (Kasai and McLeod 1969, Milligan and Jacox 1969). Clearly, formation of "isolated" ions requires ionization energy for the subject molecule from an external source and a receptor atom or molecule to trap the removed electron elsewhere in the matrix. This chapter describes two cation systems studied extensively in the infrared and the ultraviolet and visible regions and one anion system studied in the infrared. Recent ESR studies of isolated cations and anions will be described in the chapter by Knight (ch. 7, this volume).

Molecular complexes and clusters are receiving extensive attention in supersonic nozzle beams backed by a high pressure of inert gas. These complexes reveal important information on intermolecular interactions, which are fundamental to many branches of science. The matrix sample provides a similar environment for molecular complexes, except for the lack of rotational freedom for most molecules, and such samples can be prepared in a straightforward manner by co-depositing gas mixtures at 5 or 12 K from two separate vacuum systems. In addition new evidence suggests that the matrix-isolated complex may be more relaxed than those prepared in supersonic expansions (Süzer and Andrews 1987b). A variety of base $\cdots$ HF complexes have been described by the author (Andrews, 1983b, Andrews et al. 1984a), base···hydrogen halide complexes have been reviewed by Barnes (1983), and Lewis acid···base complexes in matrices have been discussed (Machara and Ault 1987, Ault 1988). Hydrogen fluoride complexes to be discussed here include amine bases that involve strong hydrogen bonding, and methane and hydrogen bases that involve weak hydrogen bonding. The molecular complexes $NH_3···F_2$ and $PH_3···O_3$ are of interest in their own right, but they are of more interest as photochemical precursors to $NH_2F···HF$ and H_3PO. This chapter will also describe infrared spectra and structures of ammonia dimer and the Li complex with C_2H_2.

2. Experimental methods

Molecular ions and complexes are produced and trapped concurrently with sample deposition using the apparatus shown in fig. 1. The cold window W is cooled by one of several closed-cycle helium cryogenic refrigerators. For 10–12 K refrigeration, two-stage CTI Cryogenics Model 21 and 22 cryocoolers, and earlier models, have been used in our laboratory for over 20 years with absolutely outstanding performance records. In these mechanical devices, compressed helium at about room temperature is passed through a regenerator and expanded from an enclosed volume; the expanding gas is warmed but the remaining gas and the enclosure are cooled, since the helium works on itself in an adiabatic process. The cooler helium is forced through and cools the regenerator, and the entire process is repeated continuously. For 4–5 K refrigeration, a three-stage Air Products Heliplex is used; this system adds a Joule–Thomson expansion valve onto a two-stage mechanical cryocooler

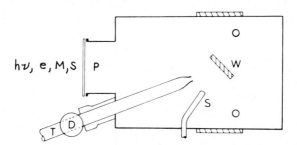

Fig. 1. Vacuum vessel base cross-section used for absorption spectroscopy of molecular ions and complexes. The diagram shows the position of the rotatable 4–12 K cold window W, optical windows O, gas sample deposition line S, open discharge tube vacuum ultraviolet source T, microwave discharge cavity D, and interchangeable plate P that can be fitted with a quartz photolysis port, a Knudsen effusion metal atom source, a low-energy electron source, or another gas deposition line. For HF, F_2 or O_3 experiments the discharge tube T is replaced by a stainless-steel gas deposition tube.

using two compressors for the helium gas employed. This three-stage system is obviously more complicated than the two-stage systems and more sensitive to impurities in the helium gas, but it can operate quite reliably for matrix infrared experiments.

Several sample preparation methods employed for molecular ions and complexes are illustrated in fig. 1. The alkali-metal reaction, photolysis and radiolysis methods have been described in earlier reviews (Andrews 1979, 1983a). The vacuum ultraviolet photo-ionization technique developed in our laboratory employed open microwave-powered discharge tubes (T) like the one shown in fig. 1 (Prochaska and Andrews 1977). The argon discharge gas was subsequently collected with the matrix, but neon and helium discharge gases were evacuated through an auxiliary pumping system. Similar direct-current powered discharge resonance lamps were used for photoelectron spectroscopy (Andrews et al. 1984a). Resonance two-photon ionization of aromatic precursors has been done by irradiation with an intense high-pressure mercury arc through the photolysis port (P) (Kelsall and Andrews 1982). Recently a thermionic electron source capable of co-depositing 10–200 µA of low-energy electrons (30–400 eV) with the matrix sample was employed to produce molecular ions during sample condensation (Süzer and Andrews 1987c). The electron source, a Knudsen cell for effusion of alkali metal atoms, or another gas sample deposition tube was connected to the matrix apparatus by interchangeable O-ring mounted plates (P) for the particular matrix reaction of interest.

3. Molecular ions

Infrared spectroscopic studies of trifluoromethyl cations and polyfluoride anions and visible–ultraviolet spectra of naphthalene cations will be described next.

3.1. Trihalomethyl cations

Our objective for investigation of infrared spectra of positive ions in solid argon was CF_3^+. Development of techniques for producing and trapping CF_3^+ led first to CCl_3^+ and the mixed chlorofluoromethyl cations. Jacox and Milligan (1971) performed hydrogen resonance photolysis of argon/chloroform samples and observed a large yield of CCl_3 radical at 898 cm^{-1} and a new 1037 cm^{-1} absorption. The latter band showed the 3:1 doublet splitting characteristic of three equivalent chlorine atoms and a carbon-13 shift to 1003 cm^{-1} which supported its identification as CCl_3^+ isolated in the matrix. Subsequent radiolysis studies of CCl_4, CCl_3Br, CCl_2Br_2, $CClBr_3$ and CBr_4 in our laboratory produced the 1037 cm^{-1} band, absorptions at 1019 and 957 cm^{-1} for CCl_2Br^+, absorptions at 978 and 894 cm^{-1} for CClBr$_2^+$ and a new band at 874 cm^{-1} for CBr$_3^+$, each from two of the above precursors, which confirmed the trichloro-methyl cation identification of the 1037 cm^{-1} band (Andrews et al. 1975). Infrared absorptions for trihalomethyl cations are listed in table 1.

Table 1
Carbon–halogen stretching vibrations (cm^{-1}) observed for trihalomethyl cations isolated in solid argon.

Cation	C–F	C–Cl	C–Br
CF$_3^+$	1665		
CF$_2$Cl$^+$	1514, 1414		
CFCl$_2^+$	1351	1142	
CCl$_3^+$		1037	
CCl$_2$Br$^+$		1019, 957	
CClBr$_2^+$		978	894
CBr$_3^+$			874
CF$_2$Br$^+$	1483, 1367		
CF$_2$I$^+$	1432, 1320		

In a subsequent argon resonance photo-ionization study of CCl$_4$, different filtered mercury-arc photolysis behavior for each new absorption was attributed to the photochemical stability of the cation in question. The photosensitive 927 and 374 cm^{-1} bands were assigned to an asymmetric CCl$_4^+$ species and the 502 cm^{-1} absorption to Cl$_3^+$. The 1037 cm^{-1} band was decreased slightly by prolonged mercury arc photolysis, which required photodetachment from chloride electron traps in the matrix, since CCl$_3^+$ itself probably does not dissociate in the mercury-arc energy range (Prochaska and Andrews 1977). The small shift for the 1037 cm^{-1} CCl$_3^+$ band in solid argon to 1035 cm^{-1} in solid krypton and the low ionization energy of CCl$_3$ (8.3 eV) (Lossing 1972) suggest that CCl$_3^+$ does not interact strongly with either matrix. An important conclusion from the mercury arc photolysis studies is that isolated cations in matrices will be of two types:

(1) cations which photodissociate with mercury-arc light; and

(2) cations photochemically stable in the mercury-arc range.

The latter must exhibit a decrease in intensity from photoneutralization by detachment of electrons from counteranions in the matrix. In this regard, appearance potential data are useful for predicting the photochemical behavior of cations. Clearly, photodetachment depends on the electron-trapping species, as will be discussed below for CF$_3^+$.

In order to bridge the gap from CCl$_3^+$ at 1037 cm^{-1} to CF$_3^+$, photo-ionization and radiolysis studies were performed in our laboratory on the Freon series CFCl$_3$, CF$_2$Cl$_2$ and CF$_3$Cl (Prochaska and Andrews 1978a,b). Sharp new 1352 and 1142 cm^{-1} bands in CFCl$_3$ experiments exhibited appropriate carbon-13 shifts and photolysis behavior for assignment to CFCl$_2^+$. Analogous bands in CF$_3$Cl experiments at 1415 and 1515 cm^{-1} showed large carbon-13 shifts and slight photolysis with the full mercury arc which indicated assignment to CF$_2$Cl$^+$. These vibrations for CFCl$_2^+$ and CF$_2$Cl$^+$ were 200–300 cm^{-1} above the corresponding free-radical values, which predict that CF$_3^+$ may absorb above 1600 cm^{-1}.

Matrix photo-ionization studies have been performed on the trifluoromethyl compounds CF$_3$Cl, CF$_3$Br, CF$_3$I and CHF$_3$ (Prochaska and Andrews 1978c) using the apparatus shown in fig. 1. The CF$_2$Cl$^+$ absorptions at 1514 and 1414 cm^{-1} were shifted to 1483 and 1367 cm^{-1} for CF$_2$Br$^+$, and to 1432 and 1320 cm^{-1} for CF$_2$I$^+$. The C–F stretching modes in the CF$_2$X$^+$ ions exhibited a pronounced heavy-halogen effect. A weak 1665 cm^{-1} band in the CF$_3$Cl study was produced with greater intensity in the CF$_3$Br, CF$_3$I and CHF$_3$ experiments at the same frequency. This 1665 cm^{-1} absorption was reduced 10% by full high-pressure mercury-arc photolysis for 2 h in fluoroform experiments. The production of the same 1665.2 cm^{-1} band from four different trifluoromethyl precursors, and the photolysis behavior, indicate assignment of the 1665 cm^{-1} band to CF$_3^+$. The trifluoromethyl cation was formed by photo-ionization of CF$_3$ radicals produced in the matrix photolysis process.

Infrared spectra for CF$_3^+$ and ^{13}CF$_3^+$ in fluoroform experiments are compared in fig. 2. The strong absorption produced from ^{13}CHF$_3$ at 1599.2 cm^{-1} (absorbance $A = 0.20$) is appropriate for v_3 of ^{13}CF$_3^+$; note the weak ^{12}CF$_3^+$ absorption ($A = 0.02$) at 1665.2 cm^{-1} due to 10% ^{12}C present in the enriched fluoroform precursor. The ^{12}C–^{13}C shift, 66.0 cm^{-1}, is, however, greater than expected for the antisymmetric C–F vibration, v_3, of a planar centrosymmetric species. This unexpectedly large carbon-13 shift can be explained by Fermi resonance between v_3 and the combination band $v_1 + v_4$ for the ^{13}CF$_3^+$ species, since the new 1641.7 cm^{-1} product band in the ^{13}CF$_3^+$ spectrum can be assigned to the combination band. The calculated position of v_3 for ^{13}CF$_3^+$, 1620 cm^{-1}, appears to coincide with the apparent position of $v_1 + v_4$, and these modes strongly interact and shift v_3 down to 1599 cm^{-1} and $v_1 + v_4$ up to 1642 cm^{-1}. In the absence of Fermi resonance, the $v_1 + v_4$ combination band for ^{12}CF$_3^+$ is expected to be 2–4 cm^{-1} above the ^{13}CF$_3^+$ counterpart, in the region of the 1624 cm^{-1} water absorption; the spectrum in fig. 2b shows the 1624 cm^{-1} band, which may contain additional absorption. The markedly increased CF$_3^+$ yield in the electric discharge experiment, fig. 2d, reveals a very strong 1665.2 cm^{-1} absorption ($A = 0.84$) and 1624 cm^{-1} absorption ($A = 0.40$) clearly in excess of the other water absorptions present (Kelsall and Andrews 1981); the latter absorption is assigned to the combination band $v_1 + v_4$ for ^{12}CF$_3^+$. The 1624 cm^{-1} combination band for ^{12}CF$_3^+$ provides a basis for determining the infrared inactive symmetric C–F bond stretching mode v_1 of ^{12}CF$_3^+$. The intensity of the v_2 and v_4 modes of BF$_3$ are approximately a factor of ten weaker

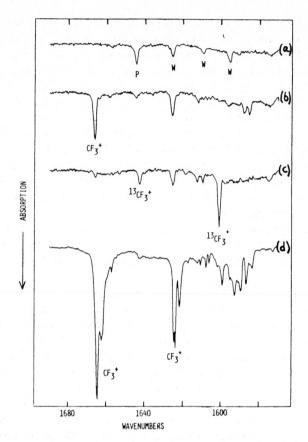

Fig. 2. Infrared spectra in the 1580–1680 cm^{-1} region for fluoroform samples condensed with excess argon at 15 K. (a) Ar:CHF$_3$ = 200:1 sample, P denotes precursor and W denotes water absorptions; (b) Ar:CH$_3$ = 800:1 with concurrent argon-resonance photo-ionization; (c) Ar:^{13}CHF$_3$ = 800:1 with photo-ionization; and (d) Ar:CHF$_3$ = 700:1 with electric discharge on matrix during condensation.

than that of v_3, and the failure to observe these weaker fundamentals of CF$_3^+$ in the present study is not surprising. However, v_4 may be estimated at 500 ± 30 cm^{-1} since $v_4 = 512$ cm^{-1} for CF$_3$ (Milligan and Jacox 1968) and 480 cm^{-1} for BF$_3$ (Dows 1959), which predicts $v_1 = 1125 \pm 30$ cm^{-1} for ^{12}CF$_3^+$.

Two similar neon resonance photo-ionization experiments have been performed in our laboratory co-depositing Ne:CHF$_3$ = 300:1 sample at 5 K with neon from a 3 mm microwave discharge tube (Andrews and Hunt 1987). New bands were observed at 1254 cm^{-1} ($A = 0.06$) for CF$_3$, 1222 cm^{-1} (weak) for CF$_2$, and 1670 and 1664 cm^{-1} ($A = 0.012$) for CF$_3^+$. All of these bands decreased by $\frac{1}{4}$ on annealing at 5–12 K. Spectra were scanned after 1–2 h periods, and product band growth rates decreased with time, limiting the effective sample deposition period for good matrix isolation to about 4 h. The neon condensation rate decreased rapidly as sample thickness increased thus limiting the amount of neon that can be condensed quickly enough to trap reactive species at 5 K. Although neon is more inert than argon, its use for ions and small molecules in infrared experiments, *which require more sample* than the ESR and fluorescence spectroscopies described in the chapters by Knight (ch. 7) and Bondybey (ch. 5), is limited. Accordingly, it is possible to produce and trap an order of magnitude more CF_3^+ in solid argon than in solid neon.

The observation of CF_3^+ in solid neon is, however, important. The split absorptions at 1670 and 1664 cm^{-1} are characteristic of many species in this smaller matrix cavity and bracket the 1665 cm^{-1} argon value. The near agreement between neon- and argon-matrix absorptions for CF_3^+ and the low ionization energy for the CF_3 radical (9.2 eV) (Lossing 1972) suggest that CF_3^+ does not interact strongly with these matrix environments. This agreement of neon- and argon-matrix values predicts the gas-phase v_3 fundamental of CF_3^+ at 1675 ± 20 cm^{-1}.

It is interesting to consider the bonding in CF_3^+ in view of its substantially increased antisymmetric stretching frequency and the increased symmetric frequency deduced from the $v_1 + v_4$ band. This increase is considerable relative to the pyramidal CF_3 species ($v_1 = 1086$ cm^{-1}, $v_3 = 1251$ cm^{-1} and the planar $^{11}BF_3$ molecule ($v_1 = 888$ cm^{-1}, $v_3 = 1454$ cm^{-1}). From the well-known back-donation of fluorine 2p electron density to the positive carbon, it is readily apparent that CF_3^+ should exhibit extensive π-bonding, and the markedly increased v_3 and v_1 modes are consistent with this bonding model.

The photolysis behavior of the 1665 cm^{-1} CF_3^+ absorption in CHF_3, CF_3Cl, CF_3Br and CF_3I experiments provides evidence for the halide counterion in these studies (Prochaska and Andrews 1978c). Since dissociation of CF_3^+ to $CF_2^+ + F$ requires at least 5 eV, which is above the mercury-arc range, a photoneutralization mechanism is required for the slight decrease of CF_3^+ on mercury-arc photolysis. The relative decrease in the CF_3^+ band on 290–1000 and 220–1000 nm photolysis in these experiments is more pronounced in the order $I^- > Br^- > Cl^- > F^-$ where these halide ions are most likely counterions in photo-ionization studies with CF_3I, CF_3Br, CF_3Cl and CHF_3, respectively. This photolysis behavior parallels the expected photodetachment cross-section in the halide series and supports the photoneutralization model for the photolysis of CF_3^+ isolated in a matrix containing halide counterions.

3.2. Bifluoride and trifluoride anions

The bifluoride ion is the best documented example of a very strong and symmetrical hydrogen bond (Larson and McMahon 1982, Ibers 1964). Bifluoride anion has been characterized in crystalline solids (Dawson et al. 1975) and as $Cs^+ HF_2^-$ ion pairs in solid argon (Ault 1978). In order to examine the HF_2^- ion free of crystal-lattice and counterion effects, the isolated HF_2^- ion was prepared by vacuum-ultraviolet irradiation of Ar/HF samples, and a strong, sharp product band was observed at $1377\,cm^{-1}$ (McDonald and Andrews 1979). The analogous Ar/DF experiments gave a new $965\,cm^{-1}$ band and the strong $Ar_n D^+$ band at $644\,cm^{-1}$ (Milligan and Jacox 1973, Wight et al. 1976). The observed $HF:DF = 1377:965 = 1.427$ ratio substantiated the v_3 assignment to a linear centrosymmetric species that can have only quartic anharmonic terms in its vibrational potential function, in agreement with the solid and ion-pair studies. However, the gaseous anion has been detected by diode laser spectroscopy in a hollow cathode discharge, and a band at $1848\,cm^{-1}$ has been assigned to the v_3 fundamental (Kawaguchi and Hirota 1986). In order to obtain new results on polyfluoride anion systems, Ar/HF samples have been subjected to low-energy electron impact during condensation at 12 K.

A thermionic electron source capable of delivering $10-200\,\mu A$ of low-energy electrons has been connected to a matrix-isolation apparatus to study electron impact on small molecules during condensation. In the first application of this technique, OH radical, OH^- anion, and matrix-solvated proton $Ar_n H^+$ have been produced by low-energy electron impact on water during condensation (Süzer and Andrews 1988). The OH^- anion absorption at $3549\,cm^{-1}$ in solid argon has been assigned to the $R(0)$ vibration–rotation band observed at $3591.5\,cm^{-1}$ in the gas phase (Owrutsky et al. 1986). This assignment is based in part on comparison with the infrared spectrum of the isoelectronic HF molecule in solid argon (Andrews and Johnson 1984), which is dominated by the $R(0)$ band shifted $38\,cm^{-1}$ to the red by the argon matrix.

The spectrum of an $Ar:HF = 100:1$ sample subjected to irradiation with $10\,\mu A$ of electrons at 150 eV for 4 h during condensation at 12 K is shown in fig. 3a (Hunt and Andrews 1987b). This sample contained strong HF cluster bands, which have been described earlier (Andrews and Johnson 1984), and strong new product bands at 1815 and $1377\,cm^{-1}$ labeled T and B, respectively, in the figure. Sample annealing decreased band B more than T and produced a broad new band at $1715\,cm^{-1}$ as shown in fig. 3b. Decreasing the HF concentration decreased T relative to B. Experiments with 40, 75 and 90% deuterium-enriched hydrogen fluoride produced a $965\,cm^{-1}$ counterpart for the $1377\,cm^{-1}$ band and two counterparts for the $1815\,cm^{-1}$ band, the first appearing at $1707\,cm^{-1}$ with partial deuterium enrichment, and the second following at $1391\,cm^{-1}$ with high deuterium enrichment.

Assignment of the $1377\,cm^{-1}$ band to v_3 of HF_2^- is reaffirmed based on the high $H:D = 1377:965 = 1.427$ ratio, which is higher than the 1.396 harmonic value and can arise for a linear, centrosymmetric species 1 (scheme 1) owing to the absence of cubic and the presence of quartic anharmonic terms in the vibrational potential. The results of recent high-quality ab initio calculations (Janssen et al. 1987) also substantiate this assignment. The higher gas-phase

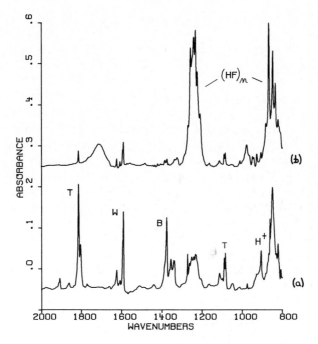

Fig. 3. Infrared spectra of Ar : HF = 100 : 1 sample: (a) co-deposited at 12 K with 10 μA of 150 eV electrons for 4 h, and (b) after warming to 25 K and recooling to 12 K.

Scheme 1.

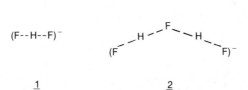

$\underline{1}$ $\underline{2}$

band exhibits a much lower $H:D = 1848:1397 = 1.323$ ratio, which is not appropriate for the v_3 fundamental of centrosymmetric HF_2^-. It is suggested that the 1848 cm^{-1} band is due to the $v_1 + v_3$ combination band of gaseous HF_2^- and that a careful search of the 1300 cm^{-1} region with other diodes will reveal the stronger v_3 fundamental (see Kawaguchi and Hirota 1987).

The 1815 cm^{-1} band is assigned to the out-of-phase H–F stretching fundamental of $H_2F_3^-$ based on agreement with solid-state spectra (Azman et al. 1967, Harmon and Gennick 1977), the concentration dependence relative to HF_2^-, the $H:D = 1815:1391 = 1.305$ ratio, which is appropriate for a C_{2v} species $\underline{2}$ (scheme 1) whose anharmonicity is dominated by cubic terms, and the appearance of a mixed H:D band at 1715 cm^{-1}. The 1090 cm^{-1} band is due to an F--H--F bending mode in $H_2F_3^-$. The $H_2F_3^-$ anion is of interest as it contains the second strongest hydrogen bond based on displacement of the H–F fundamental in $(FHFDF)^-$ to $1707 \, cm^{-1}$; HF_2^- contains the strongest hydrogen bond with the average H–F stretching fundamental at $1016 \, cm^{-1}$ (Dawson et al. 1975). These values characterize stronger hydrogen bonds than other strong complexes such as H_a–F at $1870 \, cm^{-1}$ in $(CH_3)_3N\cdots H_a\cdots F\cdots H_b$–$F$ (Andrews et al. 1986) and at $3041 \, cm^{-1}$ in $NH_3\cdots HF$ (Johnson and Andrews 1982).

The broad $1715 \, cm^{-1}$ band is in excellent agreement with solid-state observation for $K^+H_3F_4^-$ (Harmon and Gennick 1977) and it is assigned to isolated $H_3F_4^-$ in solid argon. This higher polyfluoride anion is formed at the expense of HF_2^- and $H_2F_3^-$ on diffusion and further association of HF in the matrix.

Finally, charge balance in these experiments is maintained by the matrix-solvated proton species Ar_nH^+ (Milligan and Jacox 1973, Wight et al. 1976), which is probably formed on ionization of atomic hydrogen librated in the dissociative capture of low-energy electrons by HF molecules, according to

$$HF + \text{low-energy electrons} \rightarrow H + F^-,$$

$$n\text{Ar} + H + \text{low-energy electrons} \rightarrow Ar_nH^+ + 2e,$$

$$HF + F^- \rightarrow HF_2^-,$$

$$HF_2^- + HF \rightarrow H_2F_3^-,$$

$$H_2F_3^- + HF \rightarrow H_3F_4^-.$$

The fluoride anion has an exceptionally high proton affinity, and it readily clusters with one, two or more HF submolecules in these matrix-isolation experiments as is found in solid polyfluorides.

3.3. Naphthalene cation

The naphthalene cation is a particularly interesting case for matrix study since a wealth of data are available from photoelectron spectroscopy (Eland and Danby 1968, Clark et al. 1972), absorption spectra in glassy matrices (Shida and Iwata 1973), and multi-photon dissociation spectra (Kim and Dunbar 1980). Argon discharge photo-ionization of naphthalene (hereafter denoted by N) vapor during condensation with excess argon produced new absorption band systems at $675.2 \, nm$ ($A = 0.17$), $461 \, nm$ ($A = 0.01$) and $381 \, nm$ ($A = 0.03$). The spectrum of naphthalene cation from matrix two-photon ionization of naphthalene is illustrated in fig. 4: trace a shows the spectrum of an $Ar:CCl_4:N = 3000:4:1$ sample deposited at $20 \, K$, trace b illustrates the spectrum after photolysis for $15 \, s$ with 220–$1000 \, nm$ high-pressure mercury-arc radiation. The above absorption bands were observed with a three-fold intensity

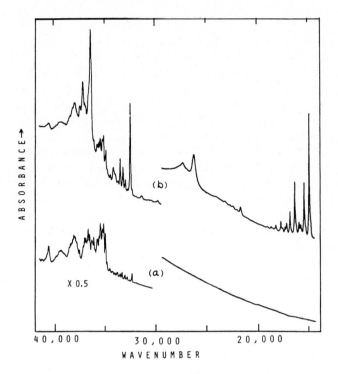

Fig. 4. Visible and ultraviolet absorption spectra of naphthalene samples in solid argon at 20 K: (a) Ar:CCl$_4$:N = 3000:4:1, and (b) same sample after 15 s exposure to 220–1000 nm high-pressure mercury-arc radiation.

increase over the one-photon production method, and two strong new systems were produced at 307.6 nm ($A = 0.78$) and 274.6 nm ($A = 0.9$) at the expense of N precursor absorption systems beginning at 313.3 and 272 nm, which were reduced by 70% (Kelsall and Andrews 1982, Andrews et al. 1982a).

The five band systems are assigned to the naphthalene radical cation (N$^+$) for the following reasons:

(a) Three of these absorption band systems are observed with two different techniques capable of forming positive ions. Argon resonance radiation ionizes N molecules during matrix condensation and the N$^+$ product is trapped in solid argon; in these studies the electron removed in ionization is probably trapped by a molecular fragment or another parent molecule. Intense mercury-arc radiation performed two-color resonance photo-ionization of N molecules isolated in solid argon with the relatively long lived S$_1$ state serving as an intermediate step in the photo-ionization process. In the latter experiments, CCl$_4$ was added as an electron trap, and positive-ion yields increased by a factor of 3 over single-photon ionization experiments, giving further support to the product indentification as a positive ion. The two highest energy absorption systems were obscured by precursor absorption in the first experiments, but the latter studies used one-tenth as much N, and the N^+ absorptions were produced at the expense of N bands (fig. 4) (Kelsall and Andrews 1982).

(b) The two visible band origin positions are in excellent agreement with photoelectron spectra (PES). The red absorption provides an excellent basis for comparison between PES and the sharp matrix absorption spectrum, since the difference between sharp vertical PES band origins can be measured to ± 0.02 eV. Sharp vertical ionization energies have been observed for the first (8.11 eV) and third (9.96 eV) ionic states of N^+. The difference between the first and third PES band origins is 1.85 ± 0.02 eV, which predicts an absorption band origin at $14\,920 \pm 160$ cm^{-1} in the gas phase, assuming no change in structure between the neutral molecule and the two ionic states. The argon matrix origin at $14\,810 \pm 3$ cm^{-1} is in agreement within measurement error, which confirms the identification of N^+ and indicates very little perturbation of N^+ by the argon matrix. The blue absorption at $21\,697 \pm 10$ cm^{-1} is also in excellent agreement with the $21\,859 \pm 160$ cm^{-1} difference between sharp origins of the first and fourth PES bands. The three ultraviolet bands are attributed to electron-promotion transitions with upper states not reached by PES, hence a comparison cannot be made.

(c) The first four transitions are in excellent agreement with the 690, 467, 387 and 308 nm absorptions assigned to N^+ produced by radiolysis of N in a freon glass at 77 K (Shida and Iwata 1973). The small red-shifts (42–400 cm^{-1}) are due to greater interaction with the more polarizable freon medium. Vibronic structure is similar in solid-argon and solid-freon samples; the bands are sharper in solid argon which facilitates more accurate vibronic measurements.

(d) The first three transitions are predicted reasonably well by simple Hückel molecular orbital theory with the molecular orbital energies calculated using $\beta = 2.77$ eV from PES (Eland and Danby 1968).

(e) The sharp UV band origins at $32\,510$ and $36\,417$ cm^{-1} are in agreement with the broad four-photon dissociation bands at $33\,200$ and $38\,200$ cm^{-1} for N^+; the agreement would probably be better if origins were resolved for the broad PDS bands (Kim and Dunbar 1980).

(f) The vibronic structure in the absorption systems, particularly the red band, and the N-d_8^+ counterpart spacings correspond closely with N and N-d_8 vibrational intervals (Mitra and Bernstein 1959). The small blue shifts in the origin bands upon deuteration range from 75 to 145 cm^{-1}; the weak S_1 origin of N at 313.2 nm exhibits a similar 102 cm^{-1} blue-shift upon deuteration.

Two regularly repeating vibrational intervals were observed in the red N^+ absorptions. The first interval, 1422 ± 6 cm^{-1}, was unchanged upon deuterium substitution. This is in agreement with the 1420 ± 40 cm^{-1} interval in the first PES band, which is probably due to the C(9)–C(10) stretching mode $\nu_4(a_g)$. The strong fundamental at 1376 cm^{-1} in the Raman spectrum of $C_{10}H_8$ is complicated by Fermi resonance, but the strong fundamental for $C_{10}D_8$ is at

1380 cm^{-1} (Mitra and Bernstein 1959). The second repeated interval in the spectra was approximately 505 cm^{-1} for N-h$_8^+$ and 485 cm^{-1} for N-d$_8^+$; this corresponds to the $v_9(a_g)$ skeletal distortion observed in Raman spectra for N-h$_8$ at 512 cm^{-1} and N-d$_8$ at 491 cm^{-1}. The 754 cm^{-1} interval corresponds to $v_8(a_g)$, the skeletal breathing mode observed at 758 cm^{-1} in the Raman spectrum of N. These observations show that ionization has a relatively small effect on the vibrational potential function of the N molecule.

4. Base···HF complexes

The argon···HF matrix system must be examined before base···HF spectra can be analyzed. Hydrogen fluoride matrix spectra have been studied extensively (Andrews 1983b, 1984, Andrews and Johnson 1984), and detailed concentration and annealing experiments have made possible the identification of HF dimer, trimer and higher clusters. The bands labeled D at 3826, 2809 and 2803 cm^{-1} in the DF/HF sample shown in fig. 5 are due to $(HF)_2$, $(HF)(DF)$ and $(DF)_2$ and the bands labeled T at 3702, 3670, 2719, 2717, 2698 and 2674 cm^{-1} are due to cyclic isotopic trimers. Dimers have also been prepared selectively by photolysis of $(H_2)(F_2)$ complexes which limits the yield of trimer (Hunt and Andrews 1985). Similar photolysis experiments with $(HD)(F_2)$ complexes trap preferentially $(H-F)(D-F)$ and show that deuterium prefers the bonding role. We turn our attention now to complexes of HF with other molecules.

4.1. Amine···HF complexes

The NH_3···HF complex was one of the first hydrogen-bonded complexes studied in our group and it remains one of the most interesting (Johnson and Andrews 1982, Andrews 1984). The $v_s(HF)$ fundamental and microwave spectrum have been observed in the gas phase (Thomas 1980, Clements et al. 1981), and the complex has three-fold symmetry. A series of matrix experiments was done with isotopic ammonia and hydrogen fluoride samples in argon at 12 K, and the new absorptions are collected in table 2. The spectrum illustrated in fig. 5 recorded for DF/HF and NH_3 identifies the HF submolecule modes due to the appearance of DF counterparts and the NH_3 submolecule mode (v_2^c) which shows only a very small effect for the acid isotope. The strong $v_s(DF)$ band at 2278 cm^{-1} (fwhm = 4 cm^{-1}) has a $v_s(HF)$ counterpart at 3041 cm^{-1} (fwhm = 16 cm^{-1}). Sharp librational v_ℓ bands were observed for HF and DF at 916.0 and 696.7 cm^{-1}, with their overtones at 1679.4 and 1315.9 cm^{-1}. The HF submolecule fundamentals observed for the complex are the $v_s(HF)$ stretching fundamental and the $v_\ell(HF)$ librational mode derived from rotational degrees of freedom from the isolated diatomic. The doubly degenerate nature of this librational motion is maintained by the C_{3v} symmetry of the complex.

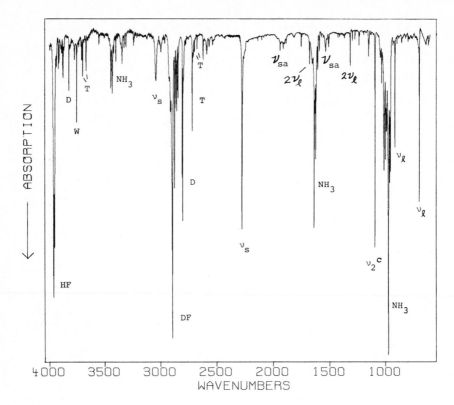

Fig. 5. FTIR spectrum of matrix formed by co-deposition of $Ar:NH_3 = 200:1$ and $Ar:DF:HF$
$= 500:2:1$ samples at 12 K.

The $v_s(HF):v_s(DF) = 3041:2278 = 1.334$ ratio is characteristic of the H–F
stretching mode in a hydrogen-bonded complex. The overtone ratio
$2v_\ell(HF):2v_\ell(DF) = 1679:1316 = 1.276$ is less than the fundamental ratio
$v_\ell(HF):v_\ell(DF) = 916:697 = 1.314$ owing to increased anharmonicity in the
overtone, which of course, makes the overtone observable in the infrared. The
positions of the v_s and v_ℓ modes provide a measure of the interaction between
acid and base as the v_s mode is decreased from the 3919 cm^{-1} stretching
fundamental and the v_ℓ mode is increased from the 44 cm^{-1} rotational
fundamental of the diatomic.

The ammonia submolecule mode v_2^c in the complex follows $^{15}NH_3$ and ND_3
substitution as does v_2 of ammonia. The v_2^c mode is not split by inversion as v_2
for NH_3 is split; this shows that the acid ligand prevents inversion of the
ammonia submolecule in the complex. Notice the effect of $^{15}NH_3$ and ND_3
substitution on the HF submolecule modes; this verifies that a reagent complex
has been formed in the matrix.

Table 2

Absorptions (cm^{-1}) for ammonia–hydrogen fluoride complexes in solid neon and argon.

Neon		Argon						Assignment	
NH$_3$ + HF	NH$_3$ + DF	NH$_3$ + HF	NH$_3$ + DF	^{15}NH$_3$ + HF	^{15}NH$_3$ + DF	ND$_3$ + HF	ND$_3$ + DF		
3106	2309	3041	2278	3042	2277	3017	2269	ν_s	$\underline{3}$
1676	1317	1679	1316	1678	1314	1664	1272	$2\nu_\ell$	$\underline{3}$
1090	1091	1093	1094	1087	1088	849	849	ν_2^c	$\underline{3}$
912	695	916	697	915	695	906	675	ν_ℓ	$\underline{3}$
1990	1560	1920	1536	1919	1531	1862	1484	ν_{sa}	$\underline{4}$
1142	1140	1155	1152	1148	1145	–	–	ν_2^c	$\underline{4}$

The $NH_3 \cdots HF$ complex has been prepared in solid neon for comparison with the argon-matrix data (table 2). First, the strong v_s band at 3106 cm^{-1} (fwhm = 10 cm^{-1}) is sharper than in solid argon and it falls approximately midway between the gas-phase (3215 cm^{-1}) and argon-matrix (3041 cm^{-1}) values. The $2v_\ell$ and v_ℓ modes are shifted only slightly and the ratio is essentially unchanged from the solid-argon value. The v_2^c band at 1090 cm^{-1} is also shifted only slightly from the solid-argon value, but the shift above NH_3 in solid neon, 122 cm^{-1}, is essentially the same as in solid argon, 120 cm^{-1}. The HF:DF ratios in solid neon are also essentially unchanged from solid-argon values. The neon medium is more inert, and the argon-matrix effect is clearly seen to be greater on v_s than on v_ℓ and v_2^c. This probably arises from the greater polarizability of argon and the large dipole change in the v_s vibration.

The weak bands labeled v_{sa} at 1920 and 1536 cm^{-1} are due to H_a–F and D_a–F motions in the 1:2 complexes based on their enhancement over the above 1:1 complex bands on annealing and on increase in DF/HF concentration. The effect of H_b–F on the H_a–F mode is a substantial shift from 3041 cm^{-1} in $\underline{3}$ to 1920 cm^{-1} in $\underline{4}$ (scheme 2); owing to the fluoride affinity of H_b–F, the H_a–F bond is substantially weakened and the stretching fundamental is decreased accordingly. The 1155 cm^{-1} band shows the proper $^{15}NH_3$ shift to be a perturbed v_2^c motion in $\underline{4}$ and it has been assigned accordingly. The neon-matrix effect on $\underline{3}$ absorptions is analogous to the effect on $\underline{1}$ absorptions. Ab initio calculations showed that the H–F bond length increases about 2% in $\underline{3}$ and about 2% more for H_a–F in $\underline{4}$ and further suggest a weak interaction between the terminal fluorine and one amine hydrogen (Kurnig et al. 1986).

Scheme 2.

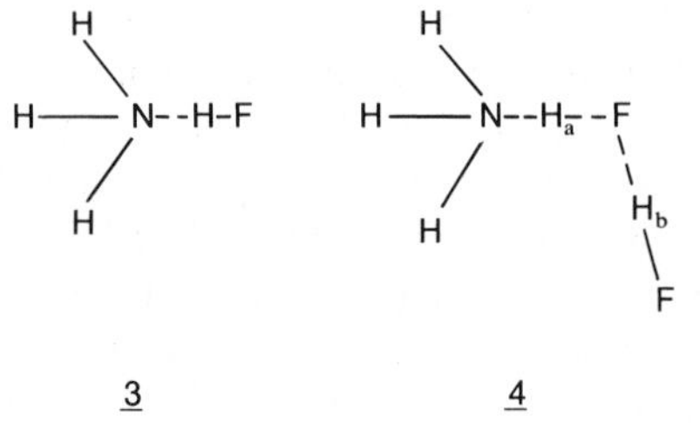

$\underline{3}$　　　　　　　　　　　　　　　$\underline{4}$

The effect of substituents on the basicity of NH_3 is of considerable chemical interest, and the spectra of HF complexes also reflect this change. Methyl substituents are electron donating and increase the proton affinity, and fluorine substituents are electron withdrawing and decrease the proton affinity. Accordingly, the v_s modes in $(CH_3)_nH_{3-n}N \cdots HF$ complexes decreased stepwise with methyl substitution (Andrews et al. 1986) and the v_s modes in $F_nH_{3-n}N \cdots HF$ complexes increased stepwise with fluorine substitution (Lascola et al. 1988).

Thus, displacement from the isolated HF fundamental value (3919 cm^{-1}) is a measure of the base$\cdots$HF interaction, and the more basic methylamines interact more strongly with HF than the less basic fluoroamines.

The trimethylamine–HF system is also important because of the $1:2$ complex. The very strong v_{sa} band at 1870 cm^{-1} was accompanied by a progression of satellite bands at 2048, 2218, 2290 and 2464 cm^{-1} (Andrews et al. 1986). The intervals above 1870 cm^{-1} are, respectively, 178, $178 + 170$, 420 and $420 + 175 \text{ cm}^{-1}$. These intervals increase slightly with DF, although the v_{sa} band shifts to 1397 cm^{-1}, and they show little effect with $(CD_3)_3N$. The isotopic data indicate assignment of the 178 cm^{-1} interval to $v_{\sigma b}$ and the 420 cm^{-1} spacing to $v_{\sigma a}$, the two hydrogen-bond stretching modes. These v_σ observations are of particular interest for two reasons. First, v_σ modes are weak infrared absorptions and their observation in combination bands provides further information about the bonding in these $1:2$ complexes. The 420 cm^{-1} $v_{\sigma a}$ fundamental characterizes a strong hydrogen bond approaching the very strong hydrogen bond in HF_2^-, which exhibits $v_1 = v_\sigma = 600 \text{ cm}^{-1}$ (Dawson et al. 1975). The 178 cm^{-1} $v_{\sigma b}$ value is comparable to the $168 \pm 3 \text{ cm}^{-1}$ value for the relatively strong CH_3CN–HF complex (Thomas 1971). Second, the observation of combination mode progressions in v_{sa} and $v_{\sigma a}$ and $v_{\sigma b}$ clearly demonstrates anharmonic coupling between the acid bond (H–F) and the hydrogen bond (N$\cdots$H$_a$) and provides support for the "frequency modulation" theory of band width in hydrogen bonding.

4.2. $NH_3 \cdots F_2$ and $NH_2F \cdots HF$ complexes

The fluorine$\cdots$ammonia matrix system yielded two interesting new complexes (Andrews and Lascola 1987). On co-deposition of Ar/NH_3 and Ar/F_2 samples, new bands appeared at 966 and 781 cm^{-1} which are shown in fig. 6a. These bands increased three-fold on sample annealing and decreased markedly on photolysis that produced the NH_2F–HF complex to be described below, hence assignment of the strong 966 and weak 781 cm^{-1} bands to the $NH_3 \cdots F_2$ complex 5 (scheme 3) follows. The strong 966 cm^{-1} band is within 1 cm^{-1} of the median of the inversion split Q-branches of NH_3 in solid argon (Abouaf-Marguin et al. 1977) which demonstrates a weak $NH_3 \cdots F_2$ interaction that

Scheme 3.

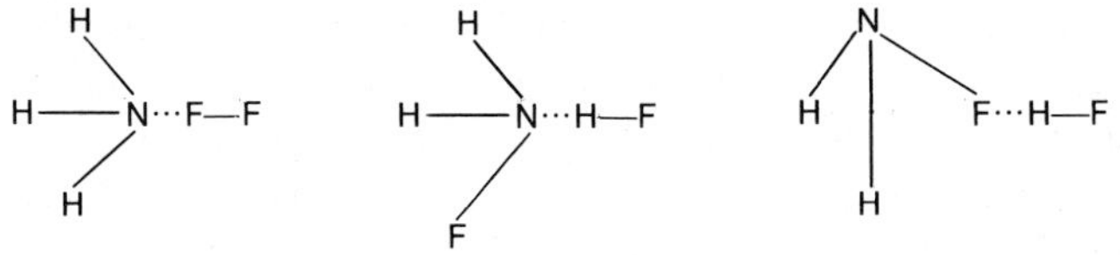

5 6 7

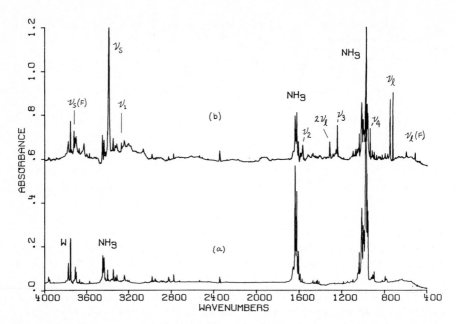

Fig. 6. FTIR spectra of ammonia and fluorine in solid argon: (a) $Ar:NH_3 = 400:1$ and $Ar:F_2$ $= 200:1$ samples co-deposited at 12 K, and (b) after temperature cycling to 27 K and photolysis for 30 min.

nevertheless prevents inversion of the NH_3 submolecule in the complex. The weak 781 cm^{-1} band is below the 892 cm^{-1} Raman fundamental of F_2 in solid argon (Howard and Andrews 1973) which suggests a weak charge-transfer interaction between NH_3 and F_2. Finally, ab initio calculations predict an increasing interaction between NH_3 and Cl_2, and NH_3 and ClF (Lucchese and Schaefer 1975). The ammonia v_2 submolecule mode increased to 988 and 1054 cm^{-1} for these complexes, respectively, demonstrating the decreased interaction and Lewis acid character of $F_2 < Cl_2 < ClF$ (Andrews and Lascola 1987, Fredin et al. 1976).

The major photolysis product bands in fig. 6b appeared at 3389 cm^{-1} (v_s), 3269 cm^{-1} (v_1), 1568 cm^{-1} (v_2), 1314 cm^{-1} ($2v_\ell$), 1244 cm^{-1} (v_3), 934 cm^{-1} (v_4) and 750, 723 cm^{-1} (v_ℓ) which are listed in table 3 along with their ND_3 and $^{15}NH_3$ counterparts. The v_s and v_ℓ bands are clearly due to acid-submolecule modes and the v_1, v_2, v_3 and v_4 bands to base-submolecule modes in a new complex with the structure proposed in $\underline{6}$ (scheme 3) based on the spectrum. This nitrogen lone-pair complex $H_2FN \cdots HF$ is weaker than $H_3N \cdots HF$ as v_s is displaced less, to 3389 cm^{-1}, and the v_ℓ modes are increased less, to 750 and 723 cm^{-1}. The base has lower than three-fold symmetry based on splitting of the librational mode into two bands. The other four bands show isotopic shifts and components for the NH_2F vibrations as assigned in table 3. Unfortunately,

Table 3
Infrared absorptions (cm^{-1}) for the amine bonded H$_2$FN$\cdots$HF complex in solid argon.

H$_2$FN$\cdots$HF	H$_2$F^{15}N$\cdots$HF	HDFN$\cdots$HF	D$_2$FN$\cdots$DF	Assignment
3389	3388	3384	2505	v_s
3269	3264	–	2399	v_1 (N–H$_2$ str)
1568	1566	1429	1151	v_2 (N–H$_2$ bend)
1314	1314	1312	998	$2v_\ell$
1244	1238	–	968	v_3 (NH$_2$ wag)
934	916	933	924	v_4 (N–F str)
750	750	–	561	v_ℓ (in-plane)
723	723	–	540	v_ℓ (out-plane)

isolated NH$_2$F has not been characterized owing to its extremely great reactivity, and inferred shifts in the NH$_2$F submolecule band positions can be used to predict the isolated-molecule values. The largest perturbation is expected to blue-shift the v_3 wag and the isolated NH$_2$F value is predicted at 1160 $\pm$ 20 cm^{-1}. A smaller blue-shift in the v_4 N–F mode is expected owing to reduction of interaction between v_3 and v_4 in the complex, and the isolated NH$_2$F value is predicted at 920 $\pm$ 10 cm^{-1}. Finally, matrix photolysis of NH$_3 \cdots$F$_2$ has produced and trapped the elusive and highly reactive NH$_2$F species as the H$_2$FN$\cdots$HF complex.

The minor product bands at 3722 cm^{-1} (v_s) and 596, 515 (v_ℓ) are assigned to the fluorine lone-pair complex NH$_2$F$\cdots$HF **7** based on comparison with CH$_3$F$\cdots$HF v_s (3774 cm^{-1}) and v_ℓ (453, 435 cm^{-1}) values (Johnson and Andrews 1980). Here the –NH$_2$ group is seen to be more electron donating than –CH$_3$ as far as basicity of the fluorine lone-pairs is concerned.

4.3. HF complexes with CH$_4$ and SiH$_4$

The contrasting chemistry and bond polarity in CH$_4$ and SiH$_4$ plays a major role in the structure of their HF complexes as deduced from matrix infrared spectra. Argon:CH$_4$:HF samples revealed the absorptions illustrated in fig. 7: a sharp, strong 3896 cm^{-1} band (v_s) just below the HF-induced Q-branch at 3919 cm^{-1}, a sharp new 3810 cm^{-1} band (v_{sa}), a weak new 2914 cm^{-1} band (v_1^c), and a new 130 cm^{-1} band (v_ℓ). In addition (HF)$_2$ bands were observed at 3826 and 189 cm^{-1} (D), and (HF)$_3$ absorptions appeared at 3702 and 153 cm^{-1} (T). Sample annealing increased the v_{sa} and (HF)$_3$ bands with little effect on v_s, v_1^c and (HF)$_2$ bands; this shows that the v_{sa} band is due to a higher-order complex than the v_s band. DF counterparts of v_s, v_{sa} and v_ℓ were observed in separate experiments at 2856, 2792 and 103 cm^{-1}. Analogous silane experiments gave a strong 3854 cm^{-1} v_s band, v_3^c at 2228 cm^{-1} above v_3 of SiH$_4$ at 2177 cm^{-1}, and v_ℓ at 165 cm^{-1} (Davis and Andrews 1987).

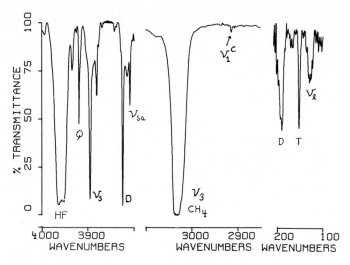

Fig. 7. Infrared spectrum of Ar:CH_4 = 200:1 and Ar:HF = 200:1 samples co-deposited at 12 K.

The v_s band for the CH_4 complex exhibits the expected ratio 3896:2856 = 1.355 for a H–F stretching vibration in a hydrogen-bonded complex, and the v_ℓ band shows an appropriate ratio 130:103 = 1.262 for a low-frequency librational motion. The weak 2914 cm^{-1} absorption is due to the v_1 fundamental of CH_4 in the complex; this mode is not active for CH_4 but the HF ligand produces enough asymmetry in the CH_4 submolecule in the complex to make the v_1 mode weakly active. Assignments for the silane–HF complex modes follow in like order.

The small shift in v_s for the methane complex, the very low v_ℓ mode and the $H^{\delta+}$ polarization suggest the reverse complex structure <u>8</u> (scheme 4) for $CH_4 \cdots FH$ while the larger shift in v_s, the higher v_ℓ mode and the $H^{\delta-}$ polarization suggest the normal complex structure for $SiH_4 \cdots HF$. The infrared spectra indicated a smaller perturbation on HF in the methane complex and the reverse structure leaves the more vulnerable acid proton free from association. Ab initio GAUSSIAN 82 calculations with the 6–31G** basis set confirmed the structures deduced from the matrix infrared spectra and gave a 2.60 Å hydrogen

Scheme 4.

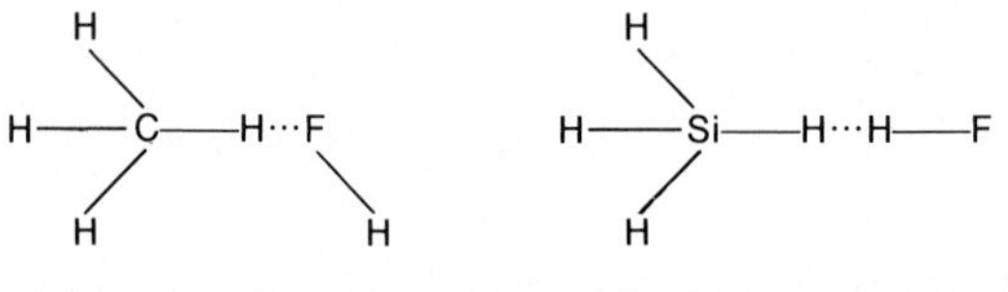

bond distance and 0.93 kcal/mol association energy for <u>8</u> and a 2.22 Å hydrogen bond length and 0.59 kcal/mol association energy for <u>9</u>. The shorter length for the weaker bond is due to the smaller covalent radius of the "hydride" base (Davis and Andrews 1987).

4.4. Hydrogen molecule · · · HF complex

Very weak, hydrogen-bonded complexes of molecular hydrogen and HF have been prepared by condensing the neon-diluted reagents at 5 K (Hunt and Andrews 1987a). Infrared spectra of a Ne:H_2:HF sample in fig. 8 reveal HF monomer, dimer (D) and trimer (T) bands at 3992, 3953, 3919, 3848 and 3706 cm^{-1} (Andrews et al 1984b), a strong, sharp v_s band at 3838 cm^{-1}, a weak v^c band at 4155 cm^{-1}, a weaker v^{cc} band at 4146 cm^{-1} and split v_{sa} and v_{sb}

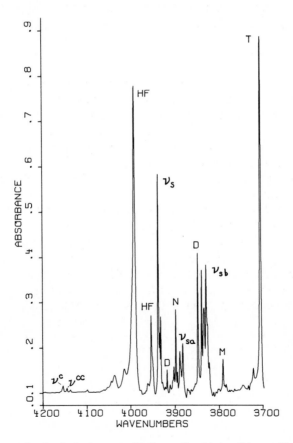

Fig. 8. FTIR spectra of molecular hydrogen and hydrogen fluoride in solid neon. Ne:H_2 = 50:1 and Ne:HF = 200:1 (11 mmol each) co-deposited at 5 K.

bands at 3889 and 3840 cm^{-1}. Changes in reagent concentration and sample annealing show that the v^{cc}, v_{sa} and v_{sb} bands are due to a higher-order complex than v^c and v_s. Accordingly, the primary product complex is identified as $H_2 \cdots HF$ 10 and the higher order complex as $H_2 \cdots (HF)_2$ 11 (scheme 5).

Scheme 5.

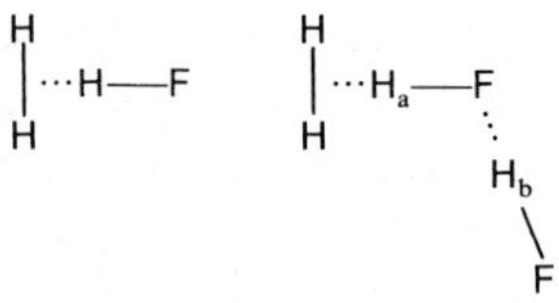

10   11

Vibrational assignments are aided by the isotopic data given in table 4. The v_s band is displaced 15 cm^{-1} below isolated HF in solid neon (3953 cm^{-1}) and exhibits an $HF:DF = 1.364$ ratio, which is appropriate for an HF complex; the v^c band is shifted 6 cm^{-1} below the H_2 fundamental in the gas phase and exhibits a $H_2:D_2 = 1.389$ ratio which is near that for the gaseous diatomics (1.390). Accordingly, the v_s and v^c bands are assigned to the H–F and H–H fundamentals in the $H_2 \cdots HF$ complex, and, in like fashion, the v^{cc}, v_{sa} and v_{sb} assignments follow as well. Comparison of fundamentals for 10, 11 and $(HF)_2$ is of chemical interest. The H_a–F and H_b–F fundamentals of $(HF)_2$ at 3919 and 3848 cm^{-1} are displaced to 3889 and 3840 cm^{-1} by the addition of H_2 to form 11; notice the greater effect for the H_a–F mode where H_2 is attached. Likewise, when H_b–F binds to 10, the H–F mode is shifted from 3938 to 3889 cm^{-1} owing to the fluoride affinity of H_b–F.

Notice the relatively larger effect of DF on H_2 and D_2 than HF listed in table 4. This is due to the shorter time-averaged distance of D from H_2 than H owing to the smaller librational amplitude of DF. In the $D_2 \cdots HF$ complex the v^c mode was red shifted only 3 cm^{-1} below the gas-phase fundamental for D_2. Clearly,

Table 4
Absorptions (cm^{-1}) for molecular hydrogen and hydrogen fluoride complexes in solid neon at 5 K.

$H_2 + HF$	$H_2 + DF$	$D_2 + HF$	$D_2 + DF$	Assignment
4155	4152	2991	2988	v^c 10
4146	4143	2983	2980	v^{cc} 11
3938	2887	3938	2886	v_s 10
3889	2850	3887	2848	v_{sa} 11
3840	2814	3838	2813	v_{sb} 11

the D_2 oscillator is less susceptible to electrical interactions than H_2 since D_2 is substantially lower in the potential well.

Initial studies on $H_2 \cdots HF$ were done in solid argon. The $H_2 \cdots HF$ complex gave a v_s mode blue shifted 8–17 cm^{-1} above the HF fundamental in solid argon. However, the value of these results was complicated by interaction with the argon matrix. Replacing one argon atom by H_2 provided a repulsive contribution to the HF intramolecular potential function and a blue-shift resulted since H_2 is only slightly more physically inert than argon. Neon is, however, still more inert and neon-matrix studies of very weak complexes are more representative of gas-phase studies. Accordingly, the neon-matrix work predicted a gas-phase band between 3945 and 3950 cm^{-1} (Hunt and Andrews 1987) which has subsequently been observed for $H_2 \cdots HF$ (Lovejoy et al. 1987, Jucks and Miller 1987).

5. Molecular complexes

Weak intermolecular interactions can be characterized by determining the perturbation one submolecule has on the other submolecule in a molecular complex. This effect is considered below for $Li \cdots C_2H_2$, $(NH_3)_2$ and ozone–small-molecule complexes.

5.1. Alkali metal atom complexes

Alkali metal atoms form complexes with a variety of molecules (e.g., C_2H_2, C_2H_4, C_6H_6, NH_3, H_2O), and in the case of lithium with π-systems the spectroscopic evidence supports a covalent bonding interaction (Manceron and Andrews 1985a,b, 1986, 1988, Süzer and Andrews 1987a,b).

Acetylene provides the most interesting complex chemistry. Lithium gives a 1:1 complex with six diagnostic infrared bands that are listed in table 5. Isotopic substitution at all positions reveals shifts that characterize the normal modes and the structure of the complex. Two C–H stretching modes were observed at

Table 5
Infrared absorptions (cm^{-1}) for acetylene, $C_2H_2 \cdots HF$ and LiC_2H_2 isotopes in solid argon.

C_2H_2	$C_2H_2 \cdots HF$	$^6LiC_2H_2$	$^7LiC_2H_2$	$^7LiC_2D_2$	$^6Li^{13}C_2H_2$	Assignment	
3374	–	2953	2953	2272	2939	v_s	(CH)
3289	3281	2908	2908	2161	2900	v_a	(CH)
1974	1973	1655	1655	1561	1597	v	(CC)
737	759	715	715	577	708	δ_a	(CH)
–	–	635	601	–	631	v	(LiC)
737	740	480	479	367	478	δ_a	(CH)

2953 and 2908 cm^{-1} shifted below acetylene values, and considerable weakening of the carbon–carbon bond is demonstrated by the position of this stretching mode at 1655 cm^{-1} near the ethylene value. The degenerate bending mode of acetylene is split into red-shifted in-plane and out-of-plane values. A strong Li–C stretching mode was observed at 601 cm^{-1} (^{7}LiC$_2$H$_2$). The structure of LiC$_2$H$_2$ is proposed to be planar with Li bridging the π-system and *cis* C–H groups with CCH angles of $140 \pm 10°$ (Manceron and Andrews 1985a). This *cis*-bridged structure 12 (scheme 6) and spectrum for LiC$_2$H$_2$ have been confirmed by ab initio molecular orbital calculations (Pouchan 1987, Nguyen 1988). Although a Li-vinylidene structure is calculated to be more stable (5 kcal/mol relative to isolated Li and C$_2$H$_2$, as compared to 3 kcal/mol for the *cis*-bridged structure), an appreciable barrier to rearrangement must separate the observed simple-bridged adduct and the calculated vinylidene structure.

Scheme 6.

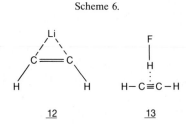

12   13

On the other hand, Na, K and Cs give only broad absorptions slightly shifted from acetylene fundamentals and new absorptions due to ethylene product (Manceron and Andrews 1987). It is clear that only a weak complex is formed between Na, K or Cs and C$_2$H$_2$, and that this complex and C$_2$H$_2$ dimer play a role in C$_2$H$_4$ formation during sample preparation. The contrasting LiC$_2$H$_2$ and CsC$_2$H$_2$ species and their spectra show a significant covalent interaction with lithium involving 2p orbitals.

The C$_2$H$_2\cdots$HF complex is a completely different situation. This complex was first described as a "T-shaped" π-complex 13 from matrix infrared spectra (McDonald et al. 1980, Andrews et al. 1982b), which was confirmed by microwave spectra (Read and Flygare 1982). The association energy of C$_2$H$_2\cdots$HF is calculated to be about 3 kcal/mol (Frisch et al. 1983), which is essentially the same as LiC$_2$H$_2$ (Pouchan 1987) although the spectral perturbations of LiC$_2$H$_2$ are more substantial than for C$_2$H$_2\cdots$HF (see table 5). The C$_2$H$_2\cdots$HF complex exhibits a v_s mode at 3746 cm^{-1}, shifted from isolated HF at 3919 cm^{-1}, and two v_ℓ modes at 426 and 381 cm^{-1} for oscillation of HF parallel and perpendicular to the π-bond. The C–H and C$\equiv$C stretching modes are red-shifted relatively little in this complex, and the bending mode is split into blue-shifted in-plane and out-of-plane counterparts at 740 and 759 cm^{-1}. In summary, the matrix infrared spectrum of the C$_2$H$_2\cdots$HF complex, both acid and base submolecule parts, provides information on structure and bonding in the complex.

5.2. Ammonia dimer

Although ammonia and ammonia dimer have been studied in matrices since the development of the technique (Pimentel et al. 1962, Fredin et al. 1976, Abouaf-Marguin et al. 1977, Barnes et al. 1980, Nishiya et al. 1985), the spectrum and structure of $(NH_3)_2$ are not completely understood. On the basis of dipole moment and rotational structure observed using the molecular-beam electric resonance method, Nelson et al. (1985) determined an asymmetric cyclic structure for $(NH_3)_2$.

The ammonia matrix system has been re-examined in our laboratory using FTIR to perform detailed concentration and annealing studies (Süzer and Andrews 1987b). These experiments revealed a group of bands at 3401, 3311, 3242, and $1000 \ cm^{-1}$ which show concentration and annealing behavior intermediate between monomer and the next set of cluster bands. This behavior characterizes the dimer species and identifies the dimer bands labeled D in fig. 9. Noteworthy is the fact that these bands exhibited ^{15}N shifts similar to the monomer fundamentals in the same region; this indicates that these dimer modes retain their monomer normal-mode character in the dimer.

Figures 9a and 9b show spectra of NH_3 and $(NH_3)_2$ in solid argon at 5 K. The very strong monomer band labeled v_2 and strong monomer lines labeled v_3 and v_1 are $R(0)$ absorptions from the ground rotational state; the much weaker R, Q and P bands in the v_2 region from higher rotational levels make only a minor contribution to the spectrum. Since NH_3 is rotationally relaxed in solid argon at 5 K, $(NH_3)_2$ must surely be relaxed. This contrasts $(NH_3)_2$ in a nozzle beam where rotational emissions from $J' = 2$ and $J' = 1$ states have been observed (Nelson et al. 1985). The spectrum in fig. 9c was recorded at 12 K; note the increased intensity of the R, Q and P monomer absorptions and the decreased intensity of monomer v_3 and v_1 bands that are reversed on recooling to 5 K as shown in fig. 9d. It is important to note that no temperature-dependent changes were observed for the D bands, thus supporting rotational relaxation for $(NH_3)_2$ in solid argon.

The observation of three dimer bands in the N–H stretching region has important structural implications for $(NH_3)_2$. The D band in the v_3 monomer region is red-shifted $46 \ cm^{-1}$ from monomer, and two D bands in the monomer v_1 region are shifted 34 and $103 \ cm^{-1}$, respectively, from the monomer fundamental. This FTIR spectrum indicates that the dimer involves one NH_2 group (the 3401 and $3311 \ cm^{-1}$ bands) and one N–H group (the $3242 \ cm^{-1}$ band) in hydrogen bonding. The observation of three distinctly different N–H stretching modes for the dimer infers sufficient structural rigidity to give distinct minima in the intermolecular potential function and allow three non-linear hydrogen bonding interactions thus perturbing and intensifying three N–H stretching modes. The dimer structure deduced from the matrix infrared spectrum is in disagreement with classical one-bond structures deduced from theoretical calculations (Frisch et al. 1986, Liu et al. 1986), and is in agreement with the model based on the microwave spectrum (Nelson et al. 1985).

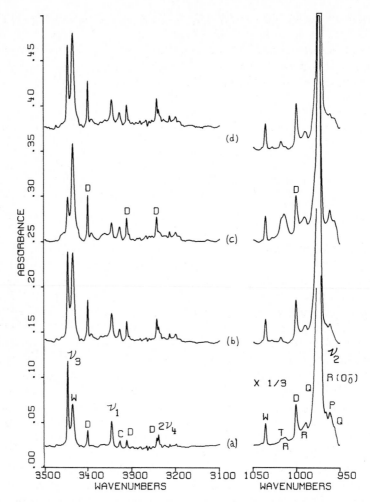

Fig. 9. Infrared spectra of NH_3 in solid argon deposited at 5 K. (a) $Ar:NH_3 = 500:1$ sample deposited for 6 h at 2 mmol/h, (b) $Ar:NH_3 = 300:1$ sample deposited for 3 h at 2 mmol/h, (c) spectrum of same 300:1 sample recorded after warming to 12 K, and (d) spectrum after recooling same 300:1 sample back to 5 K and holding for 20 min. Ordinate scale must be multiplied by 3 for the symmetric bending region. (From Süzer and Andrews 1987b.)

5.3. Ozone complexes and photochemistry

Ozone complexes and their photochemistry cover the whole range of chemistry from organic $(C_2H_4 \cdots O_3)$ to inorganic $(PH_3 \cdots O_3)$ to physical $(O_3 \cdots HF)$ and provide a synthetic source of new species as well as subjects for understanding weak intermolecular interactions. In the case of $C_2H_4 \cdots O_3$, the complex was evidenced by the photolysis products acetaldehyde and vinyl alcohol (Hawkins and Andrews 1983) whereas for $O_3\cdots HF$, three perturbed ozone and three perturbed HF motions were observed (Andrews et al. 1988). Table 6 contrasts the spectra of ozone and several ozone complexes; it is readily seen that HF perturbs ozone more strongly than phosphine and that the iodide$\cdots$ozone complexes give relatively large shifts in the ozone fundamentals.

Table 6
Vibrational fundamentals (cm^{-1}) of ozone and the ozone submolecule in complexes in solid argon.

O_3	Mode	$O_3\cdots HF$	$PH_3\cdots O_3$	$ClI\cdots O_3$	$FCl\cdots O_3$	$CH_3I\cdots O_3$	$CF_3I\cdots O_3$
1103.7	v_1	1115.7	1102.4	1110.8	–	1102.8	1106
1039.9	v_3	1026.7	1037.2	1019.5	1026	–	1032
703.6	v_2	713.2	705.2	–	–	704.7	–

The structure of these ozone complexes is symmetric (apex oxygen attachment) or asymmetric (terminal oxygen attachment). This is clearly demonstrated in fig. 10 for the $PH_3\cdots O_3$ complex with scrambled isotopic ozone. Each strong isotopic ozone doublet is accompanied by a sharp satellite band (indicated by an arrow) (Withnall et al. 1986). The important point here is that the $1:2:1:1:2:1$ relative intensity sextet matches the intensity distribution of the isotopic ozone bands, which indicates that the ozone submolecule in the complex has the same symmetry as ozone itself. In other words, the terminal oxygen atoms are equivalent giving added statistical weight in the asymmetric mixed isotopic species, and the $PH_3\cdots O_3$ complex <u>14</u> (scheme 7) binds at the central oxygen position. This is clearly not the case for $CF_3I\cdots O_3$, which gives a mixed isotopic octet for v_3 with equal intensity components (Andrews et al. 1985) indicating terminal oxygen attachment in the $CF_3I\cdots O_3$ complex <u>15</u>. The $O_3\cdots HF$ complex <u>16</u> exhibits a mixed isotopic octet for v_1 and substantial intensity enhancement in the v_1 ozone submolecule mode (Andrews et al. 1988), which demonstrates HF interaction with a terminal oxygen atom.

Photochemistry of the ozone complexes varies with the partner submolecule. The $PH_3\cdots O_3$, $P_4\cdots O_3$, $PCl_3\cdots O_3$, and $AsH_3\cdots O_3$ complexes were photolysed with high cross-section by red light, which does not dissociate isolated ozone in solid argon (Withnall et al. 1986, Andrews and Withnall 1988, Moores and Andrews 1989, Andrews et al. 1989). It is suggested that bonding

Scheme 7.

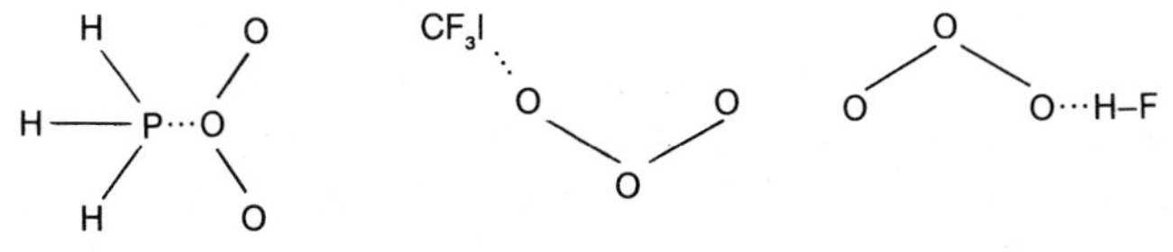

| 14 | 15 | 16 |

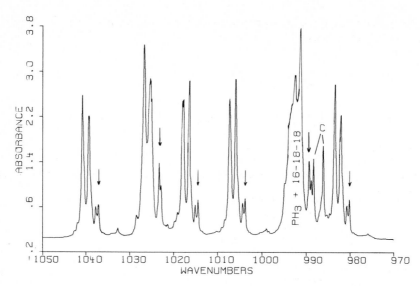

Fig. 10. Infrared spectrum for $Ar:PH_3 = 300:1$ and $Ar:^{16,18}O_3 = 200:1$ (50% ^{18}O enriched) samples co-deposited at 16 K. The PH_3 submolecule bands are labeled C and the ozone submolecule bands in the product complex are indicated by arrows. (From Withnall et al. 1986.)

interaction between these phosphorus compounds and oxygen from ozone facilitates dissociation. On the other hand, ultraviolet light in the strong ozone absorption was required for $SiH_4 \cdots O_3$ dissociation (Withnall et al. 1986, Withnall and Andrews 1985). In either case, the net effect is O-atom transfer to give a new molecular species such as FClO (Andrews et al. 1974), ClIO (Hawkins et al. 1984), CH_3IO (Hawkins and Andrews 1985), CF_3IO (Andrews et al. 1985), H_2SiO (Withnall and Andrews 1985), H_3PO (Withnall and Andrews 1987), P_4O (Andrews and Withnall 1988), and H_2NOH (Withnall and Andrews 1988).

The second-row hydrides PH_3 and SiH_4 also provided evidence for extensive reaction with the O_2 byproduct to give new species with three oxygen atoms such as O_2POH and $OSi(OH)_2$. The $PH_3 \cdots O_3$ photochemistry was extensive; three primary products were characterized – H_3PO, H_2POH and HPO. The latter reacted with O_2 upon photoexcitation to give three different HPO_3 species (Withnall and Andrews 1987). In contrast, the $NH_3 \cdots O_3$ complex photolysed to give only a single product, hydroxylamine (Withnall and Andrews 1988). The importance of the expanded valence chemistry of phosphorus is clearly indicated by the matrix photochemistry.

Acknowledgments

The author gratefully acknowledges financial support from the US National Science Foundation for the research described in this chapter and the valuable contributions of associates whose work is described in more detail in many of the following references.

References

Abouaf-Marguin, L., M.E. Jacox and D.E. Milligan, 1977, J. Mol. Spectrosc. **67**, 34.

Andrews, L., 1968, J. Am. Chem. Soc. **90**, 7368.

Andrews, L., 1969, J. Chem. Phys. **50**, 4288.

Andrews, L., 1979, Ann. Rev. Phys. Chem. **30**, 79.

Andrews, L., 1983a, in: Molecular Ions, eds T.A. Miller and V.E. Bondybey (North-Holland, Amsterdam) p. 91.

Andrews, L., 1983b, J. Mol. Struct. **100**, 281.

Andrews, L., 1984, J. Phys. Chem. **88**, 2940.

Andrews, L., and T.A. Blankenship, 1981, J. Am. Chem. Soc. **103**, 5977.

Andrews, L., and R.D. Hunt, 1987, unpublished results.

Andrews, L., and G.L. Johnson, 1984, J. Phys. Chem. **88**, 425.

Andrews, L., and R. Lascola, 1987, J. Am. Chem. Soc. **109**, 6243.

Andrews, L., and R. Withnall, 1988, J. Am. Chem. Soc. **110**, 5606.

Andrews, L., F.K. Chi and A. Arkell, 1974, J. Am. Chem. Soc. **96**, 1997.

Andrews, L., J.M. Grzybowski and R.O. Allen, 1975, J. Phys. Chem. **79**, 904.

Andrews, L., B.J. Kelsall and T.A. Blankenship, 1982a, J. Phys. Chem. **86**, 2916.

Andrews, L., G.L. Johnson and B.J. Kelsall, 1982b, J. Phys. Chem. **86**, 3374.

Andrews, L., J.M. Dyke, N. Jonathan, N. Keddar and A. Morris, 1984a, J. Am. Chem. Soc. **106**, 299.

Andrews, L., V.E. Bondybey and J.H. English, 1984b, J. Chem. Phys. **81**, 3452.

Andrews, L., M. Hawkins and R. Withnall, 1985, Inorg. Chem. **24**, 4234.

Andrews, L., S.R. Davis and G.L. Johnson, 1986, J. Phys. Chem. **90**, 4273.

Andrews, L., R. Withnall and R.D. Hunt, 1988, J. Phys. Chem. **92**, 78.

Andrews, L., R. Withnall and B.W. Moores, 1989, J. Phys. Chem. **93** (to be published).

Ault, B.S., 1978, J. Phys. Chem. **82**, 844.

Ault, B.S., 1988, Rev. Chem. Intermed. **9**, 233.

Azman, A., A. Ocvirk, D. Hadzi, P.A. Giguere and M. Schneider, 1967, Can. J. Chem. **45**, 1347.

Barnes, A.J., 1983, J. Mol. Struct. **100**, 259.

Barnes, A.J., K. Szczepaniak and W.J. Orville-Thomas, 1980, J. Mol. Struct. **59**, 39.

Clark, P.A., F. Brogli and E. Heilbronner, 1972, Helv. Chem. Acta **55**, 1415.

Clements, V.A., P.R.R. Langridge-Smith and B.J. Howard, 1981, unpublished results.

Davis, S.R., and L. Andrews, 1987, J. Chem. Phys. **86**, 3765.

Dawson, P., M.M. Hargreave and G.R. Wilkinson, 1975, Spectrochim. Acta A **31**, 1055.

Dows, D.A., 1959, J. Chem. Phys. **31**, 1637.

Eland, J.H.D., and C.J. Danby, 1968, Z. Naturforsch. a **23**, 355.

Fredin, L., B. Nelander and G. Ribbegard, 1976, Chem. Phys. **12**, 153.

Frisch, M.J., J.A. Pople and J.E. DelBene, 1983, J. Chem. Phys. **76**, 4063.

Frisch, M.J., J.E. DelBene, J.S. Binkley and H.F. Schaefer III, 1986, J. Chem. Phys. **84**, 2279.

Harmon, K.M., and I. Gennick, 1977, J. Mol. Struct. **38**, 97.

Hawkins, M., and L. Andrews, 1983, J. Am. Chem. Soc. **105**, 2523.

Hawkins, M., and L. Andrews, 1985, Inorg. Chem. **24**, 3285.

Hawkins, M., L. Andrews, A.J. Downs and D.J. Drury, 1984, J. Am. Chem. Soc. **106**, 3076.

Howard Jr, W.F., and L. Andrews, 1973, J. Am. Chem. Soc. **95**, 3045.

Hunt, R.D., and L. Andrews, 1985, J. Chem. Phys. **82**, 4442.

Hunt, R.D., and L. Andrews, 1987a, J. Chem. Phys. **86**, 3781.

Hunt, R.D., and L. Andrews, 1987b, J. Chem. Phys. **87**, 6819.

Ibers, J.A., 1964, J. Chem. Phys. **41**, 25.

Jacox, M.E., 1978, Rev. Chem. Intermed. **2**, 1.
Jacox, M.E., and D.E. Milligan, 1971, J. Chem. Phys. **54**, 3935.
Janssen, C.L., W.D. Allen, H.F. Schaefer III and J.M. Bowman, 1987, Chem. Phys. Lett. **131**, 351.
Johnson, G.L., and L. Andrews, 1980, J. Am. Chem. Soc. **102**, 5736.
Johnson, G.L., and L. Andrews, 1982, J. Am. Chem. Soc. **104**, 3043.
Jucks, K.W., and R.E. Miller, 1987, J. Chem. Phys. **87**, 5629.
Kasai, P.H., and D. McLeod Jr, 1969, J. Chem. Phys. **51**, 1250.
Kawaguchi, K., and E. Hirota, 1986, J. Chem. Phys. **84**, 2953.
Kawaguchi, K., and E. Hirota, 1987, J. Chem. Phys. **87**, 6838.
Kelsall, B.J., and L. Andrews, 1981, J. Phys. Chem. **85**, 2938.
Kelsall, B.J., and L. Andrews, 1982, J. Chem. Phys. **76**, 5005.
Kim, M.S., and R.C. Dunbar, 1980, J. Chem. Phys. **72**, 4405.
Kurnig, I.J., M.M. Szczesniak and S. Scheiner, 1986, J. Phys. Chem. **90**, 4253.
Larson, J.W., and T.B. McMahon, 1982, J. Am. Chem. Soc. **104**, 5848.
Lascola, R., R. Withnall and L. Andrews, 1988, J. Phys. Chem. **92**, 2145.
Liu, S., C.E. Dykstra, K. Kolenbrander and J.M. Lisy, 1986, J. Chem. Phys. **85**, 2077.
Lossing, F.P., 1972, Bull. Soc. Chim. Belges **81**, 125.
Lovejoy, C.M., D.D. Nelson Jr and D.J. Nesbitt, 1987, J. Chem. Phys. **87**, 5621.
Lucchese, R.R., and H.F. Schaefer III, 1975, J. Am. Chem. Soc. **97**, 7205.
Machara, N.P., and B.S. Ault, 1987, J. Phys. Chem. **91**, 2046.
Manceron, L., and L. Andrews, 1985a, J. Am. Chem. Soc. **107**, 563.
Manceron, L., and L. Andrews, 1985b, J. Phys. Chem. **89**, 4094.
Manceron, L., and L. Andrews, 1986, J. Phys. Chem. **90**, 4514.
Manceron, L., and L. Andrews, 1988, J. Am. Chem. Soc. **110**, 3840.
McDonald, S.A., and L. Andrews, 1979, J. Chem. Phys. **70**, 3134.
McDonald, S.A., G.L. Johnson, B.W. Keelan and L. Andrews, 1980, J. Am. Chem. Soc. **102**, 2892.
Milligan, D.E., and M.E. Jacox, 1968, J. Chem. Phys. **48**, 2265.
Milligan, D.E., and M.E. Jacox, 1969, J. Chem. Phys. **51**, 1952.
Milligan, D.E., and M.E. Jacox, 1973, J. Mol. Spectrosc. **46**, 460.
Mitra, S.S., and H.J. Bernstein, 1959, Can. J. Chem. **37**, 553.
Moores, B.W., and L. Andrews, 1989, J. Phys. Chem. **93** (to be published).
Nelson Jr, D.D., G.T. Fraser and W. Klemperer, 1985, J. Chem. Phys. **83**, 6201.
Nguyen, M.T., 1988, J. Phys. Chem. **92**, 1426.
Nishiya, T., N. Hirota, H. Shinohara and N. Nishi, 1985, J. Phys. Chem. **89**, 2260.
Owrutsky, J.C., N.H. Rosenbaum, L.M. Tack and R.J. Saykally, 1986, J. Chem. Phys. **84**, 5308.
Pimentel, G.C., M.D. Bulanin and M. Van Thiel, 1962, J. Chem. Phys. **36**, 500.
Pouchan, C., 1987, Chem. Phys. **111**, 87.
Prochaska, F.T., and L. Andrews, 1977, J. Chem. Phys. **67**, 1091.
Prochaska, F.T., and L. Andrews, 1978a, J. Chem. Phys. **68**, 5568.
Prochaska, F.T., and L. Andrews, 1978b, J. Chem. Phys. **68**, 5577.
Prochaska, F.T., and L. Andrews, 1978c, J. Am. Chem. Soc. **100**, 2102.
Read, W.G., and W.H. Flygare, 1982, J. Chem. Phys. **76**, 2238.
Shida, T., and S. Iwato, 1973, J. Am. Chem. Soc. **95**, 3473.
Süzer, S., and L. Andrews, 1987a, J. Am. Chem. Soc. **109**, 300.
Süzer, S., and L. Andrews, 1987b, J. Chem. Phys. **87**, 5131.
Süzer, S., and L. Andrews, 1988, J. Chem. Phys. **88**, 916.
Thomas, R.K., 1971, Proc. R. Soc. London A **325**, 133.
Thomas, R.K., 1980, unpublished results.
Wight, C.A., B.S. Ault and L. Andrews, 1976, J. Chem. Phys. **65**, 1244.
Withnall, R., and L. Andrews, 1985, J. Phys. Chem. **89**, 3261.
Withnall, R., and L. Andrews, 1987, J. Phys. Chem. **91**, 784.
Withnall, R., and L. Andrews, 1988, J. Phys. Chem. **92**, 2155.
Withnall, R., M. Hawkins and L. Andrews, 1986, J. Phys. Chem. **90**, 575.

CHAPTER 3

Matrix-Isolated Metal Clusters

M. MOSKOVITS

Department of Chemistry, and
Ontario Laser and Lightwave Research Centre
University of Toronto
Toronto, M5S 1A1
Canada

Chemistry and Physics of
Matrix-Isolated Species
L. Andrews and M. Moskovits, eds

Contents

1. Introduction

The study of clusters has become a minor industry. Matrix isolation, one of the first techniques used to study the spectra of clusters, has been joined by a large variety of gas-phase techniques aimed at determining the physical and chemical properties of clusters.

Several good reviews and monographs already exist outlining recent developments in the field of clusters (Jena et al. 1986, Bennemann and Koutecky 1985, Smith 1986, Träger and zu Putlitz 1986, Moskovits 1986, Morse 1986, Weltner and Van Zee 1984, Salahub 1987, Castleman and Keesee 1986, Olsen and Klabunde 1987, Gallezot 1986, Benedek et al. 1987). These references lean heavily towards metal and semiconductor clusters, a bias that will be reflected in this chapter and that should not be interpreted as a negative judgement of the work done with non-metallic clusters but rather as a reflection of the author's own interest.

The study of clusters proceeds from the desire to see how macroscopic properties develop in an atomic or molecular aggregate as its size increases progressively from the atom up. Often the question is asked "at what size does a cluster become macroscopic?". It is now understood that there is often no sharp transition to bulk properties, and that even that fuzzy interface occurs in a different size range according to the property investigated.

Some properties, such as the ionization potential, can be understood in a general way (Kappes et al. 1986) as a bulk phenomenon down almost to the diatomic, however, with deviations about the "bulk" value that offer important clues to the structure of the cluster. Other properties such as an anomalous crystal structure (Yokozeki and Stein 1978, Farges et al. 1986) can persist to crystallites containing thousands of atoms. The chemistry of an aggregate can also show some fascinating trends as a function of cluster size. Hydrogen chemisorption, for example, approaches bulk behaviour with clusters of 30–50 atoms, prior to which the chemisorption properties depend dramatically on cluster size: some sizes showing unusual affinity towards hydrogen, while others are essentially inert towards its chemisorption (Richtsmeier et al. 1985, Whetten et al. 1985, Geusic et al. 1985).

Matrix isolation has been used to study the spectroscopy and reactivity of atoms and clusters for several years. Early studies used UV visible absorption spectroscopy to detect electronic transitions in atoms and dimers (table 1) and

Table 1
UV-visible absorptions of matrix-isolated transition-metal polyatomic molecules (adapted from Morse 1986).

Molecule	Transitions observed (nm)	Matrix	Ref.*
Cr_3	477	Ar	[1, 2]
	479	Kr	[1]
Mn_x	253, 312, 400	Ar	[3]
	254, 317, 410	Kr	[3]
Mn_y	329, 345	Ar	[3]
	332, 336, 349	Kr	[3]
Co_3	287, 316	Ar	[4]
Ni_3	420, 480 ($\nu_{00} = 20\,820$ cm^{-1}, $\Delta\bar{G} = 202$ cm^{-1})	Ar	[5]
Cu_3	221, 230, 234, 262, 376, 500	Ar	[6]
	227, 260, 280, 335, 404, 533	CH_4	[6]
Cu_4	402	Ar	[6]
	272, 424	CH_4	[6]
Mo_3	534	Ar	[1]
	538	Kr	[1]
Rh_3	490	Ar	[4]
Ag_3	245, 440	Ar	[2, 7, 8]
	226, 235/240, 400	Ar	[9]
Ag_3	244, 260, 440	Ar	[9]
Ag_3	222, 232, 246, 422	Kr	[3, 10]
	233, 247, 420	Kr	[9]
Ag_3	445	Kr	[9]
Ag_3	242, 260, 422/440	Xe	[9]
Ag_3	474	Xe	[9]
Ag_4	213/217, 340/345	Ar	[9]
	273, 283, 347, 363, 426, 490	Ar	[7]
	222, 348	Kr	[9]
	286, 524	Kr	[3, 10]
	229, 360	Xe	[9]
Ag_5	333, 370, 396, 505	Ar	[7]
	349, 476	Kr	[3, 10]
Ag_6	340, 520	Ar	[7]
	366, 509	Kr	[3, 10]
Ag_7	536	Ar	[7]
Au_3	292, 471	Ar	[11]
Cr_2Mo	494	Ar	[1]
	496	Kr	[1]
$CrMo_2$	498	Ar	[1]
	496	Kr	[1]
$CrAg_2$	276	Ar	[2]
Ag_xMn_y	492, 502, 550, 592, 604	Kr	[3]

*References:

[1] Klotzbücher and Ozin (1978); [2] Ozin and Klotzbücher (1979); [3] Klotzbücher and Ozin (1980); [4] Ozin and Hanlan (1979); [5] Moskovits and Hulse (1977a); [6] Moskovits and Hulse (1977b); [7] Ozin and Huber (1978); [8] Mitchell and Ozin (1978); [9] Mitchell and Ozin (1984); [10] Schulze et al. (1978); [11] Klotzbücher and Ozin (1980).

IR spectroscopy to study vibrations of molecules bonded to them. Recent studies have employed a wide variety of spectroscopic techniques such as ESR spectroscopy, laser-induced luminescence spectroscopy, Raman spectroscopy, magnetic circular dichroism, Mössbauer spectroscopy, UV spectroscopy and photoemission spectroscopy using synchrotron sources, EXAFS and XANES and FTIR. In addition, some NMR and SIMS work has also been reported.

This chapter will emphasize the determination of ground-state geometry and vibrational structure of metal and semiconductor clusters and diatoms on the basis of matrix spectroscopy. This emphasis automatically limits the discussion to a sizeable but not overly large group of species for which this information is available with some confidence from among the very large number of species studied by matrix spectroscopy.

2. Shapes of clusters

A major goal of the study of clusters is the determination of the geometry of a cluster. Very few cluster geometries have been determined experimentally to date beyond the triatomics. This contrasts with the situation that exists with theory where the geometries of many systems have been determined (Salahub 1987, Koutecky and Fantucci 1986, Rao and Jena 1985, Martins et al. 1985). Here, except for clusters of the alkalis (Koutecky and Fantucci 1986, Rao and Jena 1985) where considerable agreement exists regarding their shapes, there is still a great deal of controversy regarding the most stable geometries of most clusters.

The predicted shapes of clusters of lithium are shown in fig. 1. There is a propensity for these clusters to be planar up to the septamer. Moreover, there is no tendency to adopt the bcc structure of the bulk solid even for fairly large clusters. Work both in matrices and in the gas phase tends to support the prediction that the alkali trimer is a bent triatomic, Jahn–Teller distorted to a larger apical angle from a D_{3h} parent structure. Matrix ESR results (Lindsay et al. 1982, Garland and Lindsay 1983a,b, 1984, Howard et al. 1984, Lindsay and Thompson 1982) indicate that in all alkali triatomics the barrier to pseudo-rotation is so small that they are all dynamic Jahn–Teller molecules. This was also found to be the case in a gas-phase study of Na_3 in one of its excited states (Delacrètaz et al. 1986). A resonance Raman study of matrix-isolated Li_3 (Moskovits et al. 1984) produced a spectrum that was also consistent with a pseudo-rotating Li_3 in one matrix site. In another site, however, the spectrum was interpreted based on isotopic fine structure to be that of a rigid triatomic with an apical angle less than 60°. Likewise there is some matrix evidence (Moskovits et al. 1984) for a rhomboid-shaped Li_4, a structure favoured by calculation (Koutecky and Fantucci 1985, Rao and Jena 1985) and most consistent with the resonance Raman results. This shape, too, is due to a

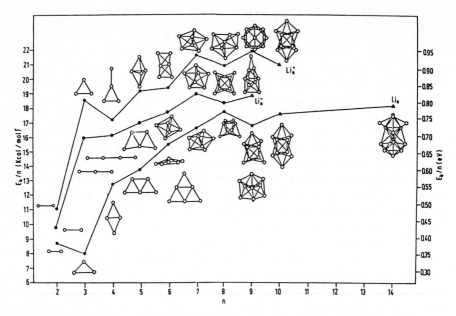

Fig. 1. Binding energy per atom of Li_n, Li_n^+ and Li_n^- as a function of n. Topologies of the SCF energy optimized clusters are shown. Used with permission from Koutecky et al. (1987).

Jahn–Teller distortion from a square planar parent molecule; but in this case the barrier to interconversion between the two equivalent rhomboidal shapes is large.

The structures of Cu_3 and Ag_3 have also been determined. The matrix Raman of Ag_3 (Schulze et al. 1978), in fact, was the first matrix result reported for a cluster although its structure was not unequivocally determined on that basis. Cu_3, Ag_3, Au_3 and Cu_2Ag (Howard et al. 1983b) were studied by matrix ESR in solid adamantane. In all cases, the molecules were determined to be bent triatomics with apical angles greater than 60°. If this is interpreted to be a result of a static Jahn–Teller distortion from a parent D_{3h} molecule, then the "obtuse" form implies a $^2B_2$ ground state. This contrasts with the findings of Kernisant et al. (1985) who report Ag_3 to be an acute angle triatomic (ground state) in solid N_2. For Cu_3, a matrix resonance Raman spectrum was also reported (DiLella et al. 1983). The prominent feature of this spectrum was a 350 cm^{-1} repeating interval identified with the symmetric stretching frequency of a dynamic Jahn–Teller Cu_3. In a gas-phase study by Morse et al. (1983), twelve vibronic lines were observed for jet-cooled Cu_3 with an origin at 5397 Å. With the exception of three hot bands, the observed spectrum provided information regarding an excited state of Cu_3, ultimately identified with a $^2E''$ state, that is Jahn–Teller unstable. Parameters associated with this state were extracted by

Morse et al. (1983), and revised by Thompson et al. (1985) after correcting an error in the Hamiltonian used by the former group.

A breakthrough in the study of Cu_3 came with the dispersed fluorescence study by Rohlfing and Valentini (1986) from which ground vibronic levels of Cu_3 were determined. Analyses by Truhlar et al. (1986) and by Zwanziger et al. (1986) assigned this spectrum to a pseudo-rotating triatomic with a Jahn–Teller stabilization energy of 221 and 305 cm^{-1}, respectively, and a barrier to pseudo-rotation of 95 and 111 cm^{-1}. Both groups assigned a 270 cm^{-1} band that was observed in the spectrum but that could not be reproduced by the quadratic Jahn–Teller calculation, to the symmetric vibration. A spin-off of this study was a reassignment by Morse (1987) of the excited state. "Extra" bands observed in a fluorescence spectrum obtained from an excited vibronic state of the excited-state manifold (Rohlfing and Valentini 1986) were best understood if the excited state were a $^2A'$ rather than a $^2E''$ state. This is also consistent with a calculation of Walch and Laskowski (1986) who predicted other parameters of the ground-state surface of Cu_3 with good accuracy. Other theoretical predictions for Cu_3 have not been as good. An extended Hückel calculation (Anderson 1978) and an SCF calculation with no CI (Bachmann et al. 1980) predicted a linear Cu_3. Most other calculations did indeed determine Cu_3 to be a D_{3h} trimer subject to Jahn–Teller distortion, but with varying barriers and rotational constants (Miyoshi et al. 1983, Richtsmeier et al. 1980).

The rich, and apparently contradictory, story of Cu_3 can be summarized as follows. In the gas phase, the molecule has a dynamic Jahn–Teller $^2E'$ ground state with a symmetric vibrational frequency of ~ 270 cm^{-1}. In the matrix the molecule may adopt an obtuse (Howard et al. 1983b) or acute rigid geometry, according to the matrix, or remain a dynamic molecule (in an argon matrix) (DiLella et al. 1983). The 350 cm^{-1} frequency observed in the latter environment is almost certainly not the symmetric frequency.

The examples of Cu_3 and Ag_3 indicate that the matrix can have a significant effect on the structures of molecules whose ground-state potential-energy surfaces are characterized by low potential barriers or shallow potential wells. This is not a surprising conclusion but one often forgotten by the matrix community.

The structures and vibrational constants of several other triatomics have been determined in rare-gas matrices, among them Sc_3, Cr_3 (Moskovits et al. 1980a, Knight et al. 1983, DiLella et al. 1982) and Ni_3 (Moskovits and DiLella 1980a); the first of which was found to have D_{3h} geometry; the next two C_{2v}.

Vibrational frequencies attributed to Cr_3 were also reported on the basis of matrix FT-IR experiments (Ozin et al. 1983). The conclusion that the molecule exists in a large number of forms in the matrix leads one to suspect this result. Likewise matrix FT-IR absorptions attributed to Fe_3 and Ni_3 (Nour et al. 1987) must be looked upon critically in view of the presence of extra bands attributed to impurities equally as intense as the assigned bands. Far IR studies have also been reported for Ag/Na clusters (Pflibsen and Huffman 1985). Infrared absorptions in the range 121–210 cm^{-1} were tentatively assigned to Ag_nNa_m ($n, m \leqslant 2$). When larger particles were generated a single, broad line was observed at 178 cm^{-1} and assigned to AgNa phonons.

Mn_3 is also found to be an unusual triatomic (Bier et al. 1988). It is a D_{3h} molecule with very weak Jahn–Teller coupling. In this limit, the vibronic eigen-energies depend on a quantum number l that results in $v_a + 1$ equally spaced levels for levels of the degenerate asymmetric vibration labeled with quantum number v_a. This is what is observed (fig. 2) yielding a Jahn–Teller parameter, $D\omega_a$, of 2.95 cm^{-1}.

Two tetratromics, Be_4 (Bondybey and English 1980a) and Sb_4 (Bondybey et al. 1981), have been reported in two elegant matrix studies that are among the earliest laser-induced fluorescence studies done in matrices. Both are tetrahedral. An ESR study of a chromium cluster tentatively assigned to Cr_4 or possibly Cr_5 has also been reported (Van Zee et al. 1985). If the species is Cr_4, then it has been speculated that it is either a distorted tetrahedron with C_{3v} symmetry or possibly Cr atom in the axial field of a Cr_3 molecule. If the molecule is Cr_5, then it is square pyramidal.

Among the pentatonics Mn_5 produces spectacular ESR spectra in solid Ar, Kr and Xe (Baumann et al. 1983, Van Zee et al. 1982). The molecule shows preferential orientation in the matrix with respect to the direction of metal

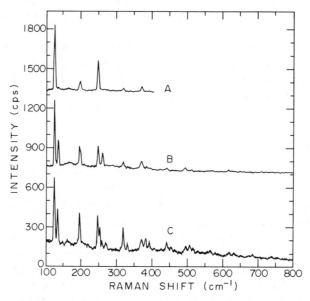

Fig. 2. Portions of the resonance Raman spectra of Mn_3 in solid Ar excited with (A) 14 550, (B) 14 750 and (C) 15 000 cm^{-1} dye laser radiation.

deposition. The spectra were consistent with either a plane-pentagonal Mn_5 species with five ferromagnetically coupled Mn atoms, or a square pyramid. The former, however, was favoured by these authors (Baumann et al. 1983, Van Zee et al. 1982).

ESR spectra attributed to Cu_5, Ag_5, $CuAg_4$ and Cu_2Ag_3 isolated in solid adamantane have been reported (Howard et al. 1983a,c, 1984). All have been interpreted as being distorted trigonal bipyramidal molecules in which two nuclei with high spin density occupy two of the equatorial positions. This structure is also one favoured in an ab initio LCAO-MO-SCF calculation (Bachmann et al. 1980).

Doubt has been cast on the proper assignment of the ESR spectrum attributed to Ag_5 by another ESR study (Bach et al. 1987) that has found additional lines belonging to the same spectrum. The revised interpretation is that the species is Ag_7. Although restricted to Ag_7, this report (Bach et al. 1987) casts doubt on all of the group 11 pentamers (Howard et al. 1983a).

The hexatomic C_6 has been found to be a linear molecule in a matrix ESR study (Van Zee et al. 1987) punctuating a long-standing debate regarding its structure. Having mentioned carbon, one cannot avoid C_{60} (Kroto et al. 1985) and the cluster $C_{60}La$ (Heath et al. 1985). These have been speculated to have the geometry of a soccer ball consisting of five- and six-sided carbon rings. The second molecule has a La atom nestled within the ball. Although there exists no direct evidence for this structure, a great deal of circumstantial clues (Yang et al. 1988) suggest that it is correct. Likewise there is some evidence that carbon clusters prefer linear chains up to C_9, and noncyclic rings in the range C_{10}–C_{29}.

Finally, a remarkable matrix ESR spectrum has been reported that has been attributed to Sc_{13} (Knight et al. 1983). The spectrum cannot be assigned to a unique geometry. Both icosahedral and cuboctahedral geometries as well as a truncated hexagonal bipyramid with D_{3h} structure are possible; although these authors prefer the icosahedral geometry.

3. Spectroscopy

In the earliest matrix cluster studies UV-visible absorption (Moskovits and Ozin 1976) spectroscopy was used to detect their presence. Bands belonging to clusters of increasing size were assigned on the basis of their relative intensities in matrices prepared with gradually increasing metal to matrix ratios. At low metal loading the matrix spectrum is dominated by absorptions due to atoms. With increasing metal/matrix ratio numerous bands appear belonging to several aggregates. Assignments are based either on the order of the sequential appearance of bands or more quantitatively, on the rate of increase of absorbance with total metal loading (Moskovits and Hulse 1977c). In both cases assignments, especially for higher clusters, can not be construed to be firm owing to the simplicity of the models used in these analyses and to the numerous band overlaps that exist in all but the simplest of cases. In addition, one often finds several bands belonging to a cluster of a given size that is isolated in different sites within the solid matrix with relative abundances that may differ according to the temperature and thermal history of the matrix. In some cases, one can convert clusters from one site to another reversibly by alternatively irradiating into the appropriate absorption (Balling et al. 1979). Such a reversible process has been observed for isolated Ag_3 and interpreted as photo-isomerization (Kettler et al. 1985, Bechthold et al. 1984).

With a few exceptions the matrix UV-visible absorption of metal clusters bands are broad and rather featureless. No definite structural information may be obtained from them. When vibrational fine structure is present as in the case of Mn_2, Fe_2 (DeVore et al. 1975), Ni_2 (Moskovits and Hulse 1977a) and Pt_2 (Jansson and Scullman 1976) it is, of course, characteristic of the excited state rather than of the ground state. The main thrust of UV-visible absorption spectroscopy is in demonstrating that a cluster is present in the matrix and in indicating at what frequency one must excite when laser-induced emission is attempted. Recently, matrix measurements on clusters have been extended to the vacuum ultraviolet (Kolb et al. 1984) by using a synchrotron source. In that frequency domain, interesting new questions regarding the nature of Rydberg states of metal atoms and clusters arise since the electron in those states will be delocalized into the region of the matrix atoms.

Some of the most refined work of this sort in the context of metal clusters and atoms has been done by Kolb and his group. In a series of papers they report the observation of facile energy transfer from excited atoms to the dimer, e.g., from Ag atom to Ag_2 (Wiggenhauser et al. 1988). They then elucidate the energy dissipation scheme for excited Ag atoms (Wiggenhauser et al. 1988), by carefully measuring the time evolution of the atom's fluorescence. Excitation of the ^2P state of Ag isolated in Ar, Kr and Xe, produced fluorescences ranging in lifetime between 5 and 60 ns (Wiggenhauser et al. 1988). Although the fluorescence lifetime varied more or less as λ^3 from band to band, it was constant within a given band, in contrast to what was claimed for Na (Balling et al. 1978).

A large body of work using matrix UV-visible spectroscopy has been assembled. Chief players in this field have been Ozin (Ozin and Mitchell 1983), Schulze (Schulze et al. 1978) and Kolb (1981). Much of this work has been reviewed by Morse (1986).

3.1. Magnetic circular dichroism (MCD)

Rather more information may be obtained from matrix UV-visible absorption spectra when it is combined with MCD. In this technique, the matrix is placed between the poles of a magnet and the light used to record the spectrum is passed through the sample coaxially with the magnetic field. Electronic states of the cluster which have non-zero spin or space angular moments will be shifted in energy and, in particular, degenerate states with non-zero spin will be split into spin–orbit terms. As the selection rule allowing transitions to these multiplets depends on the sense (right- or left-handed) of the circular polarization of the incident light, the spectrum recorded with right circularly polarized light will in those cases be shifted in frequency from that recorded with its left-handed complement. Hence, a difference spectrum $\Delta A = A_L - A_R$ (where L and R stand for right and left) will contain positive features, negative features, couplets and multiplets according to the nature of the states involved in the transition. With degenerate ground states one has the further effect of a temperature-dependent MCD spectrum as a result of the varying population of the spin–orbit states which have been split out of the ground state.

Just how powerful this technique may be in the hands of experts is illustrated by the work of Schatz (Miller et al. 1981), Vala (Rivoal et al. 1982) and Grinter (Grinter and Stern 1983). For example, Miller et al. (1981) have reported MCD spectra of Mg, Ca and Sr atoms and clusters. Based on the presence or absence of MCD couplets, unambiguous assignments of bands were at times possible. In the MCD spectrum of Mg_2, for example, weak bands can immediately be assigned to $^1\Sigma_g^+ \rightarrow {}^1\Sigma_u^+$ transitions, while intense couplets are assigned to $^1\Sigma_g^+ \rightarrow {}^2\Pi_u$ transitions. Information is not only obtained from the type of MCD spectral features observed but also from quantitative analysis of these spectral features. The powerful deductions based on these is illustrated by the analysis surrounding a band at $31\,700\ cm^{-1}$ assigned to Mg_3 (Miller et al. 1981). The MCD couplet observed is unambiguous evidence for a degenerate excited state; possible only for linear or D_{3h} geometries. This restriction together with arguments regarding the intensity of the MCD couplet relative to the integrated absorbance of the band, restricts the options further to D_{3h}.

The use of MCD coupled with varying sample temperature is illustrated by the work of Rivoal et al. (1982) on Mn_2. The MCD spectrum shows a striking temperature dependence prompting these authors to confirm the conclusion of Van Zee et al. (1981), and of Baumann et al. (1983) based on an ESR study, that Mn_2 is an antiferromagnetically coupled diatomic.

MCD [and the related technique of magnetic linear dichroism (MLD)] were key techniques in settling two important issues pertaining to matrix-isolated metal atoms. The first issue is the origin of the three-line spectrum associated with the Cu $^2P \leftarrow {}^2S$ transitions in solid rare gases (Shirk and Bass 1968). Although several explanations were suggested for the observation of three lines, rather than the two $^2P_{3/2} \leftarrow {}^2S_{1/2}$, $^2P_{1/2} \leftarrow {}^2S_{1/2}$ in the gas phase (Moore 1958), two were most plausible:

(1) the lifting of the degeneracy of the ^2P state by a static distortion of the matrix about the atom (Ruffolo et al. 1984, 1985), and

(2) a Jahn–Teller effect in the excited state due to coupling with non-cubic lattice vibrations (Belyava et al. 1973, Ozin 1980, Hulse 1979, Miller et al. 1981).

Based on a moments analysis, Vala and co-workers (Zeringue et al. 1983, Vala et al. 1984) showed that the MCD spectrum of matrix-isolated Cu is inconsistent with the static distortion explanation; simultaneous spin–orbit coupling and Jahn–Teller distortion, on the other hand, accounts for the observed results for Cu in solid Ar, Kr and Xe extremely well.

The second issue relates to the ground state of Ni atoms isolated in rare-gas matrices. The ground state of Ni in the gas-phase is $^3F_4(3d^84s^2)$; however, there is a state $^3D_3(3d^94s^1)$ only 205 cm^{-1} above this. Jacobi et al. (1980) proposed that in Ar matrices, these two states invert their order, making $^3D_3$ the ground state. Grinter and Stern (1983) concluded likewise on the basis of MCD. On the basis of further work, Barrett et al. (1984) proposed that in Ar, Kr and Xe matrices, Ni atoms reside in two sites: in one $^3D_3$, in the other $^3F_4$ being the ground state. In a compelling analysis based both on MCD and MLD, Vala et al. (1985) show that there are indeed two sites with two ground states and that on annealing the sites where the Ni ground state is $^3D_3$ tend to be favoured.

3.2. Mössbauer spectroscopy

Mössbauer spectroscopy is a form of γ-ray absorption spectroscopy applicable to atoms bound in solids, which absorb without recoiling and hence with little Doppler shift from the frequency of the emitter. The technique is restricted to atoms possessing nuclides with appropriate lifetimes. Chief among them are ^{57}Fe and ^{117}Sn; however, matrix studies with ^{151}Eu have also been reported (Montano 1982).

Mössbauer spectra yield three parameters from which useful information may be gleaned. These are the isomer shift, the electric quadrupole interaction and the magnetic hyperfine interaction.

The isomer shift (IS) is proportional to the difference in the electronic charge density at the nucleus of the absorber and the source, i.e.,

$$IS = \tfrac{2}{3}\pi Ze^2 \Delta(r^2) \left[|\psi(0)|_A^2 - |\psi(0)|_S^2 \right],$$

where $\Delta(r^2)$ is the change in the mean square radius of the nucleus on going from the ground to its excited state, and $|\psi(0)|_A^2$ and $|\psi(0)|_S^2$ are the electronic probability density at the nucleus for the absorber and source, respectively. Since only s-electrons penetrate the nucleus to any extent, the IS is a sensitive measure of the s-electron density.

The quadrupole shift is proportional to the nuclear electric quadrupole moment and the electric field gradient at the nucleus.

Using these parameters, McNab et al. (1971) showed, for example, that the effective electron configuration about an iron atom in Fe_2 is approximately $3d^64s^{1.47}$. The decrease from its single atom value of $3d^64s^2$ is attributed to the involvement of s-electrons in the Fe–Fe bond. Likewise, Montano (1980) has shown how the nuclear electric quadrupole splitting and magnetic hyperfine interaction may be used to narrow down the ground-state representation in Fe_2.

Using an external magnetic field he determined that the magnetic hyperfine interaction was large. This implies that the ground state of Fe_2 is a high-spin state. Likewise, the electric quadrupole splitting implies a negative field gradient eliminating Δ_g and Δ_u (Montano 1980) as possible candidates for the representation of the ground state. The expected magnitude of the e.q.s. for a Σ state is double that for a Π state. The observed magnitude was closer to that expected for Σ than for Π. Accordingly, a $^7\Sigma$ ground state was postulated.

While this remarkable bit of deduction might imply that Mössbauer spectroscopy is a powerful technique for determining the electronic structure of clusters, it is in fact, rather inconclusive in most cases much like absorption spectroscopy. (One should, in fact, consider even the Fe_2 result as not entirely conclusive.) Nevertheless, it has been used to detect a large number of species of the form FeM with M = Cr, Mn, Co, Ni, Cu or Pt (Dyson and Montano 1978). With ^{151}Eu the diatomic (Eu_2) and multimers, possibly Eu_3–Eu_5 have been observed using Mössbauer spectroscopy (Montano 1982). Clusters such as Fe_3, Fe_3Pt, Fe_2Cr, Fe_2Cr_2 and $FeCr_3$ (Dyson and Montano 1979, 1980, Shamai et al. 1982) have also been reported. In the case of these and other clusters the assignments must be considered tentative. Mössbauer has also been used to study tin (Bos and Howe 1974) and antimony (Nagarathna et al. 1982) clusters.

3.3. EXAFS and XANES

Two very promising related techniques for cluster-structure determination are extended X-ray absorption fine structure (EXAFS) and X-ray absorption near edge structure (XANES). In them the low-intensity interference structure observed near X-ray absorption edges is analyzed to yield an atom–atom correlation function from which atom–atom distances are determined. Using this technique the internuclear separation in Fe_2 was determined to be 1.87 ± 0.13 Å in argon and 2.02 ± 0.02 Å in Ne matrices (Montano and Shenoy 1980, Purdum et al. 1982). With matrices richer in Fe, iron–iron distances intermediate between those of Fe_2 and bulk iron were determined but for clusters of unknown nuclearity.

EXAFS has also been used to follow the progress with size of the lattice parameter in small silver microcrystallites made by the gas aggregation technique (Montano et al. 1984). Remarkably good EXAFS spectra were obtained from which it was determined that the lattice parameter decreased with decreasing cluster size, for crystallites ranging in size between 20 and 100 Å. The results were adequately explained in terms of striction due to surface tension.

3.4. Electron spin resonance (ESR) spectroscopy

ESR has been perhaps the most powerful technique in determining the geometrical and electronic structures of matrix-isolated clusters. Matrices are deposited on a cooled sapphire or metal rod placed between the poles of a magnet. Electron spin resonances are then detected by sweeping the magnetic field in order to achieve resonance between a microwave frequency applied to the sample simultaneously with the magnetic field and the energy interval between Zeeman levels split out of degenerate spin–orbit states by the magnetic field.

The measurements yield two major parameters, the g-tensor and the a-tensor. Both of these are isotropic in fluid systems where orientation is random. In matrices anisotropic g- and a-tensors are often encountered. The g-values play roughly the role of the chemical shift in NMR, determining the magnetic field strength at which the "center" of the ESR spectrum occurs. It may, therefore, be used as a rough chemical indicator. For cluster studies, it is the a-tensor which is the most revealing. It measures the magnitude of electron-spin–nuclear-spin hyperfine interactions. The size of the a-value is proportional to the Fermi contact term $|\psi(0)|^2$ mentioned earlier in the context of Mössbauer spectroscopy. This term is a measure of the "spin density" of the unpaired electrons "upon" a given atom. Since the hyperfine interaction splits the ESR spectrum into multiplets, one may use the information as an indication of the number of atoms in the cluster (from the number of hyperfine lines) and as a clue to geometry (from the spin density distribution among the atoms in a cluster).

ESR does have some drawbacks. It is, of course, only sensitive to paramagnetic species. So, for example, it will not detect molecules such as Na_2, Na_4, Na_6, etc. Depending on which species one is seeking, this limitation may be beneficial in that it simplifies the spectrum of a matrix containing a range of clusters. It is also necessary, occasionally, to work with isotopically pure metals in cases where the naturally occurring element has several isotopes each with different nuclear spin. The jumble of lines that would occur otherwise would make interpretation difficult. Finally, it is not always as easy to interpret ESR spectra as has been implied above. Occasionally, more than one size cluster may result in the same number of ESR lines. With very large clusters, one may not be sure that one has observed all the lines in a hyperfine set as illustrated previously in the case of Ag_5 (Howard et al. 1983a). Other complications in interpretation might also arise as a result of superhyperfine interactions with nuclear spins belonging to the matrix host atoms, large g- and a-tensor anisotropies, quadrupole terms and large zero-field splittings.

Occasionally, the non-observance of an ESR spectrum of a cluster whose presence in the matrix is indicated by optical data may be a telling result. In the most banal case it indicates a non-paramagnetic ground state. In certain other cases, e.g. for La_3 (Knight et al. 1983) it may indicate that the molecule is linear with a ground state of non-zero, orbital angular momentum leading to a highly broadened ESR spectrum due to the large g-tensor anisotropy.

The non-observance of Cu_3 in solid argon by ESR is somewhat more puzzling. It may indicate that in that medium the molecule is a dynamic Jahn–Teller molecule, this, too, leading to extreme broadening, in contrast with the result in the adamantane matrix (Howard et al. 1983b). Something akin to this has been observed with pseudo-rotating Na_3 (Lindsay et al. 1982) in whose ESR spectrum some hyperfine lines are broadened to the point of disappearance while others continue to be observed.

A recent triumph of matrix ESR spectroscopy is the determination of the linear structure of C_6 (Van Zee et al. 1987).

3.5. Fluorescence, Raman and resonance Raman

Two techniques that have been successful in providing structural information on matrix-isolated metal clusters are laser-induced fluorescence and resonance Raman. For diatomics, these methods can yield vibrational constants; while for polyatomics, structural information is possible either of a general sort as in cases when more than one fundamental is observed, or of a more precise variety when isotopic fine structure can be resolved in the vibrational lines. Isotopic fine structure, when it is present, is also of great use in confirming or at least corroborating the assignment of a fluorescence or resonance Raman progression.

The technique is simple. The matrix is irradiated with a laser line and the luminescence excited is dispersed spectroscopically. Alternatively, a part of or the total fluorescence is collected as a function of the frequency of the exciting laser. This is called an excitation spectrum and its vibrational structure is characteristic of the upper state, i.e., the information it provides is related to that obtained in an absorption spectrum.

A possible danger in interpreting fluorescence spectra is that one cannot identify the two states connected by the transition. So, for example, the fluorescence may represent transitions between two excited states, the lower of which may erroneously be assumed to be the ground state. Such a state of affairs indeed prevailed with some of the alkaline earth dimers and with the molecules Pb_2. In the latter case, it was matrix data (Bondybey and English 1977) that clarified the situation. In the case of Bi_2 it was shown that the wrong carrier had been identified and the published data ascribed to Bi_2 actually resulted from a spectrum of Bi_4 (Bondybey and English 1980a).

No technique is immune to error. For example, the beautiful fluorescence spectrum ascribed by Ahmed and Nixon (1979) to Ni_2 was shown by Bondybey (Rasanen et al. 1987) to be that of Se_2, presumably a contaminant from a previous experiment. The correct vibrational frequency of Ni_2 was reported recently by Lineberger (Ervin et al. 1988).

Likewise, some controversy surrounds the molecule Al_2. A gas-phase emission reported by Innes and co-workers (Ginter et al. 1964) was interpreted to be to a $^3\Sigma$ ground state. The corresponding absorption was not observed, however, in the matrix absorption spectrum (Douglas et al. 1983, Abe and Kolb 1983) suggesting that the $^3\Sigma$ state involved as the lower state in the gas phase transition is not the ground state. Of the possible terms for the ground state of Al_2, $^3\Sigma_g^-$, $^1\Sigma_g^+$ and $^3\Pi_u$ are the most likely (Fu et al. 1988). A magnetic deflection experiment (Cox et al. 1986) indicated that ground state Al_2 is a triplet; and most probably $^3\Pi_u$.

The problem of ensuring that the lower state involved in an emission is the ground state can be resolved in most cases where Raman or resonance Raman emissions are observed. These spectra are distinguished from (relaxed) fluorescence by using various exciting frequencies and noting that the positions of the spectra *relative* to the frequency of the exciting radiation remains constant. By contrast, in fluorescence the *absolute* frequencies of the emissions are constant.

Raman emissions have been reported for Ca_2 (Miller and Andrews 1980), Zn_2, Cd_2 (Givan and Loewenschuss 1979), Ag_2, Ag_3 (Schulze et al. 1978) and Cr_3 (Moskovits et al. 1980a), while resonance Raman has been used to study Sc_2, Sc_3 (Moskovits et al. 1980a), Knight et al. 1983, DiLella et al. 1982), Mn_2 (Bier et al. 1988), Ti_2, V_2 (Cossé et al. 1980), Cr_2, Cr_3 (Moskovits et al. 1980a), Fe_2 (Moskovits and DiLella 1980b), Ni_3 (Moskovits and DiLella 1980a), Cu_3 (DiLella et al. 1982), Mn_3 (Bier et al. 1988), Li_2, Li_3 and, possibly, Li_4 (Moskovits et al. 1984). In addition, Raman and resonance Raman emissions have been reported for Ga_2, In_2 and Tl_2 (Froben et al. 1983). These refer, of course, to matrix studies. In some cases gas-phase studies followed and in a few cases preceded the matrix work. For example, the vibrational spectrum of Fe_2 was observed in its photodetachment spectrum (Leopold et al. 1988). Fluorescence studies largely due to the work of Bondybey and Andrews have yielded ground-state vibrational constants for the group-2 dimers (Miller and Andrews 1980), and of Sn_2 (Bondybey and English 1980b), Pb_2 (Bondybey and English 1977), Sb_2, Sb_4 (Bondybey et al. 1981, Bondybey and English 1980a), Bi_2, Bi_4 (Bondybey and English 1980a) and Mo_2 (Pellin et al. 1981).

In all cases but two, the gas-phase and matrix vibrational frequencies agree to within one or two wave numbers. The exceptions are the group-2 dimers and Cr_2. With Mg_2 and Ca_2, the matrix frequencies were greater than their gas-phase counterparts (Miller and Andrews 1980). This was interpreted by those authors as being due to the matrix cage effect. In the matrix, the long-range portion of the effective interatomic potential is steepened as a result of interactions with the matrix cage. This has negligible effect on the harmonic frequency of a molecule with a deep potential and a short bond. It may, however, be important for long-bonded molecules characterized by a shallow interatomic potential, as in Mg_2 and Ca_2. Dichromium is a strongly bound molecule and hence for it this effect is expected to be minor (Moskovits et al. 1985). In this instance, however, the authors argued on the basis of the inordinately large anharmonicity observed for this molecule and the observation of a long-lived state of Cr_2 that the potential describing this molecule is somehow unusual, perhaps due to a second minimum in the ground-state potential at about 3 Å.

Although it was stated earlier that resonance Raman ensures ground-state data, the possibility exists that the same laser radiation which excites the resonance Raman emissions also pumps the molecule into a long-lived electronically excited state which then acts as the lower state in a resonance Raman process. Such observations have been made with V_2 (Cossé et al. 1980) and Pb_2 (Sontag et al. 1983). Normally, it is distinguishable from true ground-state resonance Raman by the fact that the emission intensity of the resonance Raman originating from an excited state depends quadratically on laser fluence, at least before saturation occurs. Upon saturation the fluence dependence reverts to linear, and excited-state resonance Raman cannot be distinguished from that originating in the ground state. With the large number of low lying electronically excited states that transition-metal molecules possess, the possibility that even a resonance Raman spectrum is not characteristic of the ground state cannot be totally discounted.

A summary of spectroscopic data of homonuclear metal diatoms and clusters is presented in Table 2.

3.6. Ultraviolet photoelectron spectroscopy (UPS)

In photoelectron spectroscopy, a UV source is used to photo-ionize the sample, and the current of photoelectrons is analyzed according to the values of their kinetic energies, yielding the electronic density of states as a function of energy, directly.

Matrix-isolated metal clusters are difficult to study using UPS because of the low metal/matrix ratios and the tendency of matrices to charge up. The high photon fluxes available at synchrotrons overcome the former and the use of thin matrices (150–200 Å) ameliorates the latter difficulty.

A number of UPS studies of matrix-isolated metal atoms and clusters have been reported, among them, studies involving Au and Cu atoms (Rotermund and Kolb 1985, Schmeisser et al. 1981). The UPS spectrum of Au atoms in solid neon deposited on a Ni(110) surface and excited with 16.3 eV synchrotron radiation shows two prominent peaks at 6.7 and 8.2 eV that are assigned to the spin–orbit split 5d level of the Au atom (Rotermund and Kolb 1985). Taking into account the 4.1 eV work function of the substrate, a 0.5 eV relaxation shift is estimated with respect to the corresponding UPS transitions in the gas phase. This is smaller than what is observed for Cu atoms in solid Ar, Kr or Xe (~ 1.3 eV) (Schmeisser et al. 1981). When matrices richer in Cu were deposited new bands occur at 5.5 and 9.5 eV, signalling the formation of dimers and clusters.

For larger Au clusters, the UP spectrum is dominated by a broad peak, 4–6 eV, probably resulting from the overlapping d-band emissions of various gold clusters. The emission shifts to lower binding energies as the cluster size increases.

It is clear that UPS provides only a rough gauge of the trend in the development of the d-band in the solid from atomic states and is not a refined tool for

Table 2
Selected ground-state constants of homonuclear metal clusters*.

Cluster	r_e (Å)	ω_e (cm^{-1})	$\omega_e X_e$ (cm^{-1})	D_e (eV)	Ground state
Diatomics					
^{7}Li$_2$	2.6729	351.43	2.610	1.07	$^1\Sigma_g^+$
^{7}Be$_2$ [1]	2.45	275.8	26.0	0.098	$^1\Sigma_g^-$
^{11}B$_2$	1.59	1051.3	9.35	3.08	$^3\Sigma_g^+$
^{23}Na$_2$	3.0788	159.124	0.7254	0.730	$^1\Sigma_g^+$
^{24}Mg$_2$ (gas)	3.890	51.12	1.645	0.0526	$^1\Sigma_g^+$
Ar matrix [2]		90.8	0.60	0.10–0.4	
^{27}Al$_2$	2.466	350.01	2.022	1.57	$^3\Pi_u$?
^{39}K$_2$	3.9051	92.021	0.2829	0.520	$^1\Sigma_g^+$
^{40}Ca$_2$ (gas) [3]	4.277	65.07	1.09	0.135	$^1\Sigma_g^+$
Ar matrix		81.7 [2]	0.52 [2]	0.40	
		74.0 [4]	0.61 [4]		
^{45}Sc$_2$		239.9 [5]	0.93 [5]	1.3 [6]	$^5\Sigma$ [7]
^{48}Ti$_2$		407.9 [8]	1.08 [8]	1.3 [9]	
^{51}V$_2$	1.76	537.5 [8]	4.3 [8]	1.85 [10]	$^3\Sigma_g^-$ [10]
^{52}Cr$_2$ (gas) [11]	1.68	$-\Delta G_{1/2} = 452.34$		1.56 [9]	$^1\Sigma_g^-$
matrix [12]		427.5	15.75		
^{55}Mn$_2$ Kr matrix	3.4 [13]	76.4 [14]	0.53 [14]	0.44 [9]	
^{56}Fe$_2$ matrix	1.87–2.02 [15]	300.26 [16]	1.45 [16]	1.3 [9]	$^1\Sigma_g^+$
				1.2 [17]	
				1.7 [9]	
^{59}Co$_2$		290 [18]			
^{58}Ni$_2$	2.20 [19]	290 [20]		2.07 [19]	$^1\Gamma_g$ or $^3\Gamma_u$
^{63}Cu$_2$	2.2197	264.55	1.025	2.05	$^1\Sigma_g^+$
^{64}Zn$_2$		80 [21]		0.19 [21]	$^1\Sigma_g^+$
^{69}Ga$_2$		180 [22]	1 [22]		$^3\Sigma_u^+$
^{85}Rb$_2$		57.31	0.105	0.49	$^1\Sigma_g^+$
^{88}Sr$_2$ (gas)		39.6 [23]		0.14 [23]	$^1\Sigma_g^+$
(Ar matrix)		~47 [2]			
^{89}Y$_2$				1.6 [9]	
^{93}Nb$_2$ (Kr matrix)		421 [24]	1.4 [24]	5.2 [9]	

Species		Frequency			State
^{98}Mo$_2$ (gas)	1.939 [25]	477.1 [25]		4.2 [25]	$^1\Sigma_g^+$
matrix		476 [26]			
^{103}Rh$_2$				2.9 [9]	
^{106}Pd$_2$				1.1 [9]	$^1\Sigma_g^+$
^{107,109}Ag$_2$	2.482	192.4	0.643	1.66	$^1\Sigma_g^+$
^{114}Cd$_2$		~58 [21]	0.8 [22]	0.09 [9]	
^{115}In$_2$		118 [22]	0.5 [27]	1.0 [9]	
^{120}Sn$_2$		118 [27]		1.9 [9]	$^1\Sigma_g^+$
^{121}Sb$_2$ (gas)	2.34	269.9	0.55	3.1	
Ne matrix		270.4 [5]	0.57 [5]		
^{133}Cs$_2$	4.47	42.022	0.0823	0.397	$^1\Sigma_g^+$
^{139}La$_2$				2.50 [28]	$^1\Sigma_g^+$
^{184}W$_2$		297 [29]			
^{186}Re$_2$		340 [30]			
^{197}Au$_2$	2.4719	190.9	0.420	2.31	$^1\Sigma_g^+$
^{202}Hg$_2$		36 [31]		0.08 [9]	$^1\Sigma_g^+$
^{205}Tl$_2$		80 [27]	0.5 [22]	0.6 [9]	
^{208}Pb$_2$ (gas) [32]	2.93	109.7	0.19	0.85	0_g^+
Ar matrix [33] [34]		112.2	0.35		
^{209}Bi$_2$ (Kr matrix)		172.7 [35]	0.34 [35]	2.0 [9]	$^1\Sigma_g^+$
^{210}Po$_2$		155.715	0.3353		

	Geometry	Frequency
Triatomics		
^{7}Li$_3$	$^2E'$ (dynamic Jahn–Teller); Atomization energy: 1.8 eV [36]; Symmetric stretching frequency = 310 [37] (Xe matrix); A rigid form was found in solid Kr and in another site of solid Xe [37]	
^{27}Al$_3$ [38]	Near D$_{3h}$	$\nu_1 = 353$; $\nu_2 = 231$ (in Xe) $\nu_1 = 351$; $\nu_2 = 247$ (in Kr)
^{45}Sc$_3$ [5]	Close to equilateral triangle	$\nu_1 (?) = 273$; $\nu_2 (?) = 133$
^{52}Cr$_3$ [12]	Bent triatomic	$\nu_1 = 247$; $\nu_2 = 145$; $\nu_3 = 151$
^{55}Mn$_3$ [14]	Weak Jahn–Teller distorted D$_{3h}$	$\nu_1 = 308$; $\nu_2 = 123$; $\nu_3 = 226$
^{58}Ni$_3$	Bent triatomic	$\nu_1 = 197$; $\nu_{2,3} = 130$ $\nu_1 (39) = 232$; $\nu_2 (40) = 85\text{--}95$

Table 2 (continued)

	Geometry	Frequency
^{63}Cu$_3$	$^2E'$; Dynamic Jahn–Teller in gas phase [41]; Bent triatomic in adamantane matrix [42]; Atomization energy: 3.05 eV [43]; Symmetric vibration: 270 [41]	$\nu_1 = 120.5$ [44]
Y$_3$ ^{107}Ag$_3$	$^2B_2$ [7] Bent triatomic $^2A_1$ [11], $^2B_2$ [42]; Atomization energy: 2.62 eV [43]	
Au$_3$	Atomization energy: 3.80 eV [43] [45]	
Tetratomics		
Be$_4$	T_d	$\nu_1(a_1) = 151$ (in Ar [46]). $\nu_1(a_1) = 149.7$; $\nu_2(e) = 89.8$; $\nu_3(t_2) = 120.4$ (in Ne [46])
Sb$_4$	T_d	$\nu_1(a_1) = 241$; $\nu_2(e) = 140?$; $\nu_3(t_2) = 179?$ (in Ne matrix [47])

*This table draws heavily on the compilation of Weltner and Van Zee (1984). Only numbers based on experimental data are included. Speculative assignments are marked with a (?). When no reference is given, data were obtained either from Weltner and Van Zee (1984) or from Huber and Herzberg (1979).

References: [1] Bondybey (1984); [2] Miller and Andrews (1980); [3] Balfour and Whitlock (1975); [4] Bondybey and Albiston (1978); [5] Moskovits et al. (1980a); [6] Verhaegen et al. (1964); [7] Knight et al. (1983); [8] Cossé et al. (1980); [9] Gingerich (1980); [10] Langridge-Smith et al. (1984); [11] Riley et al. (1983); Michalopoulos et al. (1982); [12] DiLella et al. (1982); [13] Van Zee et al. (1981); Baumann et al. (1983); [14] Bier et al. (1988); [15] Montano and Shenoy (1980); Purdum et al. (1982); [16] Moskovits and DiLella (1980b); [17] Moskovits and DiLella (1982); [18] Moskovits et al. (1980b); [19] Morse et al. (1984); [20] Ervin et al. (1988); [21] Givan and Loewenschuss (1979); [22] Froben et al. (1983); [23] Bergman and Liao (1980); [24] Moskovits and Limm (1986); [25] Hopkins et al. (1983); [26] Pellin et al. (1981); [27] Bondybry and English (1980b); [28] Verhaegen et al. (1964); [29] Pellin et al. (1985); [30] Leopold et al. (1986); [31] Epstein and Powers (1953); [32] Bondybey and English (1982); [33] Bondybey and English (1977); [34] Sontag et al. (1983); [35] Manzel et al. (1981); [36] Wu (1976); [37] Moskovits et al. (1984); [38] Fu et al. (1988); [39] Moskovits and DiLella (1980a); [40] Woodward et al. (1988); [41] Rohlfing and Valentini (1986); [42] Howard et al. (1983b); [43] Hilpert and Gingerich (1980); [44] Schulze et al. (1978); [45] Kant et al. (1968); [46] Bondybey and English (1980a); [47] Bondybey et al. (1981); Bondybey and English (1980a).

studying details of the electronic structure of clusters, at least not when cluster distributions are present and with the poor energy resolution normally available.

4. Size selected clusters

One impediment to the study of clusters in matrices has been the presence of clusters of different size in the same matrix. At times, this has seriously impeded spectral assignments and reduced markedly the upper limit in the size of cluster that could be studied. This is because in most cases the matrix-isolated clusters are formed from atomic precursors during the formation of the matrix. As the mean cluster size in the distribution increases, so does the breadth of the distribution. Hence a matrix containing substantial amounts of the septamer, say, also contains appreciable quantities of eight or ten other species. Of course from time to time a cluster is formed that is considerably stabler than its immediate neighbours, a so-called magic number cluster, and the matrix may become particularly rich in that size cluster. For most metals, clusters smaller than M_8 do not show such behaviour although there is anecdotal evidence that septamers of the alkalis and of Ag are unusually plentiful in matrices. In gas-phase alkali clusters, magic numbers are well established (De Heer and Knight 1987) and successfully explained in terms of electronic shell closings.

Attempts have been made in various ways to create matrices that contain either a single cluster species or, at least, a much narrower distribution than that formed by sequential polymerization of metal atoms. Schumacher and co-workers (Schulze et al. 1980), for example, used a Stern–Gerlach magnet to deflect sodium clusters of various spin and hence various size and in this way, generated a matrix considerably enriched in Na_2. Likewise Wöste and co-workers (Delacrètaz et al. 1987) used a quadrupole mass filter to select iron-cluster ions of various size and attempted to matrix-isolate them.

Ozin and co-workers (Ozin and Huber 1978) proposed an ingenious scheme based on two photoprocesses in an effort to narrow the cluster size distribution in a matrix. In the first step, irradiation into atomic absorptions produced diffusion of atoms and their subsequent aggregation to clusters. In the second step, selective photodissociation of a chosen cluster caused them to fragment into smaller clusters.

For example, irradiation of Ag atoms in $^2S \rightarrow {}^2P$ absorption resulted in Ag_n ($n = 2$–5) (Ozin and Huber 1978). The analysis of the steps involved in this process is heavily based on the work of DeMore and Davidson (1959). Ozin and Huber (1978) do not make a clear distinction in this paper between matrix softening, i.e., vibrational relaxation of the atom/cage excited complex resulting in matrix heating, and a direct mobilization of the atom, i.e., an electronic to translational energy conversion. The former would be a thermal effect and, although more localized than that obtained with external heating of the matrix, it would nevertheless promote the mobility of all species within a certain radius of the excited atom. The latter would initially mainly involve linear motion of the excited atom, although ultimately it would result in matrix heating as momentum transfer to the matrix occurs.

The notion of photoaggregation was implicitly refined to the second process in a paper by Klotzbücher and Ozin (1978) in which Cr and Mo atoms seemed to be mobilized independently by irradiating into their respective absorption bands. This work was not conclusive, however, since the two metals were deposited from different sources resulting in a matrix that was rich in one metal on one side and in the other on the other. The spectrum, however, was taken of the entire matrix suggesting that processes occurring in one half of the matrix occurred throughout it.

Attempts by other groups to reproduce the selective mobilization claimed for photoaggregation produced mixed results. Welker and Martin (1979) report the dimerization of Na upon irradiating an atomic absorption. Their result however, is not as clear-cut as that of Ozin and Klotzbücher (1978). A number of photolysis processes seem to interfere with the photomobilization. Moreover, those authors do not distinguish clearly in their analysis between the disappearance of a species due to photodissociation as opposed to photoaggregation, referring to both processes as "phototolysis". Other groups (Gruen 1980) have had more difficulty in repeating the photoaggregation results of Ozin and Huber (1978).

Schwentner (1988) has attempted laser-induced photoaggregation of matrix-isolated Au and subsequent dissociation of Au_2 and Au_3, and was likewise unable to promote atomic diffusion with the power densities used by Ozin and Huber (1978).

Although controversy still exists regarding the nature of photoaggregation, a reasonable conclusion is that the process is largely thermal and that it does not entail direct electronic to translational energy conversion. It is nevertheless a valiant attempt at producing a matrix that is monodispersed in a cluster species, a goal still vibrant in current research.

5. Exotic matrix environments for clusters

The desire to produce monodispersed matrix-isolated clusters, larger clusters and clusters that are stable to further aggregation even at moderately high temperatures has encouraged experimentation with unusual matrix materials including hydrocarbon (Hulse and Moskovits 1981, Ozin 1985) and frozen water (Ozin et al. 1980) matrices.

Ingenious environments such as phenylated siloxane polymers that would enclose small metal clusters through the simultaneous interaction of several phenyl groups have been attempted largely by Ozin and his group (Ozin 1980). Similar directions were established by Klabunde and his group (Klabunde 1983).

The use of phenylated siloxanes has produced some noteworthy results. On increasing the metal loading in a molybdenum/siloxane system, one observes a sequential appearance of five UV-visible absorption bands that are assigned to Mo_x ($x = 1–5$) (Ozin 1985). A variation on this theme has been to use other polymers such as polyolefins to achieve similar results (Ozin 1985).

A most exciting environment for cluster growth, possessing admirably many of the properties desired for the preparation of monodispersed clusters, is the large array of natural and synthetic zeolites (Gallezot 1986). They form too large a subject to review here, despite several formal similarities (among many differences) that metal clusters encapsulated in zeolites bear to matrix-isolated clusters.

6. Conclusion

The matrix environment as a tool for studying metal clusters is being strongly challenged by the many recently developed gas-phase techniques. Nevertheless, it remains an important medium, especially, for determining the spectroscopic properties of ground-state clusters.

If sources of mass-selected clusters are developed and if cluster fragmentation or aggregation can be overcome during matrix-isolation, then this technique will once again be a key player in the cluster drama.

References

Abe, H., and D.M. Kolb, 1983, Ber. Bunsenges. Phys. Chem. **87**, 523.
Abe, H., W. Schulze and B. Tesche, 1980, Chem. Phys. **47**, 95.
Ahmed, F., and E.R. Nixon, 1979, J. Chem. Phys. **71**, 3547.
Anderson, A.B., 1978, J. Chem. Phys. **68**, 1744.
Bach, S.B.H., D.A. Garland, R.J. Van Zee and W. Weltner Jr, 1987, J. Chem. Phys. **87**, 869.
Bachmann, C., J. Demuynck and J. Veillard, 1980, Faraday Symp. Chem. Soc. **14**, 1970.
Balfour, W.J., and R.F. Whittlock, 1975, Can. J. Phys. **53**, 472.
Balling, L.C., M.D. Harvey and J.F. Dawson, 1978, J. Chem. Phys. **69**, 1670.
Balling, L.C., M.D. Harvey and J.J. Wright, 1979, J. Chem. Phys. **70**, 2404.
Barrett, C.P., R.G. Graham and R. Grinter, 1984, Chem. Phys. **86**, 199.
Baumann, C.A., R.J. Van Zee, S.V. Bhat and W. Weltner Jr, 1983, J. Chem. Phys. **78**, 190.
Bechthold, P.S., U. Kettler and W. Krasser, 1984, Solid State Commun. **52**, 347.
Belyava, A.A., Y.B. Predtchenskii and L.D. Schcherba, 1973, Opt. Spectr. (USSR) **34**, 21.
Benedek, G., T.P. Martin and G. Pacchioni, eds, 1987, Elemental and Molecular Clusters (Springer, Berlin).
Bennemann, K.H., and J. Koutecky, eds, 1985, Surface Science Vol. 156, entire volume.
Bergman, T., and P.F. Liao, 1980, J. Chem. Phys. **72**, 886.

Bier, K.D., T.L. Haslett, A.D. Kirkwood and M. Moskovits, 1988, J. Chem. Phys. **89**, 6.

Bondybey, V.E., 1984, Chem. Phys. Lett. **109**, 436.

Bondybey, V.E., and C. Albiston, 1978, J. Chem. Phys. **68**, 3172.

Bondybey, V.E., and J.H. English, 1977, J. Chem. Phys. **67**, 3405.

Bondybey, V.E., and J.H. English, 1980a, J. Chem. Phys. **73**, 42.

Bondybey, V.E., and J.H. English, 1980b, J. Mol. Spectrosc. **84**, 388.

Bondybey, V.E., and J.H. English, 1982, J. Chem. Phys. **76**, 2165.

Bondybey, V.E., and J.H. English, 1983, Chem. Phys. Lett. **94**, 443.

Bondybey, V.E., G.P. Schwartz and J.E. Griffiths, 1981, J. Mol. Spectrosc. **84**, 328.

Bos, A., and A.T. Howe, 1974, J. Chem. Soc. Faraday Trans. II, **70**, 440.

Castleman Jr, A.W., and R.G. Keesee, 1986, Chem. Rev. **86**, 589.

Cossé, C., M. Fouassier, T. Mejean, M. Tranquille, D.P. DiLella and M. Moskovits, 1980, J. Chem. Phys. **73**, 6076.

Cox, D.M., D.E. Trevor, R.L. Whetten, E.A. Rohlfing and A. Kaldor, 1986, J. Chem. Phys. **84**, 4651.

deHeer, W.A., and W.D. Knight, 1987, in: Elemental and Molecular Clusters, eds G. Benedek, T.P. Martin and G. Pacchioni (Springer, Berlin) p. 45.

Delacrètaz, G., E.R. Grant, R.L. Whetten, L. Wöste and J.W. Zwanziger, 1986, Phys. Rev. Lett. **56**, 2598.

Delacrètaz, G., P. Fayet, J.P. Wolf and L. Wöste, 1987, in: Elemental and Molecular Clusters, eds G. Benedek, T.P. Martin and G. Pacchioni (Springer, Berlin) p. 64.

DeMore, W.B., and N. Davidson, 1959, J. Am. Chem. Soc. **81**, 5869.

DeVore, T.C., A. Ewing, H.F. Franzen and V. Calder, 1975, Chem. Phys. Lett. **35**, 78.

DiLella, D.P., W. Limm, R.H. Lipson, M. Moskovits and K.V. Taylor, 1982, J. Chem. Phys. **77**, 5263.

DiLella, D.P., K.V. Taylor and M. Moskovits, 1983, J. Phys. Chem. **87**, 524.

Douglas, M.A., R.H. Hauge and J.L. Margrave, 1983, J. Phys. Chem. **87**, 2945.

Dyson, W., and P.A. Montano, 1978, J. Am. Chem. Soc. **100**, 7439.

Dyson, W., and P.A. Montano, 1979, Phys. Rev. B **20**, 3619.

Dyson, W., and P.A. Montano, 1980, Solid State Commun. **33**, 191.

Epstein, L.R., and M.D. Powers, 1953, J. Phys. Chem. **57**, 336.

Ervin, K.M., J. Ho and W.C. Lineberger, 1988, J. Chem. Phys., to be submitted.

Farges, J., M.F. deFeraudy, B. Raoult and G. Torchet, 1986, in: Physics and Chemistry of Small Clusters, NATO ASI Series, Vol. 158, eds P. Jena, B.K. Rao and S.N. Khanna (Plenum Press, New York) p. 15.

Froben, F.W., W. Schulze and U. Kloss, 1983, Chem. Phys. Lett. **99**, 500.

Fu, Z.W., G.W. Lemire, Y.M. Hamrick, S. Taylor, J.C. Shui and M.D. Morse, 1988, J. Chem. Phys. **88**, 3524.

Gallezot, P., 1986, in: Metal Clusters, ed. M. Moskovits (Wiley, New York) p. 219.

Garland, D.A., and D.M. Lindsay, 1983a, J. Chem. Phys. **78**, 2813.

Garland, D.A., and D.M. Lindsay, 1983b, J. Chem. Phys. **78**, 5946.

Garland, D.A., and D.M. Lindsay, 1984, J. Chem. Phys. **80**, 4761.

Geusic, M.E., M.D. Morse and R.E. Smalley, 1985, J. Chem. Phys. **82**, 590.

Gingerich, K.A., 1980, Faraday Disc. **14**, 109 (compilation).

Ginter, D.S., M.L. Ginter and K.K. Innes, 1964, Astrophys. J. **139**, 365.

Givan, A., and A. Loewenschuss, 1979, Chem. Phys. Lett. **62**, 592.

Grinter, R., and D.R. Stern, 1983, J. Chem. Soc. Faraday Trans. II, **79**, 1011.

Gruen, D.M., 1980, personal communication.

Heath, J.R., S.C. O'Brien, Q. Zhang, Y. Liu, R.F. Curl, F.K. Tittel and R.E. Smalley, 1985, J. Am. Chem. Soc. **107**, 7779.

Hilpert, K., and K.A. Gingerich, 1980, Ber. Bunsenges. Phys. Chem. **84**, 739.

Hopkins, J.B., P.R.R. Langridge-Smith, M.D. Morse and R.E. Smalley, 1983, J. Chem. Phys. **78**, 1627.

Howard, J.A., R. Sutcliffe, J.S. Tse and B. Mile, 1983a, Chem. Phys. Lett. **94**, 56.

Howard, J.A., K.F. Preston, R. Sutcliffe and B. Mile, 1983b, J. Phys. Chem. **87**, 536.

Howard, J.A., R. Sutcliffe and B. Mile, 1983c, J. Phys. Chem. **87**, 2268.

Howard, J.A., R. Sutcliffe and B. Mile, 1984, Chem. Phys. Lett. **112**, 84.

Hulse, J.E., 1979, Ph.D. Thesis (University Microfilms, Ann Arbor, MI).

Hulse, J.E., and M. Moskovits, 1976, Surf. Sci. **57**, 125.

Hulse, J.E., and M. Moskovits, 1981, J. Phys. Chem. **85**, 109.

Jacobi, K., D. Schmeisser and D.M. Kolb, 1980, Chem. Phys. Lett. **69**, 113.

Jansson, K., and R. Scullman, 1976, J. Mol. Spectrosc. **61**, 299.

Jena, P., B.K. Rao and S.N. Khanna, eds, 1986, Physics and Chemistry of Small Clusters, NATO ASI series, Vol. 158 (Plenum, New York).

Jortner, J., A. Pullman and B. Pullman, eds, 1987, Large Finite Systems (Reidel, Dordrecht, The Netherlands).

Kant, A., S.S. Lin and B. Strauss, 1968, J. Chem. Phys. **49**, 1983.

Kappes, M.M., M. Schär, P. Radi and E. Schumacher, 1986, J. Chem. Phys. **84**, 1863.

Kernisant, K., G.A. Thompson and D.M. Lindsay, 1985, J. Chem. Phys. **82**, 4739.

Kettler, U., P.S. Bechtold and W. Krasser, 1985, Surf. Sci. **156**, 867.

Klabunde, K.J., 1983, J. Mol. Catal. **21**, 57 and references therein.

Klotzbücher, W.E., and G.A. Ozin, 1978, J. Am. Chem. Soc. **100**, 2262.

Klotzbücher, W.E., and G.A. Ozin, 1980, Inorg. Chem. **19**, 3776.

Knight, L.B., M.W. Woodward, R.J. Van Zee and W. Weltner Jr, 1983, J. Chem. Phys. **79**, 5820.

Kolb, D.M., 1981, NATO Adv. Study Inst., Ser. C, **76**, 447.

Kolb, D.M., H.H. Rotermund, W. Schrittenlacher and W. Schroeder, 1984, J. Chem. Phys. **80**, 695.

Koutecky, J., and P. Fantucci, 1986, Chem. Rev. **86**, 539.

Koutecky, J., V. Bonacic-Koutecky, I. Boustani, P. Fantoucci and W. Pewestorf, 1987, in: Large Finite Systems, eds J. Jortner, A. Pullman and B. Pullman (Reidel, Dordrecht, The Netherlands) p. 303.

Kroto, H.W., J.R. Heath, S.C. O'Brien, R.F. Curl and R.E. Smalley, 1985, Nature **318**, 162.

Langridge-Smith, P.R.R., M.D. Morse, G.P. Hansen, R.E. Smalley and A.J. Jerer, 1984, J. Chem. Phys. **80**, 593.

Leopold, D.G., T.M. Mitler and W.C. Lineberger, 1986, J. Am. Chem. Soc. **108**, 178.

Leopold, D.G., J. Almlöf, W.C. Lineberger and P.R. Taylor, 1988, J. Chem. Phys. **88**, 3780.

Lindsay, D.M., and G.A. Thompson, 1982, J. Chem. Phys. **74**, 1114.

Lindsay, D.M., D. Garland, F. Tischler and G.A. Thompson, 1982, in: Metal Bonding and Interactions in High Temperature Systems, eds J.L. Gole and W.C. Stwalley (ACS Symposium no. 179) p. 69.

Manzel, K., U. Engelhardt, H. Abe, W. Schulze and F.W. Froben, 1981, Chem. Phys. Lett. **77**, 514.

Martins, J.L., J. Buttet and R. Car, 1985, Phys. Rev. B **31**, 1804.

McNab, T.K., H. Micklitz and P. Barrett, 1971, Phys. Rev. B **4**, 3787.

Michalopoulos, D.L., M.E. Geusic, S.G. Hansen, D.E. Powers and R.E. Smalley, 1982, J. Phys. Chem. **86**, 3914.

Miller, J.C., and L. Andrews, 1980, Appl. Spectrosc. Rev. **16**, 1.

Miller, J.C., R.L. Mowery, E.R. Krausz, S.M. Jacobs, H.W. Kim, P.N. Schatz and L. Andrews, 1981, J. Chem. Phys. **74**, 6349.

Mitchell, S.A., and G.A. Ozin, 1978, J. Am. Chem. Soc. **100**, 6776.

Mitchell, S.A., and G.A. Ozin, 1984, J. Phys. Chem. **88**, 1425.

Miyoshi, E., H. Tatewaki and T. Nakamura, 1983, J. Chem. Phys. **78**, 815.

Montano, P.A., 1980, Faraday Symp. Chem. Soc. **14**, 79.

Montano, P.A., 1982, J. Phys. C **15**, 565.

Montano, P.A., and G.K. Shenoy, 1980, Solid State Commun. **35**, 53.

Montano, P.A., W. Schulze, B. Tesche, G.K. Shenoy and T.I. Morrison, 1984, Phys. Rev. B **30**, 672.

Moore, C.E., 1958, NBS Circular, No. 467, Vol. 1–3.

Morse, M.D., 1986, Chem. Rev. **86**, 1049.
Morse, M.D., 1987, Chem. Phys. Lett. **133**, 8.
Morse, M.D., J.B. Hopkins, P.R.R. Langridge-Smith and R.E. Smalley, 1983, J. Chem. Phys. **79**, 5316.
Morse, M.D., G.P. Hansen, P.R.R. Langridge-Smith, L.S. Zheng, M.E. Geusic, D.L. Michalopoulos and R.E. Smalley, 1984, J. Chem. Phys. **80**, 5400.
Moskovits, M., ed., 1986, Metal Clusters (Wiley, New York).
Moskovits, M., and D.P. DiLella, 1980a, J. Chem. Phys. **72**, 2267.
Moskovits, M., and D.P. DiLella, 1980b, J. Chem. Phys. **73**, 4917.
Moskovits, M., and D.P. DiLella, 1982, in: Metal Bonding and Interaction in High Temperature Systems, eds J.L. Gole and W.C. Stwalley, ACS symposium no. 179.
Moskovits, M., and J.E. Hulse, 1977a, J. Chem. Phys. **66**, 3988.
Moskovits, M., and J.E. Hulse, 1977b, J. Chem. Phys. **67**, 4271.
Moskovits, M., and J.E. Hulse, 1977c, J. Chem. Soc. Faraday Trans. II, **73**, 471.
Moskovits, M., and W. Limm, 1986, Ultramicroscopy **20**, 83.
Moskovits, M., and G.A. Ozin, eds, 1976, Cryochemistry (Wiley, New York).
Moskovits, M., D.P. DiLella and W. Limm, 1980a, J. Chem. Phys. **80**, 626.
Moskovits, M., D.P. DiLella and A. Loewenschuss, 1980b, unpublished results.
Moskovits, M., W. Limm and T. Mejean, 1984, in: Dynamics on Surfaces, eds B. Pullman, J. Jortner, A. Nitzan and B. Gerber (Reidel, Dordrecht, The Netherlands) p. 437.
Moskovits, M., W. Limm and T. Mejean, 1985, J. Chem. Phys. **82**, 4875.
Nagarathna, H.M., J.J. Choi and P.A. Montano, 1982, J. Chem. Soc. Faraday Trans I **78**, 923.
Nour, E.M., C. Alfaro-Franco, K.A. Gingerich and J. Laane, 1987, J. Chem. Phys. **86**, 4779.
Olsen, A.W., and K.J. Klabunde, 1987, Encyclopedia of Physical Science and Technology **8**, 108.
Ozin, G.A., 1980, Faraday Symp. Chem. Soc. **14**, 7.
Ozin, G.A., 1985, Chemtech **15**, 488 (for a review and references).
Ozin, G.A., and A.J.L. Hanlan, 1979, Inorg. Chem. **18**, 1781.
Ozin, G.A., and H. Huber, 1978, Inorg. Chem. **17**, 155.
Ozin, G.A., and W.E. Klotzbücher, 1979, Inorg. Chem. **18**, 2101.
Ozin, G.A., and S.A. Mitchell, 1983, Angew. Chem. **95**, 706.
Ozin, G.A., H. Huber, P. McKenzie and S.A. Mitchell, 1980, J. Am. Chem. Soc. **102**, 1548.
Ozin, G.A., M.D. Baker, S.A. Mitchell and D.F. McIntosh, 1983, Angew. Chem. **95**, 157.
Pellin, M.J., T. Foosnaes and D.M. Gruen, 1981, J. Chem. Phys. **74**, 5547.
Pellin, M.J., M.H. Mendelsohn and D.M. Gruen, 1985, unpublished results.
Pflibsen, K.P., and D.R. Huffman, 1985, Surf. Sci. **156**, 793.
Purdum, H., P.A. Montano, G.K. Shenoy and T. Morrison, 1982, Phys. Rev. B **25**, 4412.
Rao, B.K., and P. Jena, 1985, Phys. Rev. B **32**, 2058.
Rasanen, M., L.A. Heimbrook and V.E. Bondybey, 1987, J. Mol. Struct. **157**, 129.
Richtsmeier, S.C., J.L. Gole and D.A. Dixon, 1980, Proc. Nat. Acad. Sci. USA **77**, 5611.
Richtsmeier, S.C., E.K. Parks, K. Liu, L.G. Pobo and S.J. Riley, 1985, J. Chem. Phys. **82**, 3659.
Riley, S.J., E.K. Parks, L.G. Pobo and S. Wexler, 1983, J. Chem. Phys. **79**, 2577.
Rivoal, J.C., J. Shakhs Emampour, K.J. Zeringue and M. Vala, 1982, Chem. Phys. Lett. **92**, 313.
Rohlfing, E.A., and J.J. Valentini, 1986, Chem. Phys. Lett. **126**, 113.
Rotermund, H.H., and D.M. Kolb, 1985, Surf. Sci. **156**, 785.
Ruffolo, M.G., S. Ossicini and F. Forstmann, 1984, J. Chem. Phys. **80**, 2401.
Ruffolo, M.G., S. Ossicini and F. Forstmann, 1985, J. Chem. Phys. **82**, 3988.
Salahub, D.R., 1987, Adv. Chem. Phys. **69**, 447.
Schmeisser, D., K. Jacobi and D.M. Kolb, 1981, J. Chem. Phys. **75**, 5300.
Schulze, W., H.V. Becker and H. Abe, 1978, Chem. Phys. **35**, 177.
Schulze, W., E. Schumacher and S. Leutwyler, 1980, unpublished results.
Schwenter, N., 1988, unpublished results.

Shamai, S., M. Pasternak and H. Micklitz, 1982, Phys. Rev. B **26**, 3031.

Shirk, J.S., and A.M. Bass, 1968, J. Chem. Phys. **49**, 5156.

Smith, D.J., ed., 1986, Ultramicroscopy, Vol. 20 (entire volume).

Sontag, H., B. Eberle and R. Weber, 1983, Chem. Phys. **80**, 279.

Thompson, T.C., D.G. Truhlar and C.A. Mead, 1985, J. Chem. Phys. **82**, 2392.

Träger, F., and G. Zu Putlitz, eds, 1986, Z. Phys. D **3**(2,3) entire volume.

Truhlar, D.G., T.C. Thompson and C.A. Mead, 1986, Chem. Phys. Lett. **127**, 287.

Vala, M., K.J. Zeringue, J. Shakhs Emampour, J.C. Rivoal and R. Pyzalski, 1984, J. Chem. Phys. **80**, 2401.

Vala, M., M. Eyring, J. Pyka, J.C. Rivoal and C. Grisola, 1985, J. Chem. Phys. **83**, 969.

Van Zee, R.J., C.A. Baumann and W. Weltner Jr, 1981, J. Chem. Phys. **74**, 6977.

Van Zee, R.J., C.A. Baumann, S.V. Bhat and W. Weltner Jr, 1982, J. Chem. Phys. **76**, 5636.

Van Zee, R.J., C.A. Baumann and W. Weltner Jr, 1985, J. Chem. Phys. **82**, 3912.

Van Zee, R.J., R.F. Ferrante, K.J. Zeringue, W. Weltner Jr and D.W. Ewing, 1987, J. Chem. Phys. **88**, 3465.

Verhaegen, G., S. Smoes and J. Drowart, 1964, J. Chem. Phys. **40**, 239.

Vidal, C.R., 1980, J. Chem. Phys. **72**, 1864.

Walch, S.P., and B.C. Laskowski, 1986, J. Chem. Phys. **84**, 2734.

Welker, T., and T.P. Martin, 1979, J. Chem. Phys. **70**, 5683.

Weltner Jr, W., and R.J. Van Zee, 1984, Ann. Rev. Phys. Chem. **35**, 291.

Whetten, R.L., D.M. Cox, D.J. Trevor and A. Kaldor, 1985, Phys. Rev. Lett. **54**, 1494.

Wiggenhauser, H., W. Schroeder and D.M. Kolb, 1988, J. Chem. Phys. **88**, 3434.

Woodward, J.R., S.H. Cobb and J.L. Gole, 1988, J. Phys. Chem. **92**, 1404.

Wu, C.H., 1976, J. Chem. Phys. **65**, 3181.

Yang, S., K.J. Taylor, M.J. Craycroft, J. Conceicao, C.L. Pettiette, O. Cheshnovsky and R.E. Smalley, 1988, Chem. Phys. Lett. **144**, 43.

Yokozeki, A., and G.D. Stein, 1978, J. Appl. Phys. **49**, 2229.

Zeringue, K.J., J. Shakhs Emampour, J.C. Rivoal and M. Vala, 1983, J. Chem. Phys. **78**, 2231.

Zwanziger, J.F., R.L. Whetten and E.R. Grant, 1986, J. Phys. Chem. **90**, 3298.

CHAPTER 4

The Stabilization and Spectroscopy of Free Radicals and Reactive Molecules in Inert Matrices

Marilyn E. JACOX

Molecular Spectroscopy Division
National Institute of Standards and Technology
Gaithersburg, Maryland 20899
USA

Chemistry and Physics of
Matrix-Isolated Species
L. Andrews and M. Moskovits, eds

Contents

1. Introduction

During the past thirty years, the matrix-isolation technique has proved to be a valuable tool for obtaining the spectra of free radicals trapped in inert, transparent solids. Today, it is challenged by a host of new laser-based techniques which have demonstrated their ability to obtain high-resolution vibrational and electronic spectra of transient molecules in the gas phase, free of possible perturbation by the matrix environment. Despite this challenge, there are important reasons why matrix studies of transient molecules are, in fact, complementary to laser studies and should continue to yield valuable contributions to the study of chemical reaction intermediates.

While lasers are tunable over limited spectral ranges, observations in rare-gas or nitrogen matrices provide a broad spectral survey, extending from the far infrared to the vacuum ultraviolet. Thus, all of the products can be detected. As will be seen, shifts of the matrix bands from the gas-phase band center are, typically, sufficiently small to provide a guide for the choice and tuning of lasers for more detailed gas-phase studies. While isotopic studies are crucial to positive identifications in either system, they are much more readily conducted in matrix systems, for which only a few milligrams of expensive isotopically enriched substances are usually required. Matrix absorption observations may also be helpful in spectral assignments, since contributions of "hot bands" to the spectrum are eliminated; at cryogenic temperatures all of the absorptions must originate in the molecular ground state.

Matrix observations are valuable for characterizing the early stages of chemical reactions. Although in the gas phase products of all of the stages of the reaction generally contribute to the spectrum, in a matrix the reaction is very effectively quenched after the primary reaction or, at most, the earliest stages of the overall process. Energy-rich intermediates which fragment readily in the gas phase may be stabilized in detectable concentration when they are trapped in the matrix environment. In some systems, illustrated in this chapter, weakly bound reaction complexes can be stabilized and studied. These studies can provide information on reaction mechanisms which is complementary to that obtained in sophisticated molecular-beam studies. Finally, matrix observations are uniquely well suited to studies of the photodecomposition of free radicals and molecular ions. When the role of cage recombination is taken into account, information on free-radical photodecomposition thresholds and products can be obtained which is virtually inaccessible using other techniques.

The primary purpose of this chapter is to survey the principles which govern the stabilization of transient molecules in rare-gas matrices. The factors which influence the nature and yield of the trapped reaction intermediates in a number of different sampling configurations will be considered. Emphasis will be placed on results obtained in approximately the last ten years, with some discussion of earlier studies which provide background information or are concerned with especially important systems. Material in earlier reviews (Milligan and Jacox 1972, Jacox 1978a) supplements the discussion given here. For a summary of the vibrational spectra of the ground states of transient molecules, see the tables prepared by Jacox (1984d). Tables which include rotational, vibrational, and electronic spectral data for transient molecules with from three to six atoms have recently been published (Jacox 1988b). Many new experimental data, which have been published since the vibrational tables of Jacox (1984d), have been incorporated in these electronic spectral tables. Recent vibrational data for larger molecules, as well as for three- to six-atomic molecules for which electronic transitions have not been reported will be included in a supplement which is now being prepared (Jacox 1989).

2.   Spectra of radicals in matrices

2.1. Ground-state vibrations

Most of the studies of the ground-state vibrations of molecules isolated in rare-gas matrices have been conducted at a resolution of $1–2 \text{ cm}^{-1}$. While it has long been recognized that infrared absorptions of matrix-isolated molecules are sharp, recent studies have indicated that often the true band width is of the order of $0.1–0.2 \text{ cm}^{-1}$. In a tunable diode-laser study of the v_3 band of SO_2 isolated in solid argon, Dubs and Günthard (1977) reported a structured absorption, with peaks varying in width (FWHM) from 0.06 to 0.3 cm^{-1}. Dubs and Günthard (1979) found an inhomogeneous band width of 0.21 cm^{-1} for the v_{17} absorption of t-1,2-difluoroethane in solid argon, at 1047 cm^{-1}, using a tunable diode laser. Hole burning with the diode laser operated at low power gave a hole width of $0.006 \pm 0.002 \text{ cm}^{-1}$, a value equivalent to twice the homogeneous line width. Further hole-burning studies of this molecule, including studies in the heavier rare gases and in nitrogen, were conducted by Dubs et al. (1982) and by Felder and Günthard (1984). In a Fourier-transform infrared (FTIR) study at a resolution of 0.035 cm^{-1}, Swanson and Jones (1981) observed nine peaks over a range of 3 cm^{-1} for v_3 of SF_6 isolated in solid argon. The largest FWHM band width was 0.12 cm^{-1}. In a diode-laser study of this system at a resolution of 0.001 cm^{-1}, Bristow et al. (1981) reported band widths ranging from 0.034 to

0.103 cm^{-1}, with no additional structure. Jones and Swanson (1981) found that the absorptions of SeF$_6$ in rare-gas matrices are sufficiently sharp to permit location of the peaks to within 0.005 cm^{-1}. In a study of the temperature dependence of the infrared spectra of CH$_3$F and CD$_3$F isolated in argon and krypton, Jones and Swanson (1982) reported line widths which vary from 0.2 to 7.5 cm^{-1}. Jones et al. (1984) have studied v_3 of SiF$_4$ in solid argon. Two peaks separated by 0.7 cm^{-1} were reported, each with a band width of approximately 0.06 cm^{-1}.

For most molecules, ground-state vibrational absorptions in inert matrices lie close to the gas-phase band centers. Jacox (1985c) has compared the ground-state values of $\Delta G_{1/2}$ observed for diatomic molecules in the gas phase with the corresponding values observed in rare-gas and nitrogen matrices. Matrix shifts in neon are smallest. With a few exceptions, discussed below, matrix shifts in solid argon are smaller than 2%. The matrix shift increases with increasing mass of the rare gas and increases still further for a nitrogen matrix, in which hydrogen-bonding and polarization effects become important. Shifts to longer wavelengths (red shifts) are somewhat more common than are shifts to shorter wavelengths (blue shifts). The alkali metal and group IIIa halides have very large red shifts, which may be correlated with the large dipole moments and relatively low electron density in the internuclear region typical of these molecules. Data are available for only one diatomic molecular ion, C$_2^-$, which experiences only a small matrix shift. Weakly bonded species typically have large blue shifts, possibly because of the contribution by the matrix of a potential barrier to their dissociation. A similar comparison has been made by Jacox (1984d) for the vibrational fundamentals of small polyatomic free radicals

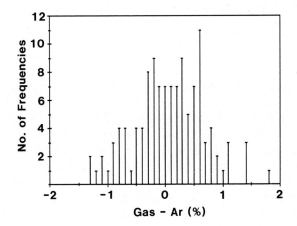

Fig. 1. Comparison of ground-state vibrational frequencies reported for transient molecules (2–16 atoms) in an argon matrix with corresponding values obtained from gas-phase measurements. Beyond scale of plot: C$_6$F$_6^+$ (v_{17}), −9.4%; C$_6$F$_6^+$ (v_{18}), −5.6%; C$_6$HF$_5^+$ (v_{11}), −2.5%; SiCl$_2$ (v_1), +2.7%.

composed of atoms no heavier than those in the first two full rows of the
Periodic Table. Similar generalizations on the relative amount of matrix shift in
various matrices hold for these species. The distribution of matrix shifts found
for solid argon is shown in fig. 1, which includes more recent observations than
those previously reported. Approximately 90% of the vibrational fundamentals
are shifted from their gas-phase values by 1% or less when the molecule is
isolated in solid argon. Shifts of more than 2% are rare.

2.2. Excited electronic states

The presence of an unfilled outer orbital, typical of free radicals, enhances the
probability that the species will possess low-lying excited electronic states, since
often there are electrons in orbitals of only slightly lower energy. The band
origins and excited-state vibrational spacings of electronic transitions of
diatomic molecules, observed in the gas phase and in rare-gas and nitrogen
matrices, have been compared by Jacox (1987a). As discussed in that review, in
observations above 4 K frequently the relatively broad maximum in the phonon
wing of the transition, rather than the sharp zero-phonon line, is detected. In
absorption studies, this phonon maximum generally lies at higher frequencies
than does the zero-phonon line, and in emission studies, at lower frequencies.
Typically, the electronic band origin of a valence transition of a diatomic
molecule isolated in solid neon shows a small blue shift but lies within about 1%
of the gas-phase value. In argon-matrix studies, a small red shift is typical, with
the band origin within about 3% of the gas-phase value. Larger red shifts are
found, successively, for krypton, xenon, and nitrogen matrices. Among the
factors which may enhance the matrix shift is charge transfer with the heavier
rare-gas matrices.

Rydberg states of molecules isolated in rare-gas matrices are much more
strongly perturbed, since the relatively large outer electron orbital generally
interacts strongly with the rare-gas lattice. In the matrix, as in the gas phase,
vibrational structure may be associated with these states. As has been noted in
earlier reviews (Jacox 1978a, 1987a), the lowest Rydberg state may be blue-
shifted from the gas-phase value by as much as several thousand cm^{-1}. On the
basis of the relatively limited data for the higher Rydberg states of molecules in
rare-gas matrices, these states have been believed to be red-shifted compared to
their gas-phase values and to be describable by Wannier-type interactions with
the rare-gas lattice. However, the recent observation by Chergui et al. (1985,
1986) of substantial blue-shifts for higher Rydberg states of NO isolated in rare-
gas matrices has called into question the validity of the Wannier model.

2.3. Methods of free-radical identification

In both gas-phase and matrix studies, free-radical identification is often
supported by chemical arguments. However, the appearance of a set of spectral

features under conditions in which a given free radical is expected – or even determined by such techniques as mass spectrometry or electron paramagnetic resonance spectroscopy – to be present or in several systems in which this species should be a common decomposition product provides less than definitive evidence for a spectroscopic identification. The existence of characteristic group frequencies may provide supporting evidence for an infrared identification. However, as was noted by Jacox (1984d), the unpaired electron typical of free radicals may lead to surprisingly large deviations of molecular vibrations from the range in which similar vibrations fall for stable molecules. Frequently, the complexity of the spectrum indicates that several unidentified products must be present. In such systems, secondary photodecomposition studies using filtered ultraviolet radiation are extremely valuable for sorting out spectral features according to the range of photolytic stability of the species which contribute them. Identification of the products of the photo-decomposition of one of the species present in the sample deposit may provide valuable clues to the identity of that species. However, the most definitive tool for positive spectroscopic identification is isotopic substitution. The sharp absorptions typical of matrix-isolated samples in the infrared spectral region are well suited to isotopic substitution studies. Analysis of the resulting data may lead to a detailed vibrational assignment and may yield information on the strengths of the chemical bonds in the molecule. Except for progressions involving a hydrogen stretching or bending vibration, it is difficult to detect isotopic shifts in the electronic spectra of matrix-isolated molecules. Where the gas-phase electronic spectrum of a free radical has not previously been reported, it is desirable to conduct infrared isotopic substitution studies on the system and then to demonstrate unique behavior (e.g., a similar photodecomposition threshold) which is common to both the vibrational and the electronic spectrum of the species.

3. Stabilization of free radicals

Techniques by which free radicals are stabilized in a matrix can be subdivided into those involving quenching of the products of a gas-phase process and those in which the free radical is formed in situ from a free-radical precursor trapped in the matrix. Gas-phase free-radical formation is characteristic of vaporization of polymeric material, high-temperature reaction systems, and pyrolysis, whereas photolysis occurs principally in the bulk of the sample deposit or on its surface layers. A variety of techniques have been used for studies of atom–molecule reactions. In some systems, the atom is formed by vaporization of a high-temperature material or by a discharge, and in others the atom is formed upon photolysis of the solid deposit. As will be considered in the following, atoms can diffuse to some extent in the matrix. Therefore, atom–molecule reactions in the solid may be important whenever atoms are generated in the system.

3.1. Vaporization of polymeric material

From the earliest days of matrix-isolation experiments, it was recognized that the technique should offer a powerful means of studying the spectra of high-temperature molecules, some of which are electron deficient and thus may also be categorized as free radicals. The first such species to be studied was C_3 (Barger and Broida 1962, Weltner and Walsh 1962, Weltner et al. 1964, Weltner and McLeod 1964a, Barger and Broida 1965), produced by the vaporization of graphite. An important factor in the success of these experiments was minimization of the heat load on the cryogenic observation surface, achieved by post-mixing the matrix gas with the high-temperature vapors, rather than subjecting the matrix material to the high temperature of the vaporization cell. The absorption studies of Weltner et al. (1964) led not only to the detection of the $\tilde{A}-\tilde{X}$ electronic transition near 410 nm, already familiar from gas-phase studies, but also to the identification of the ground-state v_3 fundamental near 2040 cm^{-1}. The emission spectroscopic studies of Weltner and McLeod (1964a) yielded the frequency of the ground-state symmetric stretching fundamental, v_1. In addition, the $\tilde{a}^3\Pi_u-\tilde{X}^1\Sigma_g^+$ emission band system was observed. Although only the $\tilde{A}^1\Pi_u$ excited state of C_3 has been seen in the gas phase, the argon-matrix absorption studies of Chang and Graham (1982b) present evidence for another excited state of C_3 near 53 000 cm^{-1}.

The isoelectronic species SiC_2 has been exceptionally difficult to study in the gas phase. Weltner and McLeod (1964b) obtained the infrared and ultraviolet absorption spectra of SiC_2 by trapping its high-temperature vapors in a neon matrix. At that time, it was presumed that SiC_2, like C_3, is linear. However, recent molecular-beam studies by Michalopoulos et al. (1984) have demonstrated that the SiC_2 molecule forms an isosceles triangle, with the silicon atom at the apex. Detailed matrix-isolation studies by Shepherd and Graham (1985), which included observations on carbon-13 enriched SiC_2 samples, support this revision of the molecular structure.

Depolymerization of low vapor pressure materials under much less extreme conditions has been used to prepare reactive molecules for subsequent trapping in matrices. For example, Jacox and Milligan (1975b) obtained a suitable yield of H_2CS for matrix-isolation studies by mixing the vapors from its trimer, s-trithiane, with an excess of argon.

A closely related technique is laser vaporization, in which the output of a pulsed YAG laser is used to vaporize material from the surface of a solid. Bondybey (1982a) used this technique to obtain SiC_2 for studies of its laser-excited fluorescence in a neon matrix. Subsequently, he conducted many studies of the spectra of diatomic species produced in this way.

3.2. High-temperature reactions

High-temperature reactions have also been used to prepare free radicals. SiF_2 (Hastie et al. 1969) and $SiCl_2$ and $SiBr_2$ (Maass et al. 1972) were stabilized by passing the tetrahalide over silicon heated to a temperature between 900 and 1150°C and co-condensing the resulting mixture with an excess of neon or argon. Kafafi et al. (1983) studied the spectrum of Si_2C prepared by heating elemental silicon in the presence of graphite and co-condensing the resulting vapors with an excess of argon. The high-temperature reaction of silver with $POCl_3$ has been used to prepare POCl for matrix-isolation sampling (Binnewies et al. 1983). The corresponding reaction with $PSCl_3$ was used to prepare PSCl (Schnöckel and Lakenbrink 1983) and with $POFBr_2$ to prepare FPO (Ahlrichs et al. 1986).

3.3. Pyrolysis

The co-condensation of the products of the pyrolysis of an organic molecule with the matrix material has been used for the stabilization of free radicals. Either because of extensive degradation of the parent molecule or because of the occurrence of secondary reactions, a complicated mixture may result, and great care must be taken in the spectroscopic identification of the products. Interpretation of the results of pyrolysis sampling experiments is greatly facilitated by detailed isotopic substitution study. Unfortunately, fluorine possesses only one stable isotope. However, because C–F bonds are exceptionally strong, when relatively simple molecules which contain them are pyrolyzed under relatively mild conditions the C–F bonds are likely not to be broken. Among the fluorocarbon radicals which have been stabilized by pyrolysis are CF_3 (Snelson 1970), C_2F_5 (Butler and Snelson 1980a), and n- and i-C_3F_7 (Butler and Snelson 1980b). By co-condensing CF_3 produced by pyrolysis with a relatively concentrated $Ar:O_2$ mixture, Butler and Snelson (1979) were able to stabilize CF_3O_2 in sufficient concentration for studies of its infrared spectrum. Recent pyrolysis studies in the laboratory of Snelson have also yielded infrared spectral data for CH_3O_2 (Ase et al. 1986) and i-C_3H_7 and i-$C_3H_7O_2$ (Chettur and Snelson 1987).

3.4. Photolysis

3.4.1. Atom diffuses from radical site

The simplest photolysis process by which free radicals may be trapped in an inert matrix is the splitting off of an atom which diffuses from the site of its photoproduction. The stabilization of HC_2 upon vacuum-ultraviolet photolysis of acetylene isolated in an argon matrix (Milligan et al. 1967, Chang and Graham 1982a, Shepherd and Graham 1987) exemplifies this process. Many other examples of this mode of free-radical stabilization have been considered in previous reviews (Milligan and Jacox 1972, Jacox 1978a).

3.4.2. Simple trapping of molecular products

Where there is a substantial barrier to the recombination of molecular fragments produced by photolysis of the matrix-isolated precursor, free radicals may be trapped in sites adjacent to other molecular fragments. Studies in the laboratory of Pacansky have led to the stabilization of alkyl radicals upon photolysis of diacyl peroxides in an argon matrix. The O–O bond splits, yielding $R-CO_2$ and $R'-CO_2$ fragments, which readily split off CO_2. The two CO_2 molecules formed in this manner suffice to inhibit the $R + R'$ reaction in the solid. Alkyl radicals stabilized in this way include CH_3 (Pacansky and Bargon 1975), C_2H_5 (Pacansky et al. 1976, Pacansky and Dupuis 1982, Pacansky and Schrader 1983), n-C_3H_7 (Pacansky et al. 1977), i-C_3H_7 (Pacansky and Coufal 1980), n-C_4H_9 (Pacansky et al. 1977, Pacansky and Gutierrez 1983), i-C_4H_9 (Pacansky et al. 1981), and t-C_4H_9 (Pacansky et al. 1981, Schrader et al. 1984). The observation of methyl benzoate, but not of phenyl acetate, on photolysis of $C_6H_5C(O)O-OC(O)CH_3$ suggests that the benzoyloxy radical is considerably more stable than the acetoxy radical (Pacansky and Brown 1983).

A variation on the photodecomposition of a stable molecule to form molecular products, one of them a transient species, is the photodecomposition of an initially formed molecular complex. Recent studies of the photodecomposition of a number of ozonides have led to the stabilization of several transient molecules, trapped in sites adjacent to the molecular oxygen product. Hawkins and Andrews (1985) have reported infrared spectral data for CH_3IO, CH_3OI, and ICH_2OH formed upon photodecomposition of $CH_3I\cdots O_3$. Andrews et al. (1985) have also reported the infrared spectra of CF_3IO and CF_3OI formed in a similar manner, and Withnall and Andrews (1987) have stabilized H_3PO, H_2POH, and HPO on photolysis of $H_3P\cdots O_3$.

3.4.3. Formation of radical in an excited state

If the free radical is formed in an excited electronic state, deactivation may be accompanied by rearrangement to a more stable ground-state structure. This process is exemplified by matrix studies of the products of the photodecomposition of methyl azide, CH_3N_3. The nitrene, CH_3N, and N_2 are formed in the primary photodecomposition. Recent studies of CH_3N in a cold molecular beam (Carrick and Engelking 1984) have demonstrated that the ground state of this species has $^3A_2$ symmetry and that its first excited state possesses ^3E symmetry. Photolysis of CH_3N_3 in a rare-gas or a nitrogen matrix leads to the stabilization of $CH_2 = NH$ (Milligan 1961, Jacox and Milligan 1975a), either because of Jahn–Teller distortion of the $\tilde{A}^3E$ state of CH_3N or because of rearrangement of an excited singlet state formed by a spin-allowed process in the primary photodecomposition. In a recent discharge-sampling study, Ferrante (1987) has been able to stabilize CH_3N in a nitrogen matrix, presumably because the radical is formed in an electronic state which is not susceptible to rearrangement.

A somewhat more complicated rearrangement is represented in the recent report by Sylwester and Dervan (1984) of the stabilization of H_2NN on photolysis of H_2NCON_3 in an argon matrix. The primary photodecomposition of the azide involves detachment of N_2. Rearrangement of the photofragment is followed by elimination of CO, which is also detected.

3.4.4. Cage recombination

Many photodecomposition processes result in the formation of fragments which have only a very low barrier to reaction. The role of cage recombination must then be considered in order to understand the processes which occur in the matrix environment. An important special case results when one of the photofragments is an atom. Early electron-spin resonance studies (Foner et al. 1960, Adrian 1960) indicated that even hydrogen atoms can be trapped to some extent in rare-gas matrices. However, there is an abundance of chemical evidence that even at 4 K atomic diffusion is to some extent possible in the rare-gas matrices. Although diffusion plays a less important role for the heavier atoms (Milligan and Jacox 1967, Jacox 1978a), the stabilization of CF_2Br on mercury-arc photolysis of $Ar:CF_2Br_2$ (Jacox 1978b) indicates that some diffusion even of Br atoms can occur. Although limited diffusion of H atoms can occur at lower temperatures, their rapid diffusion through solid argon has a sharp onset near 20 K (Willard 1982). When the H-atom concentration is relatively high, this onset is accompanied by emission in the blue spectral region.

In many systems, thermoluminescence is observed when the initial sample deposit is warmed only two or three degrees and continues while the deposit remains on the cryogenic surface. This emission, which is sometimes contributed by spectroscopically forbidden transitions, has yielded valuable information both about diffusion in matrices and about previously inaccessible molecular electronic states. Lalo et al. (1976) observed the $\tilde{a}^3B_1 - \tilde{X}^1A_1$ phosphorescence of SO_2 formed in the $S + O_2$ and $O + SO$ recombinations. Long and Pimentel (1977) identified the $\tilde{B}^3\Sigma_u^- - \tilde{X}^3\Sigma_g^-$ emission of S_2 on warming a photolyzed $Ar:OCS$ sample and determined that a thermoluminescence previously reported for this system by Brom and Lepak (1976) is contributed by SO_2. The S + S thermoluminescence was also studied by Fournier et al. (1977b) and by Lee and Pimentel (1979), who identified the previously unreported $c^1\Sigma_u^- - a^1\Delta_g$ and $A'^3\Delta_u - X^3\Sigma_g^-$ transitions of S_2. Fournier et al. (1977a) assigned to NO_2 three bands which appeared in thermoluminescence between 400 and 450 nm when O atoms and NO were present in the sample. These observations indicated that $O(^3P)$ atoms can migrate through the argon lattice even at 8 K. Both Lee and Pimentel (1978) and Tevault and Smardzewski (1978) have assigned a thermoluminescence associated with the $S + O$ reaction to the previously unreported $c^1\Sigma - a^1\Delta$ transition of SO. Tevault and Smardzewski also attributed a weaker thermoluminescence to the $A'^3\Delta - X^3\Sigma^-$ transition of SO. The $O + O$ thermoluminescence has been studied by Smardzewski (1978) and by Fournier et al.

(1979), who identified the Herzberg I ($A^3\Sigma_u^+ - X^3\Sigma_g^-$) bands of O_2. When CO was added to the system, Fournier and co-workers saw instead a thermo-luminescence which they assigned to the previously undetected $^3B_2 - \tilde{X}^1\Sigma_g^+$ transition of CO_2. There followed an interesting series of studies in their laboratory of the photodecomposition of CO_2. Mohammed et al. (1980) measured a lifetime of 570 ± 20 ms for the emission, consistent with its assignment to a forbidden transition of CO_2. They noted that the lack of structure suggests perturbation by a dissociative state of CO_2, possibly the $^1A_2$ state, calculated to lie within about 1 eV of the $^3B_2$ state. Studies by Fournier et al. (1982) produced evidence that $O(^1D)$, initially formed in the photo-decomposition of CO_2, forms an excimer with argon or the heavier rare gases, but not with neon. On warming the sample, thermal deactivation of the excimer occurs. Further details of the excimer model have been given by Maillard et al. (1982).

The occurrence of atomic diffusion in the matrix may lead to an under-estimate of the extent of decomposition in discharge-sampling experiments. Since the absorptions of SiF and SiF_2 did not appear in the deposit resulting from matrix-isolation sampling of the products of a discharge through SiF_4 (Wang et al. 1973), it was concluded that a complicated emission system observed in the discharge was contributed solely by SiF_3. The possibility of recombination of F atoms from the discharge with SiF_2 in the matrix-isolated sample was not considered. In fact, one of the Deslandres tables presented in the analysis of the "SiF_3" emission bands agrees, within experimental error, with the previously determined (Khanna et al. 1967) array for SiF_2. Mathews (1986) has since reassigned the remaining bands to SiF_2.

As molecules cannot diffuse through rare-gas or nitrogen matrices, when there is little barrier to their recombination, the molecular fragments may re-form the precursor molecule, giving an apparent inhibition of photodecomposition. The classic example is that of the recombination in the matrix of CH_2 with N_2 to form diazomethane, which undergoes irreversible photodissociation in the gas phase (Moore and Pimentel 1964).

Apparent inhibition of photodissociation has been observed in several systems in which a free radical is formed by the reaction of an atom with a molecular matrix. In studies of the reaction of photochemically produced F atoms with a CO matrix, Milligan et al. (1965a) obtained a good yield of FCO. However, only a small amount of FCO was stabilized when the F-atom precursor and CO were present in small concentration in an argon matrix. Jacox (1980a) has obtained extremely high yields of FCO in an argon matrix by using a discharge F-atom source, which has only a small photon flux associated with it. The secondary photodecomposition of FCO could then be studied using filtered mercury-arc radiation. Another example is provided by FO_2, formed in several studies (Arkell 1965, Spratley et al. 1966, Noble and Pimentel 1966) by the reaction of photochemically generated F atoms either with an O_2 matrix or

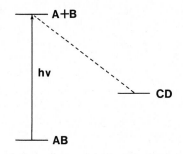

Fig. 2. Rearrangement on cage recombination.

with a relatively concentrated $Ar:O_2$ or $N_2:O_2$ sample. A maximum in the FO_2 concentration and growth in the intensity of the O_2F_2 absorptions on prolonged photolysis indicated that secondary photodecomposition was occurring. Jacox (1980c) obtained extremely high yields of FO_2 with much lower concentrations of O_2 in the argon matrix by using a discharge source of F atoms. The photodecomposition of FO_2 was also further characterized in these experiments.

A number of more complicated processes may occur as a result of cage recombination in the matrix. The first of these, illustrated in fig. 2, is rearrangement. This process is exemplified by the photodecomposition of CH_3NO_2. Gas-phase studies have indicated that under a wide range of conditions the primary process involves the formation of $CH_3 + NO_2$. However, gas-phase studies by Yamada et al. (1981) indicated that the reverse reaction leads to the formation of CH_3ONO. In an early study of the photodecomposition of CH_3NO_2 in an argon matrix, Brown and Pimentel (1958) observed many products, including *cis*- and *trans*-CH_3ONO. A recent study by Jacox (1984c) has yielded additional information regarding the identities of the other products and the processes by which they are formed.

As is illustrated in fig. 3, disproportionation can also occur on cage recombination. The photolysis of CH_3ONO in a matrix exemplifies this type of

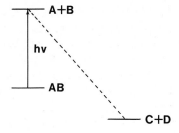

Fig. 3. Disproportionation on cage recombination.

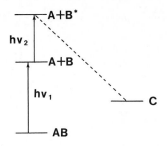

Fig. 4. Photoinduced cage recombination.

reaction. The photofragments CH_3O and NO, formed in the primary photo-decomposition, recombine to form H_2CO and HNO, rather than CH_3ONO. In the matrix, the H_2CO and HNO are trapped in adjacent sites in the lattice and form a hydrogen-bonded complex (Jacox and Rook 1982, Müller et al. 1982b). Müller et al. (1982a) have reported a similar disproportionation on photolysis of $(CH_3)_2NNO$ in a matrix, with $H_3C-N = CH_2 \cdots HNO$ the observed product.

Still another possible process is photoinduced cage recombination, illustrated in fig. 4. This process has been considered in some detail (Müller and Huber 1983, Müller et al. 1984a) for $H_2CO \cdots HNO$. When the sample is exposed to radiation of wavelength longer than 645 nm, exciting the HNO moiety, the previously unidentified species cis-nitrosomethanol is formed. On the other hand, if 345 nm radiation, which photoexcites the H_2CO moiety, is employed, trans-nitrosomethanol is instead formed. At wavelengths shorter than 510 nm, cis-nitrosomethanol is converted to the trans-structure, and at longer wavelengths trans-nitrosomethanol is converted to the cis-form. Müller et al. (1984b) have found that the $CH_3CHO \cdots HNO$ hydrogen-bonded complex, formed by the photolysis of matrix-isolated C_2H_5ONO, behaves similarly, except that only a single rotamer of nitrosoethanol is stabilized in the matrix. Photoexcitation of the HNO moiety in matrix-isolated $H_3C-N = CH_2 \cdots HNO$ leads to the stabilization of nitrosomethylmethylamine.

Chemiluminescence may also occur as a result of cage recombination, as is shown in fig. 5. This chemiluminescence differs from the thermoluminescence previously discussed in that it occurs during the photolysis process and is amenable to study in pulsed photolysis systems. The photophysics of such processes has been explored for excitation above the dissociation limit of a bound upper state by Bondybey and Brus (1975, 1976) and for excitation of a dissociative upper state by Brus and Bondybey (1975). In a number of systems, including halogen-atom recombination (Bondybey and Fletcher 1976, Beeken et al. 1983, Fournier et al. 1985, Nicolai et al. 1985), Cu-atom recombination (Bondybey 1982b), and the previously discussed $O + CO$ recombination (Mohammed et al. 1980, Fournier et al. 1982), evidence for new electronic states

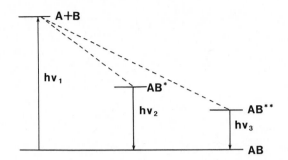

Fig. 5. Pumping of excited states by cage recombination.

which are not accessible in absorption from the ground state has been obtained in such experiments.

3.4.5. Secondary photodecomposition

Analysis of the spectrum of a sample deposit in which free radicals have been formed by photolysis is often complicated by the occurrence of secondary photodecomposition. In experiments in which acetylene was photolyzed by 122 nm radiation, leading to the first identification of the HC_2 free radical, Milligan et al. (1967) also observed the very prominent Mulliken band of C_2, formed by photodetachment of a second H atom. In addition, C_2^- was stabilized (Milligan and Jacox 1969b) by secondary photolysis processes that are not yet fully understood. 122 nm photolysis of either methylacetylene or allene in an argon matrix leads to the stabilization not only of C_3H_3 but also of C_3H_2, C_3H, and C_3 (Jacox and Milligan 1974). Secondary photodecomposition may be accompanied by rearrangement, as in the stabilization of CNC on 122 nm photolysis of CH_3CN isolated in solid argon (Jacox 1978c). It may also occur in binary reaction systems, as in the stabilization of PO and HPO on 122 nm photolysis of $Ar:PH_3:N_2O$ samples (Larzillière and Jacox 1979, 1980).

3.5. Discharge sampling

3.5.1. Free radical formed in discharge

Even at low power levels, electric and microwave discharges cause extensive fragmentation of molecules which pass through them. Therefore, only in a few relatively simple systems has it been feasible to sample free radicals formed in the discharge region. For example, in discharge-sampling experiments on $Ar:N_2:X_2$ mixtures (X = Cl, Br), Kohlmiller and Andrews (1982) have identified infrared absorptions of NCl_2 and NBr_2.

3.5.2. Role of excited rare-gas atoms

A low-power microwave discharge through a mixture of an atom precursor with an excess of argon has often been used as an atom source for studies of atom–molecule reactions in matrices. The atom precursors and almost all atoms other than argon have excited energy levels below the first excited levels of argon, which lie between 11.5 and 11.8 eV. Collisions with such species generally will rapidly degrade the argon-atom excitation. However, although H atoms can collisionally deactivate argon atoms excited in the discharge, when a discharged $Ar:D_2$ mixture was co-deposited with an $Ar:C_2H_2$ sample a good yield of HC_2 was obtained, but there was no evidence for the stabilization of deuterium-containing products (Jacox 1975). A similar yield of HC_2 was obtained when pure argon was passed through the discharge. These results suggest that deactivation of excited argon atoms upon collision with H atoms or with H_2 is relatively inefficient.

The discovery that excited argon atoms are effective in producing HC_2 from C_2H_2 outside the discharge region has led to studies on many other systems in which pure argon is passed through the discharge and the resulting beam is mixed with an $Ar:XY$ sample outside the discharge region, followed by quenching of the products on the cryogenic observation surface. The processes which can occur in this system have been considered in detail by Jacox (1978a). Often prominent absorptions of ion products have been identified in such discharge-sampling experiments. However, this discussion will be limited to consideration of the neutral free-radical products which result in excited argon-atom co-deposition studies.

An important advantage of the use of excited argon atoms for the formation of free radicals in matrix-isolation sampling systems is the reduction of the importance of secondary photodecomposition, compared to conventional photolysis systems. In 122 nm photolysis studies on matrix-isolated C_2H_2, Milligan et al. (1967) found the yield of HC_2 to be highly variable; for dilute $Ar:C_2H_2$ samples which were deposited relatively slowly the absorptions of C_2 and of C_2^- predominated. In subsequent studies using a beam of excited argon atoms as the photodecomposition source (Jacox 1975, Jacox and Olson 1987), a consistently high yield of HC_2 was obtained, leading to the first observation of the $\tilde{A}^2\Pi$–$\tilde{X}^2\Sigma^+$ absorption-band system of HC_2 by Jacox and Olson. Use of a beam of excited argon atoms, co-deposited with an $Ar:CH_3OH$ sample, led to a much higher yield of CH_2OH (Jacox 1981c) than had been obtained in earlier studies of the 147 and 122 nm photolysis of $Ar:CH_3OH$ deposits (Jacox and Milligan 1973a). Recently, Jacox (1988a) has found that the co-deposition of a beam of excited argon atoms with an $Ar:HCOOH$ sample leads to the stabilization of t-HOCO. In earlier experiments, Milligan and Jacox (1971) had obtained this important combustion reaction intermediate by subjecting $CO:H_2O$ deposits to 122 nm photolysis. However, t-HOCO can form a hydrogen bond with the CO matrix. Since OH and CO cannot diffuse through an argon matrix, the yield of t-HOCO in $Ar:CO:H_2O$ photolysis experiments was below the threshold for detection.

In some systems, the neutral products obtained in excited-argon-atom studies differ from those typical of 122 nm photolysis experiments. Studies of the co-deposition of a beam of excited argon atoms with an $Ar:CH_3CN$ sample (Jacox 1979b) led to the stabilization of a high concentration of ketenimine, $H_2C = C = NH$, in contrast to the 122 nm photolysis experiments (Jacox 1978c), in which CNC was a major product and ketenimine was not detected. Although the excited argon atoms and their resonance radiation can interact with the free-radical precursor over a path of approximately 2.5 cm, corresponding to a time of approximately 10^{-4} s for a sonic beam, before the products are frozen onto the cryogenic observation surface, usually there is little evidence for the stabilization of products which would be destroyed by cage recombination. However, in studies of the co-deposition of an $Ar:CF_3NNCF_3$ sample with a beam of excited argon atoms a high yield of isolated CF_3 was observed (Jacox 1984a).

3.6. Atom–molecule reactions

One of the most important modes by which free radicals are stabilized in inert matrices is that of atom–molecule reaction. Because atoms, unlike molecules, can diffuse through the matrix, there is a high probability for their reaction with a molecule present in small concentration. A variety of techniques have been used for introducing atoms into the deposit. Vaporization of an atomic solid has been useful for studies of the reactions of metal atoms and of atomic silicon. In situ photolysis is also frequently used for the generation of atoms; mercury-arc photolysis of HI or vacuum-ultraviolet photolysis of H_2O have been used as H-atom sources, mercury-arc photolysis of t-N_2F_2 or of F_2 leads to F-atom production, and mercury-arc photolysis of N_3CN leads to the production of C atoms. In some systems it is unnecessary to add an atom source, as an atom which is photodetached from one precursor molecule may undergo an addition reaction with another. Discharge sampling is also a valuable mode of introducing atoms into the system. It has been particularly useful in the study of the matrix-isolated products of F-atom reaction. In the following discussion, specific examples of all of these modes of atom production will be given. With few exceptions, the discussion will be limited to results obtained since an earlier review (Milligan and Jacox 1972).

3.6.1. H atoms
The first atom–molecule reaction to be studied in matrices was that of H atoms with a CO matrix. Ewing et al. (1960) identified two of the vibrational fundamentals of HCO and one of DCO formed by the mercury-arc photolysis of HBr, HI, DBr, or DI isolated in solid CO. As a result of similar experiments which also used H_2S as a photolytic H-atom source, Milligan and Jacox (1964) completed the vibrational assignment, demonstrating that the CH-stretching fundamental appeared at the exceptionally low frequency of 2488 cm^{-1} and that

the two stretching fundamentals of DCO are in Fermi resonance. Subsequently, Milligan and Jacox (1969a) were able to detect the vibrational fundamentals of HCO isolated in solid argon and to assign the $\tilde{B}-\tilde{X}$ absorption bands of HCO, and Jacox (1978e) assigned the $\tilde{C}-\tilde{X}$ absorption bands of HCO. Emission bands arising from these two transitions of HCO constitute the so-called hydrocarbon flame bands, which for many years had resisted a detailed assignment.

The first assignment of a polyatomic free radical isolated in a rare-gas matrix was that of HO_2, obtained by Milligan and Jacox (1963) in studies of the mercury-arc photolysis of $Ar:O_2:HI$ samples. The production of a significant yield of HO_2 in these experiments required the diffusion of the H atom from the site of its photoproduction and its efficient scavenging by O_2 in the matrix. Detailed deuterium and oxygen-18 isotopic-substitution studies supported the identification of all three vibrational fundamentals and determined that the molecule has a bent HOO structure. HO_2 frequently contributes to the product spectrum in photolysis or discharge-sampling experiments when oxygen impurity is present in the system; the desorption of water from the walls of the vacuum system provides an H-atom source which it is difficult to eliminate.

Bondybey and Pimentel (1972) first reported absorptions at 905 and 644 cm^{-1} in the spectra of the matrix-isolated products of $Ar:H_2$ and $Ar:D_2$ samples, respectively, that had been passed through a glow discharge. A small but definite argon-isotopic shift was noted. When krypton was substituted for argon, counterparts of these absorptions appeared at 852 and 607 cm^{-1}. Assignment of the absorptions to H or D atoms trapped in the rare-gas lattice was proposed. However, Milligan and Jacox (1973) noted that these peaks also appeared in studies of the 122 nm photolysis of a variety of hydrogen-containing compounds isolated in an argon or a krypton matrix. Common factors in all of the systems in which the peaks were relatively prominent were the presence of energy sufficient to electronically excite H atoms and the presence of a good electron acceptor. In the glow-discharge experiments, O_2 impurity may serve as an electron acceptor. Milligan and Jacox proposed the reassignment of these absorptions to HAr_2^+, DAr_2^+, HKr_2^+, and DKr_2^+, with a possibility that the cluster ion might instead contain four or six rare-gas atoms. A possible mechanism for the production of these species (illustrated for argon) is the reaction

$$H^*(2s) + Ar \rightarrow HAr^+ + e,$$

proposed by Chupka and Russell (1968), followed by cluster-ion formation with argon atoms in the matrix. Subsequent discharge sampling experiments by Wight et al. (1976) demonstrated that the absorptions assigned to HAr_2^+ and HKr_2^+ did not appear when H atoms were present but a supplementary high-energy source suitable for inducing photo-ionization or electron transfer was not. Moreover, Jacox (1976) obtained further support for the reassignment by showing that these absorptions became prominent in filtered photolysis studies of samples containing a high concentration of $HCCl_2^+$ or $DCCl_2^+$ when the threshold for photoinduced proton transfer from the dichloromethyl cation to the rare gas was exceeded.

Although most of the H-atom reactions studied in matrix-isolation experiments have been with stable molecules, Jacox (1980b) has reported the stabilization of HCF_2 in a study of the reaction of H atoms with CF_2.

A low-power microwave discharge through an $Ar:H_2$ sample provides a source of H atoms which is free of the chemical or spectral complications that may result from the presence of other molecular fragments. Furthermore, since D_2 is easily obtained and does not undergo ready isotopic exchange, the technique provides a convenient source of D atoms. In matrix sampling studies of the reaction with NO of H and D atoms formed in this way, Jacox and Milligan (1973b) obtained the spectra of five isotopic species of HNO, leading to the assignment of the NH stretching fundamental at $2716 \, cm^{-1}$, an anomalously low value, and to the reassignment of the bending fundamental at $1505 \, cm^{-1}$. These assignments have since been confirmed by gas-phase studies (Clough et al. 1973, Johns and McKellar 1977, Johns et al. 1983). Recently, a discharge source of H atoms has been used by Jacox (1987c) for studies of the $H + HCN$ reaction. Five of the six vibrational fundamentals of H_2CN isolated in solid argon have been observed, and further information on the assignment of the ultraviolet absorption bands of H_2CN has been obtained.

3.6.2. Metal atoms

Pioneering studies of lithium-atom reactions have been conducted in the laboratory of Andrews. Many free radicals have been stabilized in an argon matrix upon halogen-atom abstraction by lithium. Reviews by Andrews (1971) and by Milligan and Jacox (1972) have summarized the results of these studies.

The heavier group I metals also may abstract a halogen atom in a matrix environment. Supplementary studies of sodium-atom reactions have often been useful in determining whether lithium-atom addition products contribute to the spectra obtained for lithium-atom reaction systems.

Lithium atoms also undergo addition reactions. The formation of LiO_2 upon reaction of lithium atoms with O_2 in an argon matrix (Andrews 1969, Andrews and Smardzewski 1973) involves an appreciable amount of charge transfer. Recent studies (Manceron and Andrews 1985, 1986) have demonstrated that lithium atoms form covalently bonded complexes with C_2H_2 and C_2H_4.

Charge transfer predominates in the interactions of the heavier alkali metal atoms, which have relatively low ionization potentials, with small molecules. These systems have been considered in detail in an earlier review (Jacox 1978a).

Although charge transfer can occur for other metals, recent studies have obtained a great deal of evidence for complex formation and metal-atom insertion reactions in the matrix. The reaction with H_2O is particularly important, since its products are likely to contribute to the spectrum in virtually any other system that may be studied. This reaction has recently been studied using Mg, Ca, Sr, and Ba (Kauffman et al. 1984, Douglas et al. 1984), Sc, Ti, and V (Kauffman et al. 1985b), Cr, Mn, Fe, Co, Ni, Cu, and Zn (Kauffman et al. 1985a), and Al, Ga, and In (Hauge et al. 1980, Douglas et al. 1983). These and other metal-atom addition reactions are reviewed by Hauge et al. in ch. 10 of this volume.

3.6.3. C atoms

The discovery that cyanogen azide, N_3CN, could serve as a photolytic source of carbon atoms for matrix isolation studies (Milligan et al. 1965b, Jacox et al. 1965) led to an extensive series of carbon-atom reaction studies. These have been summarized in a previous review (Milligan and Jacox 1972). More recently, the previous investigation of the stabilization of CNN in the $C + N_2$ reaction (Milligan and Jacox 1966) has been extended, leading to the identification of new electronic band systems of CNN between 190 and 260 nm (Jacox 1978d).

3.6.4. Si atoms

The reaction of silicon atoms, produced by high-temperature vaporization, with a number of small molecules has been studied using matrix-isolation sampling. A study of the silicon-atom reactions with CO and with N_2 yielded electron-spin resonance, infrared, and ultraviolet spectra for the SiCO and SiNN free radicals (Lembke et al. 1976). Observation of the infrared spectrum of the $Si + H_2$ reaction products (Fredin et al. 1985) dictated a reassignment of the infrared spectrum of SiH_2 from that tentatively proposed as a result of the studies by Milligan and Jacox (1970) of the vacuum-ultraviolet photolysis of matrix-isolated SiH_4, in which extensive secondary photodecomposition occurred. Ismail et al. (1982a) have studied the spectrum of HSiOH, formed by the reaction of silicon atoms with H_2O in an argon matrix, and Ismail et al. (1982b) have identified HSiF, produced in the isoelectronic Si + HF reaction.

3.6.5. O atoms

In an early matrix isolation study, DeMore and Davidson (1959) found that O atoms formed by the mercury-arc photolysis of O_3 react with a nitrogen matrix, forming N_2O. When NO_2 was added to the sample, the visible absorption bands of NO_3 were also detected. Haller and Pimentel (1962) found that ketene was stabilized on 147 nm photolysis of N_2O in an argon matrix sample to which $C_2H_2-d_n$ had also been added. Hawkins and Andrews (1983) used O_3 as the photolytic O-atom source for a study of the O-atom reaction with ethylene in an argon matrix, leading to an assignment of the infrared spectrum of vinyl alcohol, $CH_2 = CHOH$. Withnall and Andrews (1985) used both ozone photolysis and a microwave discharge through an $Ar:O_2$ sample as O-atom sources for studies of the $O + SiH_4$ reaction products in an argon matrix. Although HSiOH predominated in the discharge-sampling experiments, a number of other products, resulting from secondary photodecomposition and from multiple O-atom addition, were also identified in the ozone photolysis experiments. Withnall and Andrews (1986) have also used ozone photolysis for studies of the reaction of O atoms with the methylsilanes isolated in an argon matrix, leading to the identification of infrared absorptions of methyl- and dimethylsilanone.

3.6.6. S atoms
The spectra of the products resulting from the reaction of photolytically generated S atoms with NO were studied by Hawkins and Downs (1984). Subsequently, Hawkins et al. (1985) studied the reaction of S atoms with PF_3, PCl_3, CH_4, C_2H_2, and C_2H_4 in a matrix environment.

3.6.7. F atoms
The infrared spectra of the products of the reaction of F atoms with a wide variety of molecules have recently been obtained in discharge sampling experiments. Since a detailed review of gas-phase and matrix-isolation studies of the reactions of F atoms with small molecules (Jacox 1985a) has recently appeared, only a brief summary of matrix isolation studies of the spectra of free radicals produced by F-atom reactions is given.

The first type of F-atom reaction is addition. The stabilization of FCO and of FO_2 has already been discussed. F atoms add to multiply bonded species to form fluorine-containing free radicals. The principal product of the $F + C_2H_2$ reaction is the t-2-fluorovinyl radical, for which the infrared data are sufficiently complete to permit a detailed normal coordinate analysis (Jacox 1980e). The $F + C_2H_4$ reaction yields the 2-fluoroethyl radical (Jacox 1981a). When FCD_2CH_2 or FCD_2CD_2 is irradiated with the full light of a medium-pressure mercury arc, vinyl fluoride and D atoms are formed. However, when FCH_2CH_2 or FCH_2CD_2 is similarly irradiated there is no change in the sample composition. Cage recombination of the H atom by a tunneling mechanism was proposed to explain this disparity. The $F + C_2F_4$ reaction gives a good yield of C_2F_5 (Jacox 1984b). The reaction of F atoms with C_6H_6 yields prominent absorptions of the 1-fluorocyclohexadienyl radical, with no evidence for the stabilization of an intermediate in which the F atom is bound to the π-electron system (Jacox 1982a).

F atoms form moderately strong bonds to the heavier halogen atoms in a molecule. Infrared absorptions of products of formula CF_3XF were observed in studies of the reaction of F atoms with CF_3X (X = Cl, Br, I) in an argon matrix (Jacox 1980d). Since three CF-stretching absorptions were assigned for each of these species, the CXF chain must be bent, as has been found for matrix-isolated ClF_2 and BrF_2 (Mamantov et al. 1971, Prochaska and Andrews 1977, Prochaska et al. 1978). Filtered photolysis results in the stabilization of CF_4; the relative photolytic stabilities of the three species are Cl > Br > I.

In the analogous studies on the $F + CH_3X$ reaction system, Jacox (1985b) has assigned very prominent absorptions to the X–F stretching fundamentals of the $H_3C–X–F$ addition complexes (X = Cl, Br, I). Previous gas-phase studies by Farrar and Lee (1974) had placed a lower bound of 105 kJ/mol on the IF bond strength in CH_3IF, but there were no previous studies demonstrating the stability of CH_3BrF or CH_3ClF. In the matrix studies, the X–F stretching absorptions appeared at a significantly lower frequency than that reported by Prochaska et al. (1978) for the XF diatomic molecules isolated in solid argon and at a somewhat lower frequency than that for the CF_3XF species (Jacox 1980d), suggesting that the relative order of the X–F bond strengths is $XF > CF_3XF > CH_3XF$. All three of the CH_3XF species were photolytically stable out to approximately 250 nm.

Recently, Maier and Reisenauer (1986) have reported the photo-isomerization of several haloiodomethanes isolated in an argon matrix and subjected to mercury-arc radiation, forming species with the general formula H_2CXI. Each of these species showed a very strong absorption in the visible or near ultraviolet, and irradiation in that band re-formed the initial haloiodo-methane. Maier and Reisenauer tentatively ascribed the new infrared and visible–ultraviolet absorptions to the stabilization of charge-transfer complexes, $H_2CX^+I^-$. Whether charge transfer contributes to the bonding of CF_3XF and CH_3XF remains to be established. The XF stretching modes were relatively weak for the CF_3XF species, consistent with covalent bonding. However, very strong XF-stretching absorptions appeared for CH_3ClF and CH_3BrF and a moderately intense IF stretching absorption was identified for CH_3IF, suggest-ing that charge transfer may play a significant role in the bonding of these species.

The second major type of F-atom reaction is H-atom abstraction, forming HF and a free radical. This type of reaction has been extensively studied in the gas phase, since typically it is exothermic by approximately 160 kJ/mol. Analysis of the energy distribution in the HF product has been of great interest in the development of the theory of chemical reactions. The first system studied using matrix isolation was the $F + CH_4$ reaction. Prominent absorptions of CH_3 and of HF were observed in the resulting sample deposit. The diffusion of F atoms through the matrix resulted in the stabilization of CH_3F, as well. Prominent new absorptions at 664, 1003, and 3764 cm^{-1} could not immediately be identified.

Further consideration of the processes which can occur on co-depositing a beam of F atoms with a hydrogen-containing molecule, RH, aids in the assignment of these new absorptions. These processes are illustrated in fig. 6. In its approximately 2.5 cm flight path through the "mixing region", before it collides with the cryogenic surface, the F atom can, in a typical experiment, experience two or three collisions with RH. Because the reactivity of the F atom is so great, there is a significant probability of reaction in a single collision. The free radical, R, and HF formed in such a reactive collision will, in general, be trapped in isolated sites in the solid argon. If the F atom collides with an RH

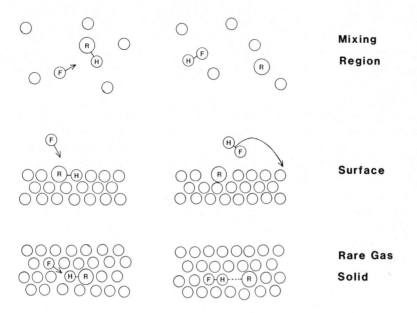

Fig. 6. H-atom abstraction by F atoms in discharge sampling matrix isolation experiments.

molecule on the surface of the solid deposit, there is a substantial probability that the HF will recoil from the surface and be trapped in an isolated site in the lattice. Once the F atom is trapped in the solid, it can continue to diffuse. However, HF formed by a reactive collision with RH in the solid cannot escape from the site of its production, but will hydrogen-bond to the free radical, R. In this process, six new vibrational absorptions will result. The first of these, the H–F stretching mode of $R \cdots HF$, generally lies in the 3000–4000 cm^{-1} spectral region. Two more absorptions, which become degenerate if the axis of the hydrogen bond has three-fold or greater symmetry, correspond to a "flexing" motion of the HF with respect to the radical moiety. These absorptions typically appear in the 300–800 cm^{-1} spectral region. The remaining three vibrations, the $R \cdots H$ stretching mode and two skeletal deformation modes, fall in the far infrared.

These arguments suggest that the hydrogen-bonded species $CH_3 \cdots HF$ should contribute to the spectrum in the $F + CH_4$ reaction studies. Furthermore, if an F atom collides with $CH_3 \cdots HF$ in the solid, $CH_3F \cdots HF$ might also be formed. Detailed isotopic substitution studies by Jacox (1979a) resulted in the assignment of the 664 cm^{-1} absorption to the out-of-plane CH_3 deformation fundamental of $CH_3 \cdots HF$, of the 3764 cm^{-1} absorption to the H–F stretching fundamental of $CH_3 \cdots HF$, and of the 1003 cm^{-1} absorption to the CF-stretching fundamental of $CH_3F \cdots HF$. Subsequent studies by Johnson and Andrews (1980) of the photolysis of $Ar:CH_4:F_2$ samples confirmed the assignment of the 1003 cm^{-1} peak to CH$_3$F$\cdots$HF. In these studies, two F atoms are formed in adjacent sites in the solid, increasing the probability of the secondary F-atom reaction. When the photolyzed sample was warmed to 25 K, enhancing the probability of F-atom diffusion, the CH$_3$$\cdots$HF absorptions appeared.

A number of other F-atom reactions which result in H-atom abstraction have also been studied using matrix-isolation sampling. Prominent absorptions of H$_2$CF and of H$_2$CF$\cdots$HF were identified in the F + CH$_3$F study (Jacox 1981b). In this study, the threshold for the photodecomposition of H$_2$CF into CF + H$_2$ was found to lie near 280 nm. The corresponding photodecomposition of H$_2$CF$\cdots$HF resulted in the appearance of a prominent absorption at 1337 cm^{-1} which was tentatively assigned to CF$\cdots$HF. In the study of the isoelectronic F + CH$_3$OH reaction, H-atom abstraction from either the CH$_3$ or the OH group could occur. However, in the matrix-isolation experiments (Jacox 1981c) high yields of both CH$_2$OH and CH$_2$OH$\cdots$HF were obtained, but CH$_3$O was not detected. On the other hand, in discharge-sampling studies of the F + CH$_3$SH reaction Jacox (1983a) has obtained evidence for H-atom abstraction from both types of H atom. CH$_2$SH was formed with sufficient energy to surmount the barrier to its rearrangement to CH$_3$S, the more stable isomer. Isomerization of the more stable CH$_3$S structure to produce CH$_2$SH occurred only on photoexcitation of CH$_3$S. H-atom abstraction from both types of CH bond was observed in studies of the F + CH$_3$CHO reaction (Jacox 1982b), leading to infrared spectroscopic data for CH$_3$CO and H$_2$CCHO, important free radicals in combustion and atmospheric pollution systems. When F atoms react with HCOOH (Jacox 1988a), the H atom may be abstracted from either site. However, even at 14 K the H–CO$_2$ product fragments into H + CO$_2$, and only CO$_2$, CO$_2$$\cdots$HF, t-HOCO and t-HOCO$\cdots$HF are detected.

H-atom abstraction by F atoms in a matrix environment provides a unique opportunity to study the spectra of free radicals hydrogen-bonded to HF. Usually, both the isolated and the hydrogen-bonded radical are detected. Independent techniques for stabilization of the free radical of interest often permit distinction of the absorptions of the isolated radical from those of the radical hydrogen-bonded to HF. The 61 cm^{-1} shift in the out-of-plane vibration of CH$_3$ observed on formation of the HF complex is exceptionally large. Andrews (1984) reviewed the infrared spectra of stable molecules hydrogen-bonded to HF in an argon matrix, and Jacox (1985a) tabulated the shifts in the molecular vibrations of both free radicals and stable molecules associated with hydrogen bonding to HF. Vibrations involving the atom to which the HF is hydrogen-bonded generally experience the largest shift, which, with few exceptions, amounts to less than about 30 cm^{-1}. The positions of the HF-stretching absorption and of the HF-"flexing" absorptions provide information about the strength of the hydrogen bond, relative to the hydrogen-bond strengths of other species isolated in an argon matrix.

Matrix-isolation studies of certain hydrogen abstraction reactions have demonstrated that F atoms can form weakly bonded complexes, information not directly accessible from previous gas-phase studies. When F atoms were co-deposited with an $Ar:CD_3NO_2$ sample (Jacox 1983b), new absorptions appeared within about $25\,cm^{-1}$ of the infrared absorptions of isolated CD_3NO_2. Upon exposure of the deposit to radiation of wavelength longer than approximately 420 nm, the new absorptions were replaced by prominent absorptions of CD_2NO_2 hydrogen-bonded to DF. This behavior contrasts with that observed in the study of the reaction of F atoms with $CH_3NO_2-d_n$ species which possess one or more unsubstituted H atoms. For these species, selective abstraction of an H atom, rather than of a D atom, occurs, and very prominent absorptions attributable to $CH_2NO_2-d_n$ hydrogen-bonded to HF are present in the spectrum of the initial sample deposit. A relatively weak, broad absorption of a species which is readily photolyzed appears near the antisymmetric NO_2 stretch of the parent molecule, but there is little change in the overall pattern of absorptions when it is destroyed by visible radiation. These observations suggest that the initial attack of the F atom on nitromethane results in the formation of an addition complex which, when H atoms are present, eliminates HF by a tunneling mechanism. A detailed normal coordinate analysis for the CH_2NO_2 moiety of the resulting $CH_2NO_2\cdots HF$ has recently been presented (Jacox 1987b).

The infrared spectra of the initial sample deposits in the $F + CH_3X$ (X = Cl, Br, I) reaction studies provide evidence not only for the CH_3XF complex, formed when the F atom attacks at the halogen end of the molecule, but also for an $F\cdots HCH_2X$ complex, formed when the F atom interacts with the methyl group. The mechanisms which account for the observed products in the reaction of F atoms with CH_3X in an argon matrix are summarized in fig. 7. In addition to the CH_3XF absorptions, the initial sample deposits of products of F atom

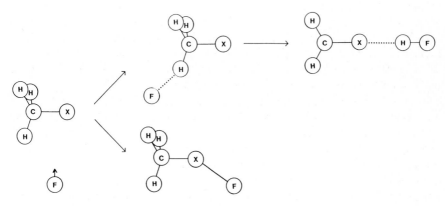

Fig. 7. Primary processes in the reaction of F atoms with CH_3X (X = Cl, Br, I) in an argon matrix.

reaction with all of the CH_3X-d_n species except CH_3Cl show new absorptions within about 25 cm^{-1} of the parent-molecule absorptions, with no absorption of isolated H_2CX and relatively weak absorptions of $H_2CX \cdots HF$. Exposure of the sample to visible light leads to the disappearance of the absorptions of the perturbed methyl halide and to the development of the $H_2CX \cdots HF$ peaks. CH_3Cl (but not CD_3Cl) differs in that the $H_2CCl \cdots HF$ absorptions are fully developed in the initial sample deposit. This behavior suggests that for all of the methyl halides a weakly bound $F \cdots HCH_2X$ complex is intermediate in the H-atom abstraction and that for CH_3Cl, but not for the other species, the barrier may be sufficiently low for tunneling to occur under the conditions of these experiments. A detailed analysis of the HF chemiluminescence in the gas-phase $F + CH_3Cl$ reaction (Wickramaaratchi et al. 1985), published at about the same time as the matrix studies, also indicated that a weakly bound complex plays a significant role in the reaction.

F atoms may undergo still other types of reaction, as is exemplified by their reaction with CH_3ONO. There was evidence for both isolated and hydrogen-bonded HF in a discharge-sampling study of this system (Jacox 1983c). However, no infrared absorptions of CH_2ONO were identified. The barrier to the fragmentation of this species into $H_2CO + NO$, which would be trapped in adjacent sites in the matrix, is expected to be small. Very prominent absorptions of FNO and of its isomer FON (Smardzewski and Fox 1974a,b,c) were present.

In discharge sampling studies of F-atom reactions, molecular fragments such as NF, NF_2, and CF_3, which are formed from the F-atom precursor, contribute to the product spectrum. However, these species are much less reactive than are F atoms and have a much smaller probability of experiencing a reactive collision in the mixing region. Furthermore, unlike F atoms, they cannot diffuse through the solid. In the many F-atom reactions which have been studied with discharge sampling, no evidence has been obtained for products which result from their reaction.

The discharge-sampling matrix-isolation studies of F atom reactions with small molecules illustrate in microcosm the complementarity of matrix-isolation experiments to gas-phase studies. The spectral survey obtained in the matrix experiments includes information on all of the primary reaction products. By adjusting the concentration of the F-atom precursor, secondary reactions can be made unimportant. The low photon flux of the discharge source permits the stabilization of high concentrations of free radicals which would be both produced and destroyed in photolysis systems. The photodecomposition thresholds of these species can then be studied at leisure. With consideration of the possible role of cage recombination, the photodecomposition products can often be determined. Energy-rich intermediates may be stabilized. The technique has permitted direct observation of weakly bound reaction complexes such as $F \cdots HCH_2X$, yielding information on the reaction mechanism which is difficult to infer from gas-phase measurements. The potential for further contributions by matrix-isolation studies to our understanding of the spectra and chemical bonding properties of free radicals and of the mechanisms by which they are formed is great.

References

Adrian, F.J., 1960, J. Chem. Phys. **32**, 972.
Ahlrichs, R., R. Becherer, M. Binnewies, H. Borrmann, M. Lakenbrink, S. Schunck and H. Schnöckel, 1986, J. Am. Chem. Soc. **108**, 7905.
Andrews, L., 1969, J. Chem. Phys. **50**, 4288.
Andrews, L., 1971, Ann. Rev. Phys. Chem. **22**, 109.
Andrews, L., 1984, J. Phys. Chem. **88**, 2940.
Andrews, L., and R.R. Smardzewski, 1973, J. Chem. Phys. **58**, 2258.
Andrews, L., M. Hawkins and R. Withnall, 1985, Inorg. Chem. **24**, 4234.
Arkell, A., 1965, J. Am. Chem. Soc. **87**, 4057.
Ase, P., W. Bock and A. Snelson, 1986, J. Phys. Chem. **90**, 2099.
Barger, R.L., and H.P. Broida, 1962, J. Chem. Phys. **37**, 1152.
Barger, R.L., and H.P. Broida, 1965, J. Chem. Phys. **43**, 2364.
Beeken, P.B., E.A. Hanson and G.W. Flynn, 1983, J. Chem. Phys. **78**, 5892.
Binnewies, M., M. Lakenbrink and H. Schnöckel, 1983, Z. Anorg. Allg. Chem. **497**, 7.
Bondybey, V.E., 1982a, J. Phys. Chem. **86**, 3396.
Bondybey, V.E., 1982b, J. Chem. Phys. **77**, 3771.
Bondybey, V.E., and L.E. Brus, 1975, J. Chem. Phys. **62**, 620.
Bondybey, V.E., and L.E. Brus, 1976, J. Chem. Phys. **64**, 3724.
Bondybey, V.E., and C. Fletcher, 1976, J. Chem. Phys. **64**, 3615.
Bondybey, V.E., and G.C. Pimentel, 1972, J. Chem. Phys. **56**, 3832.
Bristow, N.J., M. Poliakoff, P.B. Davies and P.A. Hamilton, 1981, Chem. Phys. Lett. **84**, 462.
Brom, J.M., and E.J. Lepak, 1976, Chem. Phys. Lett. **41**, 185.
Brown, H.W., and G.C. Pimentel, 1958, J. Chem. Phys. **29**, 883.
Brus, L.E., and V.E. Bondybey, 1975, Chem. Phys. Lett. **36**, 252.
Butler, R., and A. Snelson, 1979, J. Phys. Chem. **83**, 3243.
Butler, R., and A. Snelson, 1980a, J. Fluorine Chem. **15**, 89.
Butler, R., and A. Snelson, 1980b, J. Fluorine Chem. **16**, 33.
Carrick, P.G., and P.C. Engelking, 1984, J. Chem. Phys. **81**, 1661.
Chang, K.W., and W.R.M. Graham, 1982a, J. Chem. Phys. **76**, 5238.
Chang, K.W., and W.R.M. Graham, 1982b, J. Chem. Phys. **77**, 4300.
Chergui, M., N. Schwenter, W. Böhmer and R. Haensel, 1985, Phys. Rev. A **31**, 527.
Chergui, M., N. Schwenter and W. Böhmer, 1986, J. Chem. Phys. **85**, 2472.
Chettur, G., and A. Snelson, 1987, J. Phys. Chem. **91**, 913.
Chupka, W.A., and M.E. Russell, 1968, J. Chem. Phys. **49**, 5426.
Clough, P.N., B.A. Thrush, D.A. Ramsay and J.G. Stamper, 1973, Chem. Phys. Lett. **23**, 155.
DeMore, W.B., and N. Davidson, 1959, J. Am. Chem. Soc. **81**, 5869.
Douglas, M.A., R.H. Hauge and J.L. Margrave, 1983, J. Chem. Soc. Faraday Trans. I, **79**, 1533.
Douglas, M.A., R.H. Hauge and J.L. Margrave, 1984, High Temp. Sci. **17**, 201.
Dubs, M., and Hs.H. Günthard, 1977, Chem. Phys. Lett. **47**, 421.
Dubs, M., and Hs.H. Günthard, 1979, Chem. Phys. Lett. **64**, 105.
Dubs, M., L. Ermanni and Hs.H. Günthard, 1982, J. Mol. Spectrosc. **91**, 458.
Ewing, G.E., W.E. Thompson and G.C. Pimentel, 1960, J. Chem. Phys. **32**, 927.
Farrar, J.M., and Y.T. Lee, 1974, J. Am. Chem. Soc. **96**, 7570.

Felder, P., and Hs.H. Günthard, 1984, Chem. Phys. **85**, 1.

Ferrante, R.F., 1987, J. Chem. Phys. **86**, 25.

Foner, S.N., E.L. Cochran, V.A. Bowers and C.K. Jen, 1960, J. Chem. Phys. **32**, 963.

Fournier, J., J. Deson and C. Vermeil, 1977a, J. Chem. Phys. **67**, 5688.

Fournier, J., C. Lalo, J. Deson and C. Vermeil, 1977b, J. Chem. Phys. **66**, 2656.

Fournier, J., J. Deson, C. Vermeil and G.C. Pimentel, 1979, J. Chem. Phys. **70**, 5726.

Fournier, J., H.H. Mohammed, J. Deson and D. Maillard, 1982, Chem. Phys. **70**, 39.

Fournier, J., F. Salama and R.J. Le Roy, 1985, J. Phys. Chem. **89**, 3530.

Fredin, L., R.H. Hauge, Z.H. Kafafi and J.L. Margrave, 1985, J. Chem. Phys. **82**, 3542.

Haller, I., and G.C. Pimentel, 1962, J. Am. Chem. Soc. **84**, 2855.

Hastie, J.W., R.H. Hauge and J.L. Margrave, 1969, J. Am. Chem. Soc. **91**, 2536.

Hauge, R.H., J.W. Kauffman and J.L. Margrave, 1980, J. Am. Chem. Soc. **102**, 6005.

Hawkins, M., and L. Andrews, 1983, J. Am. Chem. Soc. **105**, 2523.

Hawkins, M., and L. Andrews, 1985, Inorg. Chem. **24**, 3285.

Hawkins, M., and A.J. Downs, 1984, J. Phys. Chem. **88**, 3042.

Hawkins, M., M.J. Almond and A.J. Downs, 1985, J. Phys. Chem. **89**, 3326.

Ismail, Z.K., R.H. Hauge, L. Fredin, J.W. Kauffman and J.L. Margrave, 1982a, J. Chem. Phys. **77**, 1617.

Ismail, Z.K., L. Fredin, R.H. Hauge and J.L. Margrave, 1982b, J. Chem. Phys. **77**, 1626.

Jacox, M.E., 1975, Chem. Phys. **7**, 424.

Jacox, M.E., 1976, Chem. Phys. **12**, 51.

Jacox, M.E., 1978a, Rev. Chem. Intermed. **2**, 1.

Jacox, M.E., 1978b, Chem. Phys. Lett. **53**, 192.

Jacox, M.E., 1978c, J. Mol. Spectrosc. **71**, 369.

Jacox, M.E., 1978d, J. Mol. Spectrosc. **72**, 26.

Jacox, M.E., 1978e, Chem. Phys. Lett. **56**, 43.

Jacox, M.E., 1979a, Chem. Phys. **42**, 133.

Jacox, M.E., 1979b, Chem. Phys. **43**, 157.

Jacox, M.E., 1980a, J. Mol. Spectrosc. **80**, 257.

Jacox, M.E., 1980b, J. Mol. Spectrosc. **81**, 349.

Jacox, M.E., 1980c, J. Mol. Spectrosc. **84**, 74.

Jacox, M.E., 1980d, Chem. Phys. **51**, 69.

Jacox, M.E., 1980e, Chem. Phys. **53**, 307.

Jacox, M.E., 1981a, Chem. Phys. **58**, 289.

Jacox, M.E., 1981b, Chem. Phys. **59**, 199.

Jacox, M.E., 1981c, Chem. Phys. **59**, 213.

Jacox, M.E., 1982a, J. Phys. Chem. **86**, 670.

Jacox, M.E., 1982b, Chem. Phys. **69**, 407.

Jacox, M.E., 1983a, Can. J. Chem. **61**, 1036.

Jacox, M.E., 1983b, J. Phys. Chem. **87**, 3126.

Jacox, M.E., 1983c, J. Phys. Chem. **87**, 4940.

Jacox, M.E., 1984a, Chem. Phys. **83**, 171.

Jacox, M.E., 1984b, J. Phys. Chem. **88**, 445.

Jacox, M.E., 1984c, J. Phys. Chem. **88**, 3373.

Jacox, M.E., 1984d, J. Phys. Chem. Ref. Data **13**, 945.

Jacox, M.E., 1985a, Rev. Chem. Intermed. **6**, 77.

Jacox, M.E., 1985b, J. Chem. Phys. **83**, 3255.

Jacox, M.E., 1985c, J. Mol. Spectrosc. **113**, 286.

Jacox, M.E., 1987a, J. Mol. Struct. **157**, 43.

Jacox, M.E., 1987b, J. Phys. Chem. **91**, 5038.

Jacox, M.E., 1987c, J. Phys. Chem. **91**, 6595.

Jacox, M.E., 1988a, J. Chem. Phys. **88**, 4598.
Jacox, M.E., 1988b, J. Phys. Chem. Ref. Data **17**, 269.
Jacox, M.E., 1989, J. Phys. Chem. Ref. Data, in preparation.
Jacox, M.E., and D.E. Milligan, 1973a, J. Mol. Spectrosc. **47**, 148.
Jacox, M.E., and D.E. Milligan, 1973b, J. Mol. Spectrosc. **48**, 536.
Jacox, M.E., and D.E. Milligan, 1974, Chem. Phys. **4**, 45.
Jacox, M.E., and D.E. Milligan, 1975a, J. Mol. Spectrosc. **56**, 333.
Jacox, M.E., and D.E. Milligan, 1975b, J. Mol. Spectrosc. **58**, 142.
Jacox, M.E., and W.B. Olson, 1987, J. Chem. Phys. **86**, 3134.
Jacox, M.E., and F.L. Rook, 1982, J. Phys. Chem. **86**, 2899.
Jacox, M.E., D.E. Milligan, N.G. Moll and W.E. Thompson, 1965, J. Chem. Phys. **43**, 3734.
Johns, J.W.C., and A.R.W. McKellar, 1977, J. Chem. Phys. **66**, 1217.
Johns, J.W.C., A.R.W. McKellar and E. Weinberger, 1983, Can. J. Phys. **61**, 1106.
Johnson, G.L., and L. Andrews, 1980, J. Am. Chem. Soc. **102**, 5736.
Jones, L.H., and B.I. Swanson, 1981, J. Chem. Phys. **74**, 3216.
Jones, L.H., and B.I. Swanson, 1982, J. Chem. Phys. **76**, 1634.
Jones, L.H., B.I. Swanson and S.A. Ekberg, 1984, J. Chem. Phys. **81**, 5268.
Kafafi, Z.H., R.H. Hauge, L. Fredin and J.L. Margrave, 1983, J. Phys. Chem. **87**, 797.
Kauffman, J.W., R.H. Hauge and J.L. Margrave, 1984, High Temp. Sci. **18**, 97.
Kauffman, J.W., R.H. Hauge and J.L. Margrave, 1985a, J. Phys. Chem. **89**, 3541.
Kauffman, J.W., R.H. Hauge and J.L. Margrave, 1985b, J. Phys. Chem. **89**, 3547.
Khanna, V.M., G. Besenbruch and J.L. Margrave, 1967, J. Chem. Phys. **46**, 2310.
Kohlmiller, C.K., and L. Andrews, 1982, Inorg. Chem. **21**, 1519.
Lalo, C., L. Hellner, J. Deson and C. Vermeil, 1976, J. Chim. Phys. **73**, 237.
Larzillière, M., and M.E. Jacox, 1979, in: Proc. 10th Materials Research Symposium on Characterization of High Temperature Vapors and Gases, ed. J.W. Hastie, Nat. Bur. Std., Spec. Publ. **561**, 529.
Larzillière, M., and M.E. Jacox, 1980, J. Mol. Spectrosc. **79**, 132.
Lee, Y.-P., and G.C. Pimentel, 1978, J. Chem. Phys. **69**, 3063.
Lee, Y.-P., and G.C. Pimentel, 1979, J. Chem. Phys. **70**, 692.
Lembke, R.R., R.F. Ferrante and W. Weltner Jr, 1976, J. Am. Chem. Soc. **99**, 416.
Long, S.R., and G.C. Pimentel, 1977, J. Chem. Phys. **66**, 2219.
Maass, G., R.H. Hauge and J.L. Margrave, 1972, Z. Anorg. Allg. Chem. **392**, 295.
Maier, G., and H.P. Reisenauer, 1986, Angew. Chem. **98**, 829 [Angew. Chem. Int. Ed. Engl. **25**, 819].
Maillard, D., J.P. Perchard, J. Fournier, H.H. Mohammed and C. Girardet, 1982, Chem. Phys. Lett. **86**, 420.
Mamantov, G., E.J. Vasini, M.C. Moulton, D.G. Vickroy and T. Maekawa, 1971, J. Chem. Phys. **54**, 3419.
Manceron, L., and L. Andrews, 1985, J. Am. Chem. Soc. **107**, 563.
Manceron, L., and L. Andrews, 1986, J. Phys. Chem. **90**, 4514.
Mathews, C.W., 1986, private communication.
Michalopoulos, D.L., M.E. Geusic, P.R.R. Langridge-Smith and R.E. Smalley, 1984, J. Chem. Phys. **80**, 3556.
Milligan, D.E., 1961, J. Chem. Phys. **35**, 1491.
Milligan, D.E., and M.E. Jacox, 1963, J. Chem. Phys. **38**, 2627.
Milligan, D.E., and M.E. Jacox, 1964, J. Chem. Phys. **41**, 3032.
Milligan, D.E., and M.E. Jacox, 1966, J. Chem. Phys. **44**, 2850.
Milligan, D.E., and M.E. Jacox, 1967, J. Chem. Phys. **47**, 278.
Milligan, D.E., and M.E. Jacox, 1969a, J. Chem. Phys. **51**, 277.
Milligan, D.E., and M.E. Jacox, 1969b, J. Chem. Phys. **51**, 1952.

Milligan, D.E., and M.E. Jacox, 1970, J. Chem. Phys. **52**, 2594.
Milligan, D.E., and M.E. Jacox, 1971, J. Chem. Phys. **54**, 927.
Milligan, D.E., and M.E. Jacox, 1972, in: Molecular Spectroscopy: Modern Research, eds K.N. Rao and C.W. Mathews (Academic Press, New York) pp. 259–286.
Milligan, D.E., and M.E. Jacox, 1973, J. Mol. Spectrosc. **46**, 460.
Milligan, D.E., M.E. Jacox, A.M. Bass, J.J. Comeford and D.E. Mann, 1965a, J. Chem. Phys. **42**, 3187.
Milligan, D.E., M.E. Jacox and A.M. Bass, 1965b, J. Chem. Phys. **43**, 3149.
Milligan, D.E., M.E. Jacox and L. Abouaf-Marguin, 1967, J. Chem. Phys. **46**, 4562.
Mohammed, H.H., J. Fournier, J. Deson and C. Vermeil, 1980, Chem. Phys. Lett. **73**, 315.
Moore, C.B., and G.C. Pimentel, 1964, J. Chem. Phys. **41**, 4504.
Müller, R.P., and J.R. Huber, 1983, J. Phys. Chem. **87**, 2460.
Müller, R.P., S. Murata and J.R. Huber, 1982a, Chem. Phys. **66**, 237.
Müller, R.P., P. Russegger and J.R. Huber, 1982b, Chem. Phys. **70**, 281.
Müller, R.P., J.R. Huber and H. Hollenstein, 1984a, J. Mol. Spectrosc. **104**, 209.
Müller, R.P., S. Murata, M. Nonella and J.R. Huber, 1984b, Helv. Chim. Acta **67**, 953.
Nicolai, J.-P., L.J. van de Burgt and M.C. Heaven, 1985, Chem. Phys. Lett. **115**, 496.
Noble, P.N., and G.C. Pimentel, 1966, J. Chem. Phys. **44**, 3641.
Pacansky, J., and J. Bargon, 1975, J. Am. Chem. Soc. **97**, 6896.
Pacansky, J., and D.W. Brown, 1983, J. Phys. Chem. **87**, 1553.
Pacansky, J., and J.S. Chang, 1981, J. Chem. Phys. **74**, 5539.
Pacansky, J., and H. Coufal, 1980, J. Chem. Phys. **62**, 3298.
Pacansky, J., and M. Dupuis, 1982, J. Am. Chem. Soc. **104**, 415.
Pacansky, J., and A. Gutierrez, 1983, J. Phys. Chem. **87**, 3074.
Pacansky, J., and B. Schrader, 1983, J. Chem. Phys. **78**, 1033.
Pacansky, J., G.P. Gardini and J. Bargon, 1976, J. Am. Chem. Soc. **98**, 2665.
Pacansky, J., D.E. Horne, G.P. Gardini and J. Bargon, 1977, J. Phys. Chem. **81**, 2149.
Pacansky, J., D.W. Brown and J.S. Chang, 1981, J. Phys. Chem. **85**, 2562.
Prochaska, E.S., and L. Andrews, 1977, Inorg. Chem. **16**, 339.
Prochaska, E.S., L. Andrews, N.R. Smyrl and G. Mamantov, 1978, Inorg. Chem. **17**, 970.
Schnöckel, H., and M. Lakenbrink, 1983, Z. Anorg. Allg. Chem. **507**, 70.
Schrader, B., J. Pacansky and U. Pfeiffer, 1984, J. Phys. Chem. **88**, 4069.
Shepherd, R.A., and W.R.M. Graham, 1985, J. Chem. Phys. **82**, 4788.
Shepherd, R.A., and W.R.M. Graham, 1987, J. Chem. Phys. **86**, 2600.
Smardzewski, R.R., 1978, J. Chem. Phys. **68**, 2878.
Smardzewski, R.R., and W.B. Fox, 1974a, J. Am. Chem. Soc. **96**, 304.
Smardzewski, R.R., and W.B. Fox, 1974b, J. Chem. Soc., Chem. Commun., p. 241.
Smardzewski, R.R., and W.B. Fox, 1974c, J. Chem. Phys. **60**, 2104.
Snelson, A., 1970, High Temp. Sci. **2**, 70.
Spratley, R.D., J.J. Turner and G.C. Pimentel, 1966, J. Chem. Phys. **44**, 2063.
Swanson, B.I., and L.H. Jones, 1981, J. Chem. Phys. **74**, 3205.
Sylwester, A.P., and P.B. Dervan, 1984, J. Am. Chem. Soc. **106**, 4648.
Tevault, D.E., and R.R. Smardzewski, 1978, J. Chem. Phys. **69**, 3182.
Wang, J.L.-F., C.N. Krishnan and J.L. Margrave, 1973, J. Mol. Spectrosc. **48**, 346.
Weltner Jr, W., and D. McLeod Jr, 1964a, J. Chem. Phys. **40**, 1305.
Weltner Jr, W., and D. McLeod Jr, 1964b, J. Chem. Phys. **41**, 235.
Weltner Jr, W., and P.N. Walsh, 1962, J. Chem. Phys. **37**, 1153.
Weltner Jr, W., P.N. Walsh and C.L. Angell, 1964, J. Chem. Phys. **40**, 1299.
Wickramaaratchi, M.A., D.W. Setser, H. Hildebrandt, B. Körbitzer and H. Heydtmann, 1985, Chem. Phys. **94**, 109.

Wight, C.A., B.S. Ault and L. Andrews, 1976, J. Chem. Phys. **65**, 1244.
Willard, J.E., 1982, Cryogenics **22**, 359.
Withnall, R., and L. Andrews, 1985, J. Phys. Chem. **89**, 3261.
Withnall, R., and L. Andrews, 1986, J. Am. Chem. Soc. **108**, 8118.
Withnall, R., and L. Andrews, 1987, J. Phys. Chem. **91**, 784.
Yamada, F., I.R. Slagle and D. Gutman, 1981, Chem. Phys. Lett. **83**, 409.

CHAPTER 5

Time-Resolved Laser-Induced Fluorescence Studies of the Spectroscopy and Dynamics of Matrix-Isolated Molecules

V.E. BONDYBEY

Department of Chemistry
Ohio State University
Columbus, OH 43210
USA

Chemistry and Physics of
Matrix-Isolated Species
L. Andrews and M. Moskovits, eds

Contents

1. Introduction

The matrix-isolation technique, originally developed by Pimentel (1958a) and co-workers (Whittle et al. 1954), is today well established. Over the past three decades it has been particularly useful in studies of free radicals and other transient species (Meyer 1971, Moskovits and Ozin 1976). In the early matrix studies, the emphasis was mainly on infrared or, to a lesser extent, electronic absorption spectroscopy (Jacox 1978, Andrews 1980). More recently, with the widespread availability of lasers, matrix Raman experiments became practicable, and particularly fluorescence studies became quite appealing.

The earliest laser-induced fluorescence (LIF) matrix studies used fixed frequency cw lasers, and had therefore to rely on accidental overlaps between molecular transitions and the available laser wavelengths (Shirk 1971, Bondybey and Nibler 1972). Such studies were facilitated by the various inhomogeneous and homogeneous mechanisms which broaden transitions in matrix-isolated species, and relax the exact resonance requirement. The full potential of the LIF matrix technique could, however, only be developed with the emergence of continuously tunable dye lasers permitting selective, resonant excitation of the species of interest into a particular vibrational or electronic state (Bondybey and Brus 1975a–c). Especially useful in this respect are pulsed lasers, which in addition to spectroscopic information also permit direct, time-resolved probing of molecular relaxation and dynamics (Bondybey and Brus 1980).

LIF studies offer, when compared with other spectroscopic techniques numerous advantages. Quite high among these ranks the increased sensitivity. While IR or Raman investigations are usually conducted in the $1:100-1000$ concentration range, the typical dilutions used in LIF studies are at least 1–2 orders of magnitude higher, and spectra of samples at part per million levels can often be recorded with excellent signal-to-noise ratios. Numerous spectra in our laboratory were recorded at the low part per billion levels, and while few serious efforts were reported to push the technique in this direction, it seems fairly clear that in favorable cases detection of species at the $1:10^{12}$ level is entirely feasible.

When compared with pure vibrational studies – infrared or Raman – provided the vibrational structure is clearly resolved, there is usually much less ambiguity and uncertainty in identifying the carrier of the fluorescence spectrum and assigning the observed transitions. In vibrational matrix spectroscopy just establishing which of the observed bands are due to the molecule of interest, and which are due to impurities or other species present in the matrix is often a non-trivial task requiring tedious tests and control experiments.

In LIF studies one typically observes more or less regular arrays of vibronic transitions, with lines belonging to the same electronic transition exhibiting similar line shapes (matrix analogues of the gas-phase rotational contours). Furthermore, transitions originating from the same upper state level are characterized by identical time-resolved behavior and lifetimes, and exhibit identical excitation profiles (Bondybey 1984). Very often individual isotopic species can be excited in samples with natural isotopic abundance, without the use of isotopically enriched samples. In this way, the information obtainable from the LIF spectra is often quite redundant, and many more tests and cross-checks are possible than in infrared studies.

In the present chapter we briefly review and summarize some of the available matrix fluorescence work, with emphasis upon studies employing tunable pulsed lasers. In sections 2 and 3 we discuss the general photophysics of matrix-isolated molecules, the ways in which the behavior of species in the solid differs from that of the free molecules in the gas phase, and the ways in which these differences can often be exploited to obtain information not readily available from gas-phase studies.

In sections 4–7 we then discuss several examples of the various types of chemical species and problems to which the matrix LIF techniques were applied, and compile and review some of the representative experimental data.

2. Photophysics of matrix-isolated molecules

2.1. Molecular dissociation and matrix cage effect

Some of the phenomena specific to molecules in low-temperature matrices and other condensed media are due to the cage effect. When a gaseous molecule is excited into a repulsive electronic state or, for that matter, into a dissociative continuum of a bound electronic state, it typically dissociates in a few vibrational periods, and no parent molecule fluorescence is observed. The situation in a condensed medium is, however, quite different. The relatively rigid cage of the solvent atoms surrounding the guest molecule prevents permanent separation of the atomic or molecular fragments (Schnepp and Dressler 1965, Bondybey and Brus 1976). The excess energy is rapidly dissipated into the host lattice, and recombination of the atomic or molecular fragments can take place.

Such recombination needs not necessarily produce the originally excited electronic state. It may result in the formation of any other state correlating with the particular states of the fragments, irrespective of whether they are accessible from the ground electronic state by an optically allowed process or not. As we will show later in this chapter, this can often be a very useful tool in populating directly inaccessible electronic states.

It is interesting to note that, in an adiabatic sense, the host lattice does not present a significant barrier to molecular dissociation, and, in most cases, very little energy would apparently be required to permit the solvent atoms to rearrange to allow the fragment atoms to separate. Careful measurement of the zero-phonon lines of the individual vibrational levels usually shows little medium perturbation and a smooth convergence towards the dissociation limit (Bondybey and Fletcher 1976, Bondybey et al. 1976).

In spite of this fact, one finds that, e.g., in the case of the halogens, the cage effect is almost complete. Even for excitation with more than 2 eV of excess energy above the dissociation limit, the quantum yield for permanent dissociation was found to be less than 10^{-6}. Apparently, vibrational relaxation and dissipation of the excess kinetic energy of the guest fragments occurs fast compared with the relatively slow, high-amplitude motions of the host atoms needed to permit the fragments to separate.

2.2. Vibrational relaxation

A small gas-phase molecule excited into a bound electronic state cannot relax nonradiatively, and will, in the absence of collisions, emit from the originally populated rotational and vibrational level. Quite different is the situation in a solid matrix. The rotational structure usually collapses into a single sharp line, the so-called zero-phonon line (ZPL) accompanied by a continuous phonon sideband (Rebane 1970). The lattice may also carry away some or all of the excess vibrational energy. This in many instances greatly simplifies the spectra and facilitates their assignment.

Until quite recently, there was much misunderstanding and many misconceptions regarding the vibrational relaxation of matrix-isolated molecules. On the one hand, there was the widespread intuitive expectation that relaxation in the condensed matrix systems should be very fast. This was supported by the early theoretical treatments, as well as by the experimental observation that, in most matrix studies, only the vibrationless level was evidenced in emission from excited electronic states of matrix-isolated molecules. On the other hand, several experiments (Broida and Peyron 1960, Tinti and Robinson 1968, Tinti 1968), including some of the earliest studies by Vegard (1932) on matrix-isolated N_2, seemed indicative of a remarkably slow vibrational relaxation.

Recent systematic studies, both theoretical (Nitzan et al. 1975, 1978, Diestler 1974, 1976, Lin 1974, 1976, Gerber and Berkowitz 1977a,b, Freed et al. 1977, Berkowitz and Gerber 1979) and experimental (Legay 1977, Bondybey and Brus 1980, Bondybey 1981, 1984), of molecular relaxation in matrices have led to a good qualitative understanding of the processes involved, and have shown that there is no real contradiction between the two types of experimental observations. Relaxation rates are a sensitive function of the vibrational spacing or "energy gap", polarity of the molecule, and the nature of the guest–host interaction potential. Depending on the size of the vibrational spacing and on the guest–host coupling, the rates of relaxation span more than fifteen orders of magnitude.

Laser-induced fluorescence proved to be an extremely useful tool in providing detailed experimental data and revealing the mechanisms and processes governing the molecular relaxation. Most information was obtained from two types of experiment. In the simplest experiment, laser is used to populate selectively a particular excited level, and the molecular reemission is then frequency and time resolved. The pattern of the emission occurring during the relaxation process, and the corresponding time-resolved profiles provide then quantitative information about the relaxation process (Brus and Bondybey 1975a,b). This type of experiment was particularly useful in studying relaxation involving excited electronic states.

An alternative type of experiment uses one laser to populate the level of interest, and a second independent laser to probe its population as a function of time after the initial "pump laser" pulse (Bondybey and Nibler 1972, Allamandola and Nibler 1974, Allamandola et al. 1977). The advantage of this technique is that it is readily applicable to studies of relaxation in the ground electronic states of molecules.

The available studies now provide a fairly clear and consistent picture of the relaxation of matrix-isolated molecules. In general, an overall "energy-gap law" is clearly observable: very light molecules with large vibrational spacing usually relax quite slowly, and the rates of relaxation increase exponentially with the decreasing size of the vibrational frequency. Molecules with vibrational spacing of $> 1000 \, \text{cm}^{-1}$ often have vibrational-level lifetimes in the second or millisecond range, while the lifetimes drop into the subpicosecond region as the molecular frequencies approach the phonon frequencies of the host lattice.

2.3. Vibration to rotation energy transfer

One notable exception to the above energy-gap law is observed for some relatively light molecules, in particular for small hydrides such as NH (Bondybey and Brus 1975c, Bondybey 1976a), OH (Brus and Bondybey 1975b), CH or HCl (Wiesenfeld and Moore 1979, Young and Moore 1984). In these species one observes that they relax many orders of magnitude faster than one might predict based on their vibrational frequency, and the overall energy-gap relationship discussed above. A key to understanding this exception lies in the observation that, contrary to the energy-gap law, the deuterides relax several orders of magnitude slower than the corresponding hydrides.

In vibrational relaxation of a matrix-isolated molecule it is generally assumed that the delocalized lattice phonons are accepting the vibrational energy (Nitzan et al. 1974). When the molecular frequency is increased, more phonons have to be created in the relaxation process, and the energy-gap law simply reflects the fact that higher-order processes are usually less probable, and less efficient than lower order ones.

The inclusion of an impurity molecule into the matrix introduces several degrees of freedom associated with rotation and translation of the guest. For very light molecules frequencies associated with these motions may be quite high, and in particular frequencies associated with rotational transitions of the guest are larger than those of the lattice phonons.

The anomalous behavior of the hydrides then reflects the fact that their relaxation is actually a V → R process, and the anharmonic rotational motions, rather than the delocalized and harmonic lattice phonons are the primary acceptors of the vibrational energy. While deuteration reduces the vibrational spacing in the guest molecules by roughly a factor of 1.4 ($\sqrt{2}$), the accepting rotational modes are reduced by a factor of 2. Consequently, relaxation of the deuterides is, in spite of the lower vibrational frequency, a higher order, and hence a less efficient process.

2.4. Electronic relaxation and fluorescence quantum yields

The physical principles underlying the energy-gap law and the relatively inefficient vibrational relaxation can be understood by considering the application of the Franck–Condon principle to the lattice-phonon modes. The guest–host interaction potential is rather insensitive to the particular vibrational state of the guest molecule. This is evidenced by the vibrational spectra of matrix-isolated molecules, which usually consist of sharp zero-phonon lines and show little indication of phonon sidebands.

These negligibly weak phonon sidebands are indicative of very poor Franck–Condon overlap factors for conversion of energy into the lattice phonons during vibrational excitation. The same Franck–Condon factors will, of course, be applicable to the vibrational relaxation process also, and will therefore not favor high-order processes requiring simultaneous formation of a large number of phonons.

Conversely, vibronic transitions in the electronic absorption or emission spectra often exhibit intense phonon sidebands, suggesting that solvation potentials are different for different electronic states of the guest, and that the most probable "vertical" process requires creation (or destruction) of a certain number of lattice phonons. The same Franck–Condon factors which give rise to the phonon sidebands in a radiative electronic transition will also favor phonon formation during a nonradiative relaxation.

It should be, however, noted, that the phonon Francke–Condon overlaps are only one of the factors affecting the relaxation process. In a simplified manner one can view the relaxation rate as being determined by a product of three factors:

(1) the phonon Franck–Condon factors;

(2) the intramolecular Franck–Condon factor connecting the initial and final level; and

(3) the intramolecular quantum-mechanical coupling term between the two electronic states involved.

As noted above, the "phonon Franck–Condon factors" generally favor interelectronic processes. Provided the other two factors are also favorable, internal conversions between different electronic states will prevail over direct intrastate multi-phonon vibrational relaxation. A well-studied example is matrix-isolated CN, where relaxation proceeds via a series of internal conversions between the low lying A $^2\Pi$ state and the X $^2\Sigma^+$ ground state (Bondybey 1977a–c, Bondybey and Nitzan 1977).

A similar situation prevails even in C_2 where relaxation proceeds by a sequence of intersystem crossings (Bondybey 1976a,b), even though in this case the two states involved, X $^1\Sigma_g^+$ and a $^3\Pi_u$ differ by multiplicity, and the corresponding interelectronic coupling terms are undoubtedly less favorable than in the CN case. Several beautiful examples of such interelectronic cascade processes among highly excited electronic states of CO and N_2 were reported by Schwentner and co-workers (Kühle et al. 1985a–c, 1986, Bahrdt et al. 1987).

The consequence of this rather efficient interelectronic relaxation for matrix-relaxation studies is that very often emission observed experimentally does not originate from the initially excited, optically accessible state, but from lower lying states populated by the nonradiative cascade processes.

In the diatomic molecule cases discussed above, both of the vibrational manifolds involved are sparse, and the excitation cascades back and forth between them. In polyatomic molecules, or in molecules at very high excitation energies, where the vibrational state densities are much higher, the internal conversion or intersystem crossing may be quite efficient and effectively irreversible, so that the quantum yield of fluorescence from the originally excited state may be unobservably small.

In general, electronic states which, in the gas phase, show strong coupling to lower states, as evidenced by perturbations, rarely exhibit appreciable fluorescence in the matrix. A typical example is the NH_2 radical, which is one of the free radicals most extensively studied by fluorescence in the gas phase, but whose LIF in the matrix is almost unobservably weak.

2.5. Nonradiative access to forbidden electronic states

One of the important strengths of the matrix-isolation technique is that one can take advantage of the relaxation phenomena and nonradiative cascade processes discussed in the preceding sections to access electronic states which, due to the spectroscopic selection rules, can not be directly excited from the ground state in a gas-phase experiment.

A typical example of this situation occurs in diatomic halogen molecules, exemplified by Cl_2, whose potential energy diagram is shown in fig. 1. The fluorescence spectrum of Cl_2 in matrices was first observed by excitation in the region of the B 0_u^+ state using the 4880 Å line of an argon laser, and assigned to the X $0_g^+ \leftarrow$ B 0_u^+ transition (Ault et al. 1975). A later re-examination of the system using a tunable dye laser has shown that polarization properties of

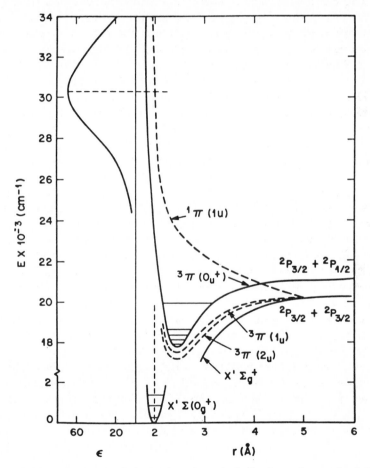

Fig. 1. Potential energy diagram of Cl_2 showing schematically the electronic states involved in the photophysical phenomena observed in rare-gas matrices. The left-hand portion of the figure shows the absorption spectrum of Cl_2. Discrete absorptions in the bound region of the 0_u state, between $\sim 18\,000$–$20\,000$ cm^{-1} are several orders of magnitude weaker than the 1_u continuum near $30\,000$ cm^{-1}.

the emission, as well as the long lifetime observed, were incompatible with the original assignment (Bondybey and Fletcher 1976). Furthermore, analysis of the emission and excitation spectrum has revealed that two separate electronic states must be involved, the absorbing $B\,0_u^+$ state, which populates by non-radiative cascade a lower lying electronic state. This emitting state, $650\,\mathrm{cm}^{-1}$ lower in energy, was identified as the $\Omega = 2$ component of the Hund's case "c" $^3\Pi$ state, $A'\,^3\Pi_u$. It was originally predicted by Mulliken (1940), but never before observed experimentally.

Rather than exciting perturbed or predissociating states, one can often also access forbidden electronic states by excitation of dissociative electronic states, or of the dissociative continua of bound electronic states. Such excitation, of course, results in a prompt dissociation of the molecule in question. However, as discussed previously, owing to the cage effect the fragments usually immediately recombine. Such a recombination may populate not only the originally excited electronic state, but also any other state correlating with the particular states of the atoms or fragments.

The Cl_2 molecule discussed above can again serve as an example. Excitation of a completely repulsive $^1\Pi_u$ electronic state near 3300 Å produces two ground-state ^2Cl atoms which, upon recombining, again populate the elusive $A'\,(2_u)$ electronic state. This excitation process via the repulsive $^1\Pi_u$ state can be distinguished from the above-discussed excitation via the bound $B\,(0_u^+)$ electronic state through the different polarization properties of the emission.

A similar situation prevails also in other halogen molecules, where the components of the $^3\Pi$ states can again be accessed nonradiatively. By now, most of the homonuclear and heteronuclear halogen molecules have been investigated, both by our group (Bondybey et al. 1976, Bondybey and Brus 1976), by Flynn and co-workers (Mandich et al. 1982, Beeken et al. 1983) and more recently Heaven and co-workers (Nicolai et al. 1985, Nicolai and Heaven 1985) and others (Langen et al. 1987).

Such nonradiative processes are, of course, not restricted to halogen but occur in numerous other molecules. For instance, the near-UV gas-phase spectroscopy of the group VI A dimers, S_2, Se_2 and Te_2, is dominated by a fully allowed $X\,^3\Sigma_g^- \rightarrow B\,^3\Sigma_u^-$ transition (Barrow and Ketteringham 1963, Gouedard and Lehmann 1976, Stone and Barrow 1975, Yee and Barrow 1972). In all of these molecules the upper $B\,^3\Sigma_u^-$ states were known to be perturbed by nearby $^3\Pi_u$ states which were, however not detected in previous gas-phase studies. In the matrix, excitation of the $B\,^3\Sigma_u^-$ states leads in each case to nonradiative population of and emission from the perturbing state (Bondybey and English 1980b,d).

While the above given examples of group VI A and VII A diatomics are typical, this list is by no means exhaustive, and, in fact, the list of forbidden electronic states which can be in the matrix populated by nonradiative pathways is becoming impressively long.

In the initial investigations of molecules which were not previously spectroscopically characterized, it is expedient, and often successful, to irradiate the matrix by high-energy UV radiation, in our laboratory usually the 1930 Å ArF excimer laser radiation. For most molecules of interest predissociated electronic states or dissociative continua are located in this energy region. The dissociation–recombination sequence, followed by a nonradiative relaxation cascade then populates lower electronic states, some of which are usually sufficiently long-lived to be detectable in emission. As an example, numerous group V A homonuclear and heteronuclear dimers were recently characterized that way (Heimbrook et al. 1985a,b, Rasanen et al. 1986). Of course, once the individual emitting states are observed using the ArF excitation, their detailed characterization with tunable dye laser excitation can be easily accomplished.

3.   Preparation of matrix samples

Detailed description of the preparation of samples for matrix-isolation studies is beyond the scope of this chapter. Furthermore, the subject has been exhaustively covered in several excellent review articles, including several chapters in this book. In this section we will only discuss several aspects of matrix preparation specific to laser-induced fluorescence studies.

Unlike infrared or optical absorption studies which usually require relatively concentrated samples and thick deposits, in laser-induced fluorescence one typically works with much higher dilutions and thinner matrices. In infrared work one also usually deposits the sample relatively slowly, with 1–2 mmol/h being typical, with overall deposition time frequently approaching or even exceeding 24 h. The slow deposition rates are dictated by the need for good optical quality samples. Rapid deposition of thick samples results in highly scattering matrices, which, furthermore, often tend to crack or peel off the substrate. In laser-induced fluorescence studies where thick deposits are not needed and where the optical quality of the sample is of less concern, more than an order of magnitude faster deposition can be used, and overall sample preparation time rarely exceeds 10–20 min.

Faster deposition is, in fact, desirable in most LIF studies. Most impurities present in the matrix are not intrinsic to the sample, but result from contamination due to wall reactions or desorption during the sample deposition. Depositing faster and cutting down on the sample preparation time reduces the concentration of impurity molecules in the sample. This is particularly important in samples which are either discharged or photolyzed, since common impurity molecules often give highly fluorescent radical fragments, which may then interfere with the spectrum of interest.

As noted above, molecules of interest in matrix-isolation investigations are often not deposited directly, but generated in situ by photolysis or photo-ionization of a suitable precursor. In vibrational studies the spectra of the precursor and those of the product of interest are of necessity in the same spectral region, and their absorptions may often overlap. Consequently, an as complete as possible conversion of the precursor into the desired product is needed, and the photolysis often proceeds for many hours, during the slow sample deposition. On the other hand, in electronic LIF studies the spectra of the precursor and the product are usually quite distinct. The radical or ionic products are open-shell molecules with spectra in the visible or near-UV range, while the stable parent compounds are quite transparent in this region, and absorb only at much shorter wavelengths. As a result, spectral interferences are rare, and the open-shell products can readily be studied in the presence of large excess of the parent compound.

The product species can, and often do, also absorb the wavelength used to photolyze the parent, and react further to form secondary products. The rate of these secondary reactions, of course, increases with the built-up of the primary product concentration. Very often, after some time a photostationary state is reached, where the formation and destruction rates of the desired product are equal, and when little benefit can be derived from continued photolysis. Unfortunately, the secondary products are much more likely to give spectral interference than the parent species.

One way of avoiding the secondary product built-up is to choose for the photolysis a wavelength which is not absorbed by the primary product. Often a simple alternative is to minimize the secondary product formation by using only short photolysis time. As a typical example may serve our studies of various molecular ions, produced by in situ photo-ionization of the parent compounds. In most of these experiments a photostationary state where the ion concentration was no longer appreciably increasing was reached after 1–2 min of 1216 Å photolysis. However, typically only 5–30 s of photolysis were needed to give in our experimental apparatus nearly 100% absorbing samples and extremely strong fluorescent signals.

4. Spectra and fluorescence of matrix-isolated atoms

The earliest observations of emission due to atoms in rare-gas solids date, in fact, from the early work of Vegard in the 1920s, who observed numerous emission lines, due mostly to forbidden transitions of oxygen and nitrogen atoms (Vegard 1924, 1930). Subsequent more systematic studies of matrix-isolated atoms relied more often on absorption work, although occasionally atomic emission was also reported. An overview of this early work demonstrated that atomic transitions are often quite broad, and considerably shifted from their gas-phase positions (Knight et al. 1974, Brom et al. 1975). Moreover, unlike in molecular electronic spectroscopy where red shifts are more common, atomic absorption bands are often blue-shifted. The availability of tunable dye laser excitation sources provided a new invaluable tool, and led to renewed activity in studies of matrix-isolated atoms.

Atomic spectra are, of course, in most cases, well understood based on extensive gas-phase work, and the possibilities of contribution from matrix studies to their spectroscopy are therefore limited. Usefulness of spectroscopic studies of matrix-isolated atoms lies in permitting easy comparisons with the free gas-phase atoms which give new information about the photophysics of matrix-isolated species and about the nature of the guest–host interactions.

In addition, time-resolved fluorescence studies following state-selective excitation provide new insights into the relaxation processes occurring in the matrix. The understanding of these phenomena derived from the studies in numerous laboratories is then invaluable in designing new experiments and interpreting spectroscopic information involving molecular species.

The broadening of the atomic transitions in the matrix can be due to inhomogeneous, or "site" effects, or to homogeneous lifetime and phonon broadening. While the source of the linewidth in the matrix spectra is usually not obvious from the absorption studies, the relative importance of the individual broadening mechanism can be established using site-selected excitation with tunable lasers.

The origin of the broadening in Ar matrix-isolated calcium atoms is clearly demonstrated in fig. 2. Dilute matrices exhibit a very strong broad and structured absorption near $24\,000\;\mathrm{cm}^{-1}$, considerably above the gas-phase atomic Ca 4s ^{1}S $\rightarrow$ 4p ^{1}P transition (Bondybey 1978). Laser excitation near the low-energy onset of this absorption at $23\,693\;\mathrm{cm}^{-1}$ produces two strong, structured emission bands, each of them over $200\;\mathrm{cm}^{-1}$ broad (FWHM). The higher-energy band extending to $23\,693\;\mathrm{cm}^{-1}$ is clearly the resonant 4p ^{1}P $\rightarrow$ 4s ^{1}S fluorescence, while the lower-energy band in the red is due to the 4p ^{3}P $\rightarrow$ 4s ^{1}S intercombination transition. When examined at higher resolution, as shown in fig. 2a, a sharp, weak zero-phonon line at $15\,438\;\mathrm{cm}^{-1}$ can be noted, somewhat blue-shifted from the gas-phase position of the ^{3}P$_0$ level at $15\,157.91\;\mathrm{cm}^{-1}$.

Figure 2b shows the excitation spectrum obtained by monitoring the weak $15\,438\;\mathrm{cm}^{-1}$ line with a $1\;\mathrm{cm}^{-1}$ bandpass, and scanning the laser. Two excitation maxima are observed, each of them exhibiting a sharp zero-phonon line. These are due to the above-mentioned resonant absorption and to a formally forbidden transition into the 3d ^{1}D level. Based on observations of this type, several fairly general conclusions can be drawn.

In the first place, homogeneous phonon broadening contributes significantly to the observed linewidths. Excitation of either the ^{1}P or the ^{3}P state requires promotion of a 4s electron into the higher lying empty 4p orbital. This apparently substantially changes the guest–host interaction potential, and requires resolution of the excited electronic state and change in the equilibrium

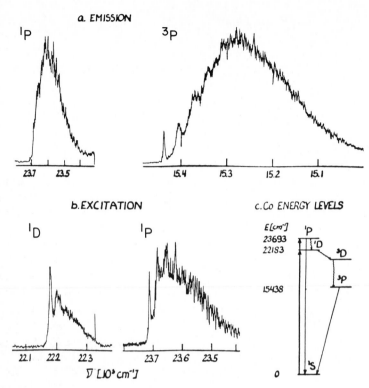

Fig. 2. Spectra of Ca atoms in solid argon matrix at 4 K. (a) Laser-induced fluorescence following excitation of the $^1S \to {}^1P$ resonant transition. Note that the scale was expanded for the inter-combination $^3P \to {}^1S$ emission band to show in more detail the phonon contour and the zero-phonon line. (b) Excitation spectrum obtained by monitoring the ZPL at 15 438 cm^{-1}. (c) Energy-level diagram explaining the radiative and nonradiative processes occurring in the matrix.

matrix cage geometry. As a result, the direct adiabatic process will have a low probability, as evidenced by the weak zero-phonon line, and the Franck–Condon factors will favor vertical excitation, requiring simultaneous excitation of lattice phonons, and shifting intensity into the phonon sideband.

In the second place, one may note that the 4s $^1S \to$ 3d ^1D transition exhibits a much less intense phonon sideband than the other transitions. Clearly, excitation of one of the two 4s electrons into the more compact "inner-shell" 3d orbital requires much less matrix-atom rearrangement and resolvation, than its promotion into the higher 4p orbital.

In the third place, the presence of the above forbidden transition demonstrates that selection rules are somewhat relaxed in the matrix. In particular, s–d type excitations occur fairly commonly with considerable intensity. One also finds that the $J = 0, \pm 1$ selection rule is not particularly important in the matrix, with the $J > 1$ transitions often appearing with comparable intensity.

Finally, the much larger width of the $^1S \rightarrow {}^1P$ transition in absorption than in emission might suggest that the solvent cage is expanded in the excited electronic state, so that in excitation one probes the steeply rising repulsive portion of the upper-state potential, while the emission terminates upon the shallow outside limb of the ground-state guest–host interaction potential.

Transition-metal atoms with their rich spectra in the easily accessible visible and near-ultraviolet region and with the variety of states differing in their orbital occupancies and spin multiplicities, provide very attractive systems for fluorescence matrix studies. Particularly interesting is the spectroscopy of matrix-isolated nickel atoms, which have been studied extensively in several laboratories (Breithaupt et al. 1983, Cellucci and Nixon 1984, 1985). They are characterized by two nearly isoenergetic groups of states arising from the orbital occupancies of $3d^8 4s^2$ and $3d^9 4s$, respectively, with $3d^8 4s^2$ $^3F_4$ level being the actual ground state in the gas-phase atoms.

Ni emission bands observed in the matrix can be classified into three separate groups (Rasanen et al. 1986). Transitions connecting states arising from the same orbital occupancy are quite sharp ($< 3\ \mathrm{cm}^{-1}$), both in emission and in excitation spectra. They also exhibit, both in solid neon and argon, very small gas–matrix shifts.

Transitions interconnecting states with different occupancies, i.e., interchanging an electron between the partially filled 3d and 4s orbitals are still sharp upon site-selective excitation, with little evidence for phonon sidebands, suggesting little change in the guest–host potential. They, however, exhibit more substantial inhomogeneous broadening of $\sim 30\ \mathrm{cm}^{-1}$ in solid neon. In solid argon the inhomogeneous linewidth is over $500\ \mathrm{cm}^{-1}$, considerably more than the gas-phase energy difference between the ground state and the lowest state of the $3d^9 4s$ electron configuration, $^3D_3$. As a result of this situation, for part of the inhomogeneous distribution the latter state is actually the ground state in the argon matrix.

Finally, transitions involving promotion of one of the electrons into the higher lying, previously empty 4p valence orbital are, like in the case of Ca, quite broad even in solid neon, and shifted from their gas-phase positions to the extent that unambiguous assignment to specific transitions becomes difficult. Excitation of these higher lying states results in emission which is relatively independent of the exact laser wavelength, indicating that homogeneous phonon broadening plays a dominant role in this case.

It should be noted that matrix absorption studies usually involve fully allowed transitions of this last category, which often require a considerable matrix rearrangement and excited-state resolution. This explains the large gas–matrix shifts and broad bands typically observed in absorption work. Observation of the usually forbidden transitions involving inner-shell orbitals requires the higher sensitivity of laser-induced fluorescence techniques.

5. Metal dimers and larger clusters

Studies of metal dimers, clusters and their chemistry and structure are an area of rapidly expanding current scientific interest. The major thrust of research in several dozen laboratories is centered in this area, with new investigators entering the field at a rapid rate. This is also reflected in the number of specialized conferences and symposia devoted exclusively or predominantly to this topic. The major motivation for this intense interest is, of course, the realization that understanding of small metal clusters has important implications for the closely related and technologically important areas such as surface science, chemical catalysis and solid-state electronic device manufacture.

It should be noted at this time that matrix studies of metal atoms, dimers, and clusters substantially predates this current interest. Matrix-isolation provides a very attractive technique for spectroscopic studies of high-temperature species, and studies of atoms of alkali metals and mercury are among its earliest applications. The early matrix-isolated metal-vapor work involved in most cases ESR and electronic absorption spectroscopy. Pioneering studies by Moskovits and Ozin (1976), Gruen and co-workers (Carstens et al. 1972, Gruen 1976), Lindsay and co-workers (Lindsay et al. 1976, Lindsay and Thompson 1982), and by numerous other investigators, have provided extensive new information, which was extremely useful in formulating subsequent more detailed studies both in the matrices and in the gas phase.

Several experimental obstacles hampered the application of laser-induced fluorescence to matrix metal-cluster studies. Metal vapors are typically monatomic, and dilute matrices prepared in a conventional fashion by vaporizing the metal in a furnace or Knudsen cell, usually contain predominantly atoms. Also, alternative sample production techniques, such as metal sputtering in electrical discharges, usually result in predominantly atomic species. In order to increase the statistical probability of atoms being sufficiently close to form a dimer or a larger cluster, one had to work with rather high overall metal concentrations.

This problem was often circumvented by extensive sample annealing to permit cluster formation following sample deposition. One of the difficulties with this approach is that light impurity molecules invariably present in the matrix often diffuse through the solid much more readily than the heavier metal containing species, and the spectra can be affected by the species which result from their reactions.

New techniques were recently developed which essentially eliminate this problem and which hold a large promise for future matrix-isolated metal-cluster studies. If the metal, or other material of interest, is vaporized by a pulsed laser in the presence of a dense inert carrier gas, the vapor is rapidly collisionally cooled, and extensive clustering can occur in the gas phase (Dietz et al. 1981, Bondybey and English 1981). By using a condensible carrier gas, such as argon or neon, and trapping the products formed in a pulsed supersonic expansion, matrices greatly enhanced in dimer and cluster concentrations can be generated (Heimbrook et al. 1987).

Enhancements in relative dimer concentration by more than an order of magnitude above that achievable in comparable, conventionally deposited matrices are easily accomplished. By doping or substituting the carrier gas with a suitable reactive gas, the technique should be potentially useful for studies of a wide range of metal compounds, including oxides, nitrides, hydrides, carbides, and similar species.

A more fundamental limitation of matrix-isolated cluster studies by laser-induced fluorescence is due to the fact that most larger metal clusters, and even many dimer species exhibit vanishingly small fluorescence quantum yields. Most metal atoms, and those of the transition metals in particular, possess many low lying atomic states, which result in a large number of molecular states in the dimers, and in particular in larger clusters. As a result of this very high density of electronic states, nonradiative processes often dominate their relaxation in the solid matrices, and limit the general applicability of the fluorescence techniques.

In spite of the experimental problems, laser-induced fluorescence studies have made a substantial contribution to our current knowledge of spectroscopy and properties of metal dimers and in some cases also larger clusters. Pioneering studies in this area were carried out by Teichman and Nixon (1975a,b, 1977) who used a cw ion laser to excite fluorescence of several metal containing molecules including Pb_2. They observed fluorescence involving several low lying electronic states of the molecule, and deduced a vibrational frequency of 220 cm^{-1} for its ground electronic state.

A great advantage of tunable dye laser work is that one can excite different states and levels at will, and the highly redundant spectroscopic information obtained permits much more unambiguous interpretation of the spectral observations. Such tunable pulsed-laser matrix studies (Bondybey and English 1977d, Sontag et al. 1983, Eberle et al. 1985) have shown that the ground state frequency of Pb_2 is actually close to 110 cm^{-1}, a value which was subsequently confirmed by gas-phase fluorescence studies (Heaven et al. 1983), as well as by the results of relativistic ab initio calculations (Balasubramanian and Pitzer 1983). An added benefit of pulsed-laser studies is, of course, the information obtainable about the lifetimes of the excited states and about the occurring relaxation processes.

Similarly for Sn_2, the initial observations of a ground-state vibrational frequency of 256 cm^{-1} by matrix fluorescence and Raman techniques (Teichman et al. 1978, Epting et al. 1980) were substantiated by subsequent theoretical (Balasubramanian and Pitzer 1983), and gas-phase experimental work (Bondybey et al. 1983).

Also bismuth, as well as its nonmetallic relatives in the main group V of the Periodic Table, were quite extensively studied (Teichman and Nixon 1977, Ahmed and Nixon 1980, 1981). An early matrix-isolation fluorescence study revealed a strong emission system in the near-infrared, and seemed to confirm a gas-phase observation of a ground electronic state of Bi_2 with a fundamental frequency of about 140 cm^{-1}. However, a more detailed study using tunable dye lasers (Bondybey and English 1980c) has revealed that the carrier responsible for this emission is, in fact, the polyatomic molecule Bi_4. Later work both in the gas phase and in the matrix has then shown the Bi_2 frequency to be near 173 cm^{-1} (Ahmed and Nixon 1981b).

The lighter, nonmetallic dimers of group V are reminiscent of N_2 in that transitions from the ground state to most of the lowest excited electronic state are forbidden, either by symmetry or by spin-selection rules or by both. Numerous of these forbidden electronic transitions were recently observed by the LIF techniques in the matrix for both homonuclear and heteronuclear species (Heimbrook et al. 1985a,b, Rasanen et al. 1986), and their lifetimes were measured.

The degree of the "forbiddenness" of the singlet–triplet transitions, of course, decreases with the increasing atomic number of the nuclei. Thus the lifetime of the lowest $a\,^3\Sigma_u^+$ state, corresponding to the Vegard–Kaplan transition of nitrogen decreases from some 30 s in neon matrix-isolated N_2 to about 0.6 ms for the corresponding state of Sb_2.

Some of the results for these molecules are listed in table 1 and compared, where available, with the gas-phase data. Quite noteworthy are the negligible matrix shifts of the neon matrix data, rarely exceeding 0.2%. This is in agreement with the general observation that the vibrational frequencies of strongly covalently bound molecules are rarely significantly perturbed in solid neon, as evidenced by the recent extensive compilations of Jacox (1985, 1987, 1988).

Somewhat different is the situation for molecules with only relatively weak Van der Waals bond, whose potential can be much more significantly affected by the solid environment, and where the observed frequency shifts may be quite substantial. An interesting series of studies of the group IIA dimers was performed by Miller and Andrews (1978a–e, 1980a–c). The ground electronic states of these molecules are characterized only by a weak Van der Waals type bond, whose strength increases from Mg_2 (~ 430 cm^{-1}) to Sr_2 (~ 1100 cm^{-1}). The laser-induced fluorescence studies have shown that the well depth deepens, and the ground-state vibrational frequencies increase substantially in the solid rare gases.

In spite of the larger than usual perturbation, very useful information can be derived from studies of this type. For instance, Ca_2 was known from the gas-phase studies to have an excited singlet state near 18 964 cm^{-1} (Balfour and Whittlock 1975), but the matrix absorption and fluorescence spectra have revealed a strongest transition at much lower energy, in the red. The long controversy which ensued has ended with the observation and characterization of the new low-lying singlet state near 14 000 cm^{-1} in the gas phase (Hofmann and Harris 1984, 1986, Bondybey and English 1984). This confirmed the earlier

Table 1

Comparison of selected molecular constants of some group V diatomics in solid neon and in the gas phase.

Molecule	Neon			Gas phase			
	T_e	ω_e	$\omega_e x_e$	T_e	ω_e	$\omega_e x_e$	t (ms)
X $^1\Sigma_g^+$ ($\Omega = 0$)							
P_2	0.0	781.0	2.91	0.0	780.8	2.83	
As_2	0.0	429.1	1.05	0.0	429.5	1.12	
Sb_2	0.0	270.4	0.57	0.0	269.9	0.55	
Bi_2	0.0	173.0	0.4	0.0	172.7	0.34	
AsP	0.0	604.0	1.88	0.0	604.0	1.98	
SbP	0.0	501.0	1.28	0.0	500.1	1.63	
a $^3\Sigma_u^+$ ($\Omega = 1$)							
P_2	18794.9	566.9	2.92	18794.5	565.1	2.7	90
As_2	14495.6	314.6	1.07	14481.6	314.3	1.17	12
Sb_2	14993.0	219.1	0.48	14991.0	217.2	0.44	0.6
AsP	16417.6	439.7	1.90				26
SbP	12796.0	370.5	1.27				12
w $^3\Delta_u$ ($\Omega = 1$)							
P_2	24030.4	591.3	2.43	23938.4	591.3	2.5	28
As_2	19929.2	330.5	0.86	19914.7	330.0	0.90	55
AsP	21714.1	461.2	1.63				4.5
SbP	21313.0	(d)					9
b' $^3\Sigma_u^-$ (0_u^+)							
P_2	28485.2	606.2	2.44	28503.4	604.4	2.2	
As_2	24659.8	336.3	0.6	24641.2	337.0	0.83	
Sb_2	19092	219.9	0.57	19068.9	218.0	0.53	

Notes:

(a) All values in cm^{-1}, except the lifetime t in ms.

(b) Gas-phase values are taken from Huber and Herzberg (1979).

(c) The g,u parity in electronic state designations should be omitted for the heteronuclear species.

(d) The w $^3\Delta_u$ state of SbP is strongly perturbed and the vibrational progression irregular.

suggestion by Miller and Andrews (1980a–c) that two different singlet states of Ca_2 are involved.

Transition-metal atoms have particularly numerous low-lying atomic states, resulting in a quite complex molecular cluster spectroscopy. As a consequence of this high density of molecular states, many transition-metal dimers exhibit negligible quantum yields of fluorescence in the matrix. Even in the case of Cu, whose ground-state atoms have a closed d electron shell, and whose dimer has therefore relatively simple spectroscopy compared with most other transition metals, the lowest dimer singlet states have negligible fluorescence quantum yields owing to predissociation resulting in two ground-state Cu atoms. However, the predissociation–recombination sequence populates the lowest a $^3\Sigma_u^+$ state, as discussed in one of the preceding sections, and a long-lived red phosphorescence is observed (Gole et al. 1982), which is assigned to the a $^3\Sigma_u^+$ → X $^1\Sigma_g^+$ intercombination transition (Bondybey 1982).

This type of process is obviously not unique to Cu_2. A very similar situation seems to prevail also in Mo_2 and Cr_2, where excitation of their strong singlet state absorptions in the visible region (Klotzbücher and Ozin 1977, 1978, Kundig et al. 1975) does not result in any observable resonant fluorescence. In Mo_2, excitation of the singlet results in long-lived phosphorescence from a low-lying triplet state (Pellin et al. 1981, 1982). In Cr_2, no emission at all was observed, but excitation of the corresponding $^1\Sigma_u$ state of Cr_2 resulted in a transient bleaching of the Cr_2 absorption, with a recovery time of about 0.5 s (Moskovits et al. 1985). It appears likely that in this case also a low-lying state, probably a triplet state, is populated, whose emission is, however, shifted into the infrared, beyond the range accessible with photomultiplier detectors.

Probably in many other dimers and larger clusters for which no fluorescence was observed in the visible, emissions from low-lying electronic states occur in the infrared, and could be observable using infrared detectors. Most of these species contain strong absorptions in the visible or near-UV, permitting very efficient pumping using tunable dye lasers. This may result in very bright IR emission, which could be conveniently studied using, e.g., Fourier transform spectroscopy. Such studies of infrared emission following UV or visible excitation might prove to be an extremely useful technique for investigation of simple metal clusters and compounds.

6. Free radicals

The matrix technique was initially developed for studies of free radicals and other transient species that, because of their reactivity and ephemeral existence, are difficult to study in the gas phase. A great variety of these species has now been stabilized and studied, and several excellent reviews of the previous work in this area are available (Perutz 1985, Andrews 1979, 1980, Jacox 1987, 1988). The exact coverage of a review will depend on the applied definition of a free radical. For example, the relatively stable NO and NO_2 species could, strictly speaking, be included in a discussion of a free-radical spectroscopy, while species such as CX_2 or SiX_2 which possess a singlet ground state might be excluded. For the purpose of this discussion we will consider only molecular species and fragments which under normal environmental conditions have only a transient existence.

There are numerous techniques for production of matrix-isolated radicals, and the reader is again referred to previous reviews. Fundamentally, they can be divided into techniques where the species of interest are generated in situ, in the matrix, and those where they are generated in the gas phase and subsequently trapped. Sampling the products of pyrolysis or discharge are typical examples of the latter category. The in situ techniques usually employ photolysis of a suitable precursor, or less frequently, bombardment of the matrix by electrons or ions. Photolytic radical production studies often take advantage of the fact that, unlike heavier species, hydrogen atoms can move through rare-gas matrices with relative ease. It is therefore often convenient to choose as precursor a molecule containing one more hydrogen than the radical of interest (Jacox and Milligan 1969, Andrews and Prochaska 1979). As previously noted, the matrix-cage effect prevents the loss of heavier atoms or fragments, but permits escape of a hydrogen atom from the trapping site.

The list of radicals studied by laser-induced fluorescence is, of course, much shorter than for absorption studies, partly due to the fact that the requirement of a stable fluorescent excited electronic states reduces the range of radicals amenable to LIF study. Main reason is, however, that while IR and visible/UV absorption spectroscopies of matrix-isolated species have been practiced since early days of the matrix technique, most LIF studies are much more recent.

In cases where fluorescence studies have been carried out, a much more complete information about the properties, spectroscopy and photophysics of the carrier of the spectrum is generally obtained than from studies of vibrational, either infrared or Raman, spectroscopy alone.

As an example, one can give the group of dihalocarbenes, CX_2. These isovalent radicals all have 15 electrons in their valence shell, and are therefore strongly bent, consistent with the rules formulated by Walsh (1953) based on a semi-empirical molecular orbital treatment. They all have readily accessible transition in the visible and near-UV (Milligan and Jacox 1968) characterized by long progressions in the bending frequency v_2. They have been studied rather extensively by laser-induced fluorescence, and their lifetimes and most of their vibrational frequencies, both in the ground state and in the excited electronic states are now rather well known.

The X $\rightarrow$ A transition energies shift from $36\,878$ cm^{-1} in CF_2 (Bass and Mann 1962, Smith et al. 1976) towards the red as the fluorine atoms are substituted by their heavier, more polarizable Cl or Br relatives (Tevault and Andrews 1975a,b, Jacox and Milligan 1970), so that in CBr_2 the lowest excited singlet state is in the near-infrared, near $14\,962$ cm^{-1}. In the same direction, one notes a lengthening of the transition lifetime from 27 ns in CF_2 to 14.5 μs in CBr_2 (Bondybey and English 1977a, 1980e, Huie et al. 1977, King et al. 1978). For reasons not fully understood, this lengthening is considerably faster than one would predict based only on the v^3 dependence of the transition energy.

Another interesting group of radicals are species with 12–16 valence electrons which are, again as predicted by Walsh, all linear. C_3 (12 electrons), CCN (13), CNN (14) and NCO (15) have all been studied. Fairly complete data about their vibrational frequencies and lifetimes are now available, again with considerable contribution from matrix studies.

The rigidity of the molecules depends on the number of electrons in the partially occupied π_g orbital. One therefore sees in all of these species a considerable increase in the bending frequency v_2 upon electronic excitation, since lowest excited states involve promotion of an additional electron into this π_g orbital. One also sees a systematic increase in the bending frequency from the ground state of C_3 where the orbital is empty ($v_2 = 63$ cm^{-1}) to the excited state of NCO, where it is completely full ($v_2 = 640$ cm^{-1}).

An interesting matrix effect involves the gas to matrix shifts in these species. The degree to which the vibrational structure of these species is perturbed also seems to scale with the number of electrons in the π_g orbital. NCO in the excited A $^2\Sigma$ state where, as noted before, the orbital is full, appears quite unperturbed, with vibrational intervals being almost identical in Ne, Ar and Kr and, where data are available, also close to the gas-phase values. On the other hand, in C_3 and CCN some vibrational frequencies appear to be strongly perturbed. These perturbations are again larger in the ground states, which have fewer electrons in the π_g orbital, and are much more pronounced in the heavier rare gases. For instance, the ground state v_1 stretching frequency of CCN which is in solid Ne very close to the gas-phase value, shifts by nearly 150 cm^{-1} to 1770 cm^{-1} in solid Ar.

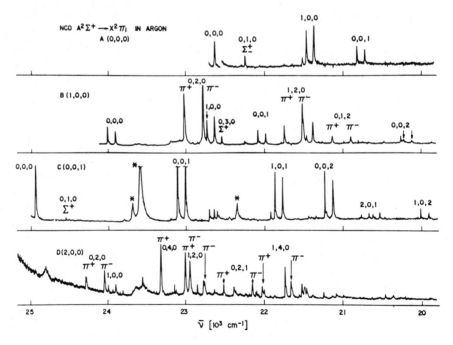

Fig. 3. Fluorescence spectra of the NCO radical in solid argon, demonstrating the lack of vibrational relaxation on the timescale of the 180 ns radiative lifetime of the excited A $^2\Sigma^+$ state. The bands denoted by an asterisk are due to the impurity CNN radical. The sharply rising background (near 25 000 cm^{-1} in trace d) is due to the CH impurity.

It is noteworthy that the rates of the vibrational relaxation in the excited states of these radicals seem to increase with increasing degree of perturbation. In NCO the rates are quite slow compared with the radiative rate, and lifetimes are in the high μs range. This can be seen in fig. 3, where in each panel only resonant emission from the level directly excited by the laser is observable. In contrast with that, no vibrationally unrelaxed fluorescence is observed in either C_3 or CCN, suggesting lifetimes in the subnanosecond range.

7. Production and fluorescence studies of ionic species

Molecular ions are a particularly elusive class of transient species. Their studies by classical gas-phase spectroscopic techniques proved to be quite difficult, and consequently a considerable effort went into developing matrix-isolation as an alternative to gas-phase studies.

Obvious prerequisites for successful laser-induced studies of molecular ions are the ability to generate the matrix-isolated ions, and the existence of fluorescent excited electronic states. Numerous techniques have been developed for generating matrix-isolated ions. If the corresponding neutral is a stable molecule, the simplest and cleanest way of generating matrix-isolated ions is to deposit the parent compound, and subsequently generate the ionic species by in situ ionization (Milligan and Jacox 1969, Bondybey et al. 1978). The matrix-cage effect discussed previously usually slows down or eliminates competing fragmentation or dissociative processes, but allows ionization to proceed unhindered. The in situ photolysis technique is, of course, limited by the range of transparency of the host material, but solid rare gases, and neon in particular, are excellent in this respect.

Electric discharges of various kinds represent a commonly used alternative technique. They have the added advantage of being also applicable to fragment ions, where the corresponding neutral entities are free radicals or otherwise unstable species. A clear disadvantage of discharge techniques is that they are quite nonselective, and one usually ends up with a complex mixture of neutral and ionic products. The same is usually true for techniques employing bombardment of the matrix by high-energy particles, such as electrons (Milligan et al. 1970, Knight et al. 1983) or protons (Andrews et al. 1975a,b).

A direct deposition of mass-selected ion beams would appear to be an almost ideal technique, but, unfortunately, experimental implementation proves to be quite difficult, and the results of numerous attempts in several laboratories were often less than spectacular. The problem here is in slowing the ions sufficiently and preventing them from reacting with each other, with impurities in the matrix, or from recombining with electrons or negatively charged species needed to maintain the overall electroneutrality of the matrix.

A new, potentially very powerful technique for trapping ions in matrices involves laser vaporization. Gas-phase studies using this technique have shown that a substantial fraction of the species produced by laser vaporization is initially ionized. Studies in our laboratory have demonstrated that, e.g., in studies of vapors of calcium and beryllium, often as much as 10% of the metal atoms were in the ionic state (Bondybey 1985). This technique should be particularly useful for highly refractory or difficult to ionize materials.

Recent studies where laser-generated graphite vapors were trapped (Rasanen et al. 1987) have revealed, in addition to C_2 and C_3 spectra, also copious amounts of C_2^- (Milligan and Jacox 1969, Allamandola and Nibler 1974, Allamandola et al. 1977, Frosch 1971), and a spectrum of an additional ion which, based on isotopic results, is almost certainly due to C_2^+.

As noted above, an obvious additional prerequisite for successful laser-induced fluorescence studies of molecular ions is the existence of sufficiently stable and long-lived excited electronic states with observable fluorescence quantum yields. This is, unfortunately, a rather serious limitation. For instance, in most simple saturated hydrocarbons and related compounds, all the valence electrons participate in providing the σ-bonds between the atoms. Removal of one of the localized σ electrons obviously weakens the overall bonding. Attempts to excite an additional bonding electron then often result in fragmentation of the ion. Even if the excited state does not fragment, the electron deficit would lead to a changed upper-state geometry and a severe broadening of the matrix spectrum.

In view of this, the best candidates for laser-induced fluorescence studies are multiply bonded, unsaturated species, where removal of one of the delocalized π electrons usually does not severely reduce the overall bonding and stability of the resulting ion, and where transitions are possible which interchange electron between partially filled π-type orbitals. Accordingly, most of the ionic species studied thus far are derived from aromatic hydrocarbons, polyolefins and acetylenic, triple-bond containing compounds (Maier 1980, Bondybey and Miller 1980, 1983).

A serious potential obstacle to successful matrix studies is the relatively strong interaction of ionic species with the rare-gas solid. Solvation energies of neutral and nonpolar guests in rare-gas solids are quite small, usually at least 1–2 orders of magnitude smaller than the bond energies in typical covalently bound molecules (200–600 kJ/mol). This large disparity between the intramolecular bonding of the guest and the guest–host interaction forces forms the basis of matrix-isolation spectroscopy.

On the other hand, it has been shown experimentally that the solvation energies of molecular ions are, even in the rare-gas solids, quite large (Gedanken et al. 1973). The reported values, of the order of 120–200 kJ/mol, are not so different from the binding forces within the guest. Such strong coupling might cause broadening of the electronic spectrum. Obviously, in the limit where the spectrum is broad and featureless, little useful information can be derived. In view of the strong interactions one also could, and should, enquire, whether spectroscopic information, even if obtained, is meaningful as far as spectroscopy of the free guest is concerned.

Fortunately, experimental studies have shown that, in spite of the relatively strong interactions, sharp and well-resolved fluorescence spectra can be obtained for many molecular ions. More importantly, it has been demonstrated that, in particular when solid-neon matrices are used, the matrix shifts and perturbations are often quite insignificant (Bondybey and English 1979a).

The importance of matrix material and the advantage of solid neon in studies of ionic species is demonstrated quite dramatically in figs. 4 and 5. The fluorescence spectrum of diacetylene radical cation, $C_4H_2^+$, in solid Ne exhibits a sharp, well-resolved structure. The origin band at 19 708 cm^{-1} is only insignificantly (16 cm^{-1}) red-shifted from its gas-phase position, and most vibrational energies are changed by less than 0.5%. On the other hand, the corresponding emission in solid argon is completely featureless and shifted some 2000 cm^{-1} to lower energies.

In solid neon, in spite of the relatively large solvation energy of the cations, the interaction largely consists of polarization of the solvent atoms, but there is little chemical charge transfer interaction. Even more importantly, the solvation is not state-specific, i.e., the guest–host interaction potential is very similar for both the electronic states involved, resulting in the sharp zero-phonon line spectrum.

While the electron affinities of the cation, 10.17 and 12.62 eV in the ground

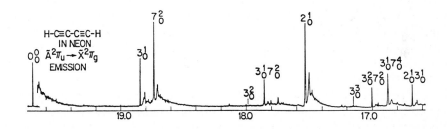

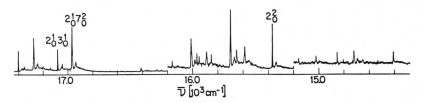

Fig. 4. Fluorescence spectrum of the diacetylene radical cation in solid neon. A sharp discrete structure is observable up to above 7000 cm^{-1} of vibrational energy, and the vibrational structure shows, where gas-phase data are available, quite insignificant matrix perturbation.

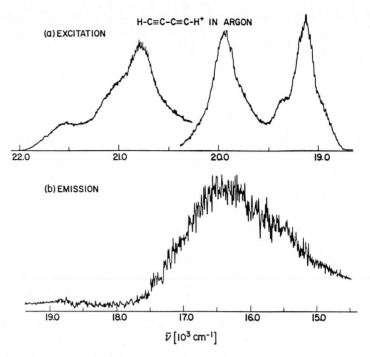

Fig. 5. Fluorescence excitation spectrum (a) and laser-induced fluorescence (b) of the diacetylene radical cation in solid argon taken under similar conditions as the solid Ne spectrum in fig. 4. The excitation shows a rather broad strongly perturbed vibrational structure. The emission is strongly red-shifted and exhibits no resolvable vibrational fine structure.

state and excited state, respectively, are much smaller than the ionization potential of the host neon, 24.6 eV, the disparity is much smaller in the argon matrix. Its ionization potential, 15.76 eV, is much lower, and much stronger charge transfer can take place, with the solvent acting as a Lewis base and the ion as a Lewis acid. Such interaction will be state-specific, and stronger in the excited electronic state, which has electron affinity higher by its excitation energy. The result of this charge transfer are the perturbed frequencies, strong red shifts in emission, and broad, and in extreme cases featureless, spectra, as seen in argon matrix for the diacetylene cation.

It should be noted that, since the interactions will, in general, be weaker in the ground electronic states, relatively unperturbed molecular ion spectra may be obtained in solid argon or krypton. Infrared studies of molecular ions have been summarized in several excellent reviews by Andrews, Jacox, and others. For some species with low ionization potential, useable optical absorption and fluorescence spectra can also be obtained in the heavier rare gases. For instance, relatively well-resolved spectra for several fluorobenzene cations have been observed in solid argon and even krypton.

In general, however, neon appears to be the best medium for studies of charged species, since, in addition to giving transparent nonscattering samples of excellent optical quality, it also gives best resolved and least perturbed molecular spectra.

8. Summary

In this chapter we have discussed the fluorescence of matrix-isolated molecules. We have attempted to describe the photophysics of matrix-isolated atoms and molecules, and the ways in which their behavior differs from that of the free gas-phase species. We have, in particular, explored the question of whether or how can these differences in behavior in matrices be turned into an advantage.

The main disadvantages of matrix studies are the absence of information obtainable in the gas phase by analysis of rotational fine structure, as well as the possible perturbation of the spectra by the solid environment. One also has to consider that alternative gas-phase low-temperature techniques are available today. We have shown, however, that the medium effects and perturbations are, fortunately, in many instances quite minor. Furthermore, we hope to have shown that by taking advantage of the differences in matrix processes and photophysics, one can obtain information otherwise not obtainable, or difficult to obtain in gas-phase studies.

While nothing can be done about the lack of rotational data in matrix work, and ultimately gas-phase studies may be needed to obtain definitive structural information, the matrix fluorescence techniques can be invaluable for quickly obtaining spectroscopic information about electronic and vibrational structures, molecular dynamics and lifetimes. It is particularly useful in studies of strongly forbidden transitions through nonradiative access to otherwise inaccessible states. We have discussed the application of the fluorescence technique to several different types of transient species, including metal atoms, clusters, free radicals and molecular ions.

While alternative techniques now, as noted above, exist, one can foresee continued important role for matrix studies, with gas-phase and matrix studies more often being complementary, rather than redundant. As shown in the present chapter, matrix laser-induced fluorescence studies have yielded large amounts of useful spectroscopic data in the past, and undoubtedly will continue being useful in a wide range of spectroscopic applications in the future as well.

Acknowledgement

This work was supported by the Chemistry Division of the National Science Foundation under contract no. CHE-8803169, and by the Ohio State University under the University Research Grant Program.

References

Ahmed, F., 1985, J. Chem. Phys. **82**, 4401.
Ahmed, F., and E.R. Nixon, 1979, J. Chem. Phys. **71**, 3547.
Ahmed, F., and E.R. Nixon, 1980, J. Mol. Spectrosc. **83**, 64.
Ahmed, F., and E.R. Nixon, 1981a, J. Chem. Phys. **74**, 2156.
Ahmed, F., and E.R. Nixon, 1981b, J. Chem. Phys. **75**, 110.
Ahmed, F., and E.R. Nixon, 1981c, J. Mol. Spectrosc. **87**, 101.
Allamandola, L.J., and J.W. Nibler, 1974, Chem. Phys. Lett. **28**, 335.
Allamandola, L.J., A.M. Rohjantalab, J.W. Nibler and T. Chappell, 1977, J. Chem. Phys. **67**, 99.
Andrews, L., 1979, Ann. Rev. Phys. Chem. **30**, 79.
Andrews, L., 1980, Adv. Infrared Raman Spectrosc. **7**, 59.
Andrews, L., and F.T. Prochaska, 1979, J. Chem. Phys. **70**, 4717.
Andrews, L., B.S. Ault, J.M. Grzybowski and R.O. Allen, 1975a, J. Chem. Phys. **62**, 2461.
Andrews, L., J.M. Grzybowski and R.O. Allen, 1975b, J. Phys. Chem. **79**, 903.
Ault, B.S., W.F. Howard and L. Andrews, 1975, J. Mol. Spectrosc. **55**, 217.
Bahrdt, J., P. Gurtler and N. Schwentner, 1987, J. Chem. Phys. **86**, 6108.
Balasubramanian, K., and K.S. Pitzer, 1983, J. Chem. Phys. **78**, 321.
Balfour, W.J., and R.F. Whittlock, 1975, Can. J. Phys. **53**, 472.
Barrow, R.F., and J.M. Ketteringham, 1963, Can. J. Phys. **41**, 419.
Bass, A.M., and D.E. Mann, 1962, J. Chem. Phys. **36**, 3501.
Beeken, P.B., M. Mandich and G.W. Flynn, 1982, J. Chem. Phys. **76**, 5995.
Beeken, P.B., E.A. Hanson and G.W. Flynn, 1983, J. Chem. Phys. **78**, 5892.
Berkowitz, M., and R.B. Gerber, 1979, Chem. Phys. **37**, 369.
Bondybey, V.E., 1976a, J. Chem. Phys. **65**, 5138.
Bondybey, V.E., 1976b, J. Chem. Phys. **65**, 2296.
Bondybey, V.E., 1977a, J. Chem. Phys. **66**, 4237.
Bondybey, V.E., 1977b, J. Chem. Phys. **66**, 995.
Bondybey, V.E., 1977c, J. Mol. Spectrosc. **64**, 180.
Bondybey, V.E., 1978, J. Chem. Phys. **68**, 1308.
Bondybey, V.E., 1981, Adv. Chem. Phys. **47**, 521.
Bondybey, V.E., 1982, J. Chem. Phys. **77**, 3771.
Bondybey, V.E., 1984, Ann. Rev. Phys. Chem. **35**, 591.
Bondybey, V.E., 1985, Science **227**, 125.
Bondybey, V.E., and L.E. Brus, 1975a, J. Chem. Phys. **62**, 620.
Bondybey, V.E., and L.E. Brus, 1975b, J. Chem. Phys. **63**, 2223.
Bondybey, V.E., and L.E. Brus, 1975c, J. Chem. Phys. **63**, 794.
Bondybey, V.E., and L.E. Brus, 1976, J. Chem. Phys. **64**, 3724.
Bondybey, V.E., and L.E. Brus, 1980, Adv. Chem. Phys. **41**, 269.
Bondybey, V.E., and J.H. English, 1977a, J. Mol. Spectrosc. **68**, 89.
Bondybey, V.E., and J.H. English, 1977b, J. Chem. Phys. **67**, 664.
Bondybey, V.E., and J.H. English, 1977c, J. Chem. Phys. **67**, 2868.
Bondybey, V.E., and J.H. English, 1977d, J. Chem. Phys. **67**, 3405.
Bondybey, V.E., and J.H. English, 1978a, J. Mol. Spectrosc. **70**, 236.
Bondybey, V.E., and J.H. English, 1978b, J. Chem. Phys. **68**, 4641.
Bondybey, V.E., and J.H. English, 1979a, J. Chem. Phys. **71**, 777.
Bondybey, V.E., and J.H. English, 1979b, J. Chem. Phys. **70**, 1621.
Bondybey, V.E., and J.H. English, 1980a, J. Chem. Phys. **84**, 388.
Bondybey, V.E., and J.H. English, 1980b, J. Chem. Phys. **72**, 3113.
Bondybey, V.E., and J.H. English, 1980c, J. Chem. Phys. **73**, 42.
Bondybey, V.E., and J.H. English, 1980d, J. Chem. Phys. **72**, 6479.

Bondybey, V.E., and J.H. English, 1980e, J. Mol. Spectrosc. **79**, 416.
Bondybey, V.E., and J.H. English, 1981, J. Chem. Phys. **74**, 6978.
Bondybey, V.E., and J.H. English, 1984, Chem. Phys. Lett. **111**, 195.
Bondybey, V.E., and C. Fletcher, 1976, J. Chem. Phys. **64**, 3615.
Bondybey, V.E., and T.A. Miller, 1980, J. Chem. Phys. **73**, 3053.
Bondybey, V.E., and T.A. Miller, 1983, in: Molecular Ions: Spectroscopy Structure and Chemistry, eds T.A. Miller and V.E. Bondybey (North-Holland, Amsterdam) p. 125.
Bondybey, V.E., and J.W. Nibler, 1972, J. Chem. Phys. **56**, 4719.
Bondybey, V.E., and A. Nitzan, 1977, Phys. Rev. Lett. **38**, 889.
Bondybey, V.E., S.S. Bearder and C. Fletcher, 1976, J. Chem. Phys. **64**, 5243.
Bondybey, V.E., J.H. English and T.A. Miller, 1978, J. Am. Chem. Soc. **100**, 5251.
Bondybey, V.E., T.A. Miller and J.H. English, 1980, J. Chem. Phys. **72**, 2193.
Bondybey, V.E., C.B. Vaughn, T.A. Miller, J.H. English and R.H. Shiley, 1981, J. Chem. Phys. **74**, 6584.
Bondybey, V.E., M.C. Heaven and T.A. Miller, 1983, J. Chem. Phys. **78**, 3593.
Bondybey, V.E., R.C. Haddon and J.H. English, 1984, J. Chem. Phys. **80**, 5432.
Breithaupt, B., J.E. Hulse, D.M. Kolb, H.H. Rotermund, W. Schroeder and W. Schrittenlacher, 1983, Chem. Phys. Lett. **95**, 513.
Brewer, L., and G.D. Brabson, 1966, J. Chem. Phys. **44**, 3274.
Brewer, L., G.D. Brabson and B. Meyer, 1965, J. Chem. Phys. **42**, 1385.
Broida, H.P., and M. Peyron, 1960, J. Chem. Phys. **32**, 1068.
Brom, J.M., and H.P. Broida, 1975, J. Chem. Phys. **63**, 3718.
Brom, J.M., W.D. Hewett and W. Weltner, 1975, J. Chem. Phys. **62**, 3122.
Brus, L.E., and V.E. Bondybey, 1975a, J. Chem. Phys. **63**, 3123.
Brus, L.E., and V.E. Bondybey, 1975b, J. Chem. Phys. **63**, 786.
Carstens, D.H., W. Brashear, D.R. Eslinger and D.M. Gruen, 1972, Appl. Spectrosc. **26**, 184.
Cellucci, T.A., and E.R. Nixon, 1984, J. Chem. Phys. **81**, 1174.
Cellucci, T.A., and E.R. Nixon, 1985, J. Phys. Chem. **89**, 1991.
Diestler, D.J., 1974, J. Chem. Phys. **60**, 2692.
Diestler, D.J., 1976, Chem. Phys. Lett. **39**, 39.
Dietz, T.G., M.A. Duncan, D.E. Powers and R.E. Smalley, 1981, J. Chem. Phys. **74**, 6511.
Eberle, B., H. Sontag and R. Weber, 1984, J. Chem. Phys. **80**, 2984.
Eberle, B., H. Sontag and R. Weber, 1985, Chem. Phys. **92**, 417.
Epting, M.A., and E.R. Nixon, 1983, J. Chem. Phys. **78**, 999.
Epting, M.A., M.T. McKenzie and E.R. Nixon, 1980, J. Chem. Phys. **73**, 134.
Freed, K.F., and H. Metiu, 1977, Chem. Phys. Lett. **48**, 262.
Freed, K.F., D.L. Yeager and H. Metiu, 1977, Chem. Phys. Lett. **49**, 19.
Frosch, R.P., 1971, J. Chem. Phys. **54**, 2660.
Gedanken, A., B. Raz and J. Jortner, 1973, J. Chem. Phys. **58**, 1178.
Gerber, R.B., and M. Berkowitz, 1977a, Chem. Phys. Lett. **49**, 260.
Gerber, R.B., and M. Berkowitz, 1977b, Phys. Rev. Lett. **39**, 1000.
Gole, J.L., J.H. English and V.E. Bondybey, 1982, J. Phys. Chem. **86**, 2560.
Gouedard, G., and J.C. Lehmann, 1976, J. Phys. B **9**, 2113.
Gruen, D.M., 1976, in: Cryochemistry, eds M. Moskovits and G.A. Ozin (Wiley, New York) p. 441.
Heaven, M.C., T.A. Miller and V.E. Bondybey, 1983, J. Phys. Chem. **87**, 2072.
Heimbrook, L.A., N. Chestnoy, M. Rasanen, G.P. Schwartz and V.E. Bondybey, 1985a, J. Chem. Phys. **83**, 6091.
Heimbrook, L.A., M. Rasanen and V.E. Bondybey, 1985b, Chem. Phys. Lett. **120**, 233.
Heimbrook, L.A., M. Rasanen and V.E. Bondybey, 1987, J. Phys. Chem. **91**, 2468.
Hofmann, R.T., and D.O. Harris, 1984, J. Chem. Phys. **81**, 1047.
Hofmann, R.T., and D.O. Harris, 1986, J. Chem. Phys. **85**, 3749.

Huber, K.P., and G. Herzberg, 1979, Constants of Diatomic Molecules (Van Nostrand, New York).
Huie, R.E., N.J.T. Long and B.A. Thrush, 1977, Chem. Phys. Lett. **2**, 197.
Jacox, M.E., 1978, Rev. Chem. Intermed. **1**, 1.
Jacox, M.E., 1985, J. Mol. Spectrosc. **113**, 286.
Jacox, M.E., 1987, J. Mol. Struct. **157**, 43.
Jacox, M.E., 1988, J. Phys. Chem. Ref. Data **17**, 269.
Jacox, M.E., and D.E. Milligan, 1969, J. Chem. Phys. **50**, 3252.
Jacox, M.E., and D.E. Milligan, 1970, J. Chem. Phys. **53**, 2688.
King, D.S., P.K. Schenck and J.C. Stephenson, 1978, in: Lasers in Chemistry, Royal Institution (Elsevier) p. 441.
Klotzbücher, W.E., and G.A. Ozin, 1977, Inorg. Chem. **16**, 984.
Klotzbücher, W.E., and G.A. Ozin, 1978, J. Am. Chem. Soc. **100**, 2262.
Knight Jr, L.B., R.D. Britain, M.A. Starr and C.H. Joyner, 1974, J. Chem. Phys. **61**, 5289.
Knight Jr, L.B., J.M. Bostick, R.W. Woodward and J. Steadman, 1983, J. Chem. Phys. **78**, 6415.
Kühle, H., J. Bahrdt, R. Fröhling and N. Schwentner, 1985a, J. Phys. (France) **46**, C7-315.
Kühle, H., J. Bahrdt, R. Fröhling, N. Schwentner and H. Wilcke, 1985b, Phys. Rev. B **31**, 4854.
Kühle, H., R. Fröhling, J. Bahrdt and N. Schwentner, 1985c, J. Chem. Phys. **84**, 666.
Kühle, H., N. Schwentner, H. Gabriel and I. Gersonde, 1986, Chem. Phys. Lett. **132**, 570.
Kundig, E.P., M. Moskovits and G.A. Ozin, 1975, Nature **254**, 503.
Langen, J., K.P. Lodermann and U. Schurath, 1987, Chem. Phys., in press.
Legay, F., 1977, Chemical and Biological Applications of Lasers, Vol. 2 (New York).
Leutloff, D., and D.M. Kolb, 1979, Ber. Bunsenges. Phys. Chem. **83**, 666.
Lin, S.H., 1974, J. Chem. Phys. **61**, 3810.
Lin, S.H., 1976, J. Chem. Phys. **65**, 1053.
Lindsay, D.M., and G.A. Thompson, 1982, J. Chem. Phys. **77**, 1114.
Lindsay, D.M., D.R. Herschbach and A.L. Kwiram, 1976, Mol. Phys. **32**, 1199.
Maier, J.P., 1980, Chimia **34**, 219.
Mandich, M., P. Beeken and G.W. Flynn, 1982, J. Chem. Phys. **77**, 702.
McCarthy, M., and G.W. Robinson, 1959, J. Chim. Phys. **56**, 723.
McKenzie, M.T., and E.R. Nixon, 1983, Chem. Phys. **75**, 115.
McKenzie, M.T., M.A. Epting and E.R. Nixon, 1982, J. Chem. Phys. **77**, 735.
Meyer, B., 1971, Low Temperature Spectroscopy (Elsevier, New York).
Miller, J.C., and L. Andrews, 1978a, J. Chem. Phys. **68**, 1701.
Miller, J.C., and L. Andrews, 1978b, J. Am. Chem. Soc. **100**, 6956.
Miller, J.C., and L. Andrews, 1978c, J. Chem. Phys. **69**, 936.
Miller, J.C., and L. Andrews, 1978d, J. Am. Chem. Soc. **100**, 2966.
Miller, J.C., and L. Andrews, 1978e, J. Chem. Phys. **69**, 2054.
Miller, J.C., and L. Andrews, 1980a, Appl. Spectrosc. Rev. **16**, 1.
Miller, J.C., and L. Andrews, 1980b, J. Mol. Spectrosc. **80**, 178.
Miller, J.C., and L. Andrews, 1980c, J. Phys. Chem. **84**, 401.
Milligan, D.E., and M.E. Jacox, 1968, J. Chem. Phys. **48**, 2265.
Milligan, D.E., and M.E. Jacox, 1969, J. Chem. Phys. **51**, 1952.
Milligan, D.E., M.E. Jacox and W.A. Guillory, 1970, J. Chem. Phys. **52**, 3864.
Moskovits, M., and G.A. Ozin, eds, 1976, Cryochemistry (Wiley, New York).
Moskovits, M., W. Limm and T. Mejean, 1985, J. Chem. Phys. **82**, 4875.
Mulliken, R.S., 1940, Phys. Rev. **57**, 500.
Nicolai, J.P., and M.C. Heaven, 1985, J. Chem. Phys. **83**, 6538.
Nicolai, J.P., L.J. Van de Burgt and M.C. Heaven, 1985, Chem. Phys. Lett. **115**, 496.
Nitzan, A., S. Mukamel and J. Jortner, 1974, J. Chem. Phys. **60**, 3929.
Nitzan, A., S. Mukamel and J. Jortner, 1975, J. Chem. Phys. **63**, 200.
Nitzan, A., M. Shugard and J.C. Tully, 1978, J. Chem. Phys. **69**, 2525.

Pellin, M.J., T. Foosnaes and D.M. Gruen, 1981, J. Chem. Phys. **74**, 5547.
Pellin, M.J., T. Foosnaes and D.M. Gruen, 1982, ACS Symp. Ser. **179**, xxx.
Perutz, R.N., 1985a, Chem. Rev. **85**, 11.
Perutz, R.N., 1985b, Chem. Rev. **85**, 91.
Pimentel, G.C., 1958a, J. Am. Chem. Soc. **80**, 62.
Pimentel, G.C., 1958b, Spectrochimica Acta **12**, 94.
Rasanen, M., L.A. Heimbrook, G.P. Schwartz and V.E. Bondybey, 1986, J. Chem. Phys. **85**, 86.
Rasanen, M., J.Y. Liu, T.P. Dzugan and V.E. Bondybey, 1987, Chem. Phys. Lett., in press.
Rebane, K., 1970, Impurity Spectra of Solids (Plenum, New York).
Schnepp, O., and K. Dressler, 1965, J. Chem. Phys. **42**, 2482.
Shirk, J.S., 1971, J. Chem. Phys. **55**, 3608.
Smith, C.E., M.E. Jacox and D.E. Milligan, 1976, J. Mol. Spectrosc. **60**, 381.
Sontag, H., B. Eberle and R. Weber, 1983, Chem. Phys. **80**, 279.
Stone, T.J., and R.F. Barrow, 1975, Can. J. Phys. **53**, 1976.
Teichman, R.A., and E.R. Nixon, 1975a, J. Mol. Spectrosc. **55**, 192.
Teichman, R.A., and E.R. Nixon, 1975b, J. Mol. Spectrosc. **57**, 14.
Teichman, R.A., and E.R. Nixon, 1977, J. Chem. Phys. **67**, 1977.
Teichman, R.A., M.A. Epting and E.R. Nixon, 1978, J. Chem. Phys. **68**, 336.
Tevault, D.E., 1982, J. Chem. Phys. **76**, 2859.
Tevault, D.E., and L. Andrews, 1975a, J. Am. Chem. Soc. **97**, 1707.
Tevault, D.E., and L. Andrews, 1975b, J. Mol. Spectrosc. **54**, 54.
Tevault, D.E., and L. Andrews, 1975c, J. Mol. Spectrosc. **54**, 110.
Tinti, D.S., 1968, J. Chem. Phys. **48**, 1459.
Tinti, D.S., and G.W. Robinson, 1968, J. Chem. Phys. **49**, 3229.
Vegard, L., 1924, Nature **114**, 357.
Vegard, L., 1930, Ann. Phys. **6**, 487.
Vegard, L., 1932, Z. Phys. **75**, 30.
Walsh, A.D., 1953, J. Chem. Soc. **x**, 2266.
Whittle, E., D.A. Dows and G.C. Pimentel, 1954, J. Chem. Phys. **22**, 1943.
Wiesenfeld, J.M., and C.B. Moore, 1979, J. Chem. Phys. **70**, 930.
Wight, C.A., B.S. Ault and L. Andrews, 1975, J. Mol. Spectrosc. **56**, 239.
Wight, C.A., B.S. Ault and L. Andrews, 1976, Inorg. Chem. **15**, 2147.
Yee, K.K., and R.F. Barrow, 1972, J. Chem. Soc. Faraday Trans. II, **68**, 1181.
Young, L., and C.B. Moore, 1984, J. Chem. Phys. **81**, 3137.

CHAPTER 6

Chemistry on Ground State and Excited Electronic Surfaces Induced by Selective Photo-excitation in Matrices

Heinz FREI and George C. PIMENTEL

Chemical Biodynamics Division
Lawrence Berkeley Laboratory
University of California
Berkeley
CA 94720, USA

Chemistry and Physics of
Matrix-Isolated Species
L. Andrews and M. Moskovits, eds

Contents

1. Introduction

Tunable laser light sources have opened new opportunities for clarification of the course of chemical reactions. Both on ground and excited electronic surfaces, it is now possible to initiate chemical reactions at precisely controlled initial energies and excitations. In principle, such possibilities bring the reaction coordinate and the activated complex closer to experimental reach.

The earliest attempts to develop these possibilities have revealed the rapidity of intramolecular energy redistribution and the necessity to understand these processes better. Some of these important but diversionary processes are either mute or inactive in cryogenic matrix environments. We shall discuss here a number of types of experiments which show that a cryogenic matrix offers a unique environment for the study of selective, photo-induced chemistry.

One special function of a matrix is to hold together a pair of potentially reactive molecules in a "sustained collisional complex". This permits convenient and leisurely infrared spectroscopic study of these reactive pairs before and after selective excitation and in the absence of diffusion and thermally induced reactions. Of course, the diagnostic power of the IR spectral regions for matrix samples derives from the narrow band widths and the fact that essentially all molecules are vibrational chromophores (with the sole exception of homonuclear diatomics). Furthermore, prolonged photolytic exposure is possible, so low extinction coefficients and/or low quantum yields can be dealt with. For systems with low activation energies, one may attempt to stimulate reactions on the electronic ground-state surface by selective excitation of fundamental, overtone, or combination vibrations, using mid- or near-IR photons. Both for vibrationally stimulated and selective electronically induced chemistry, infrared monitoring of product growth allows detailed study of quantum efficiencies to reaction, branching ratios, and sequential processes. Of course, the possibility of trapping transient intermediates and studying them by UV/visible and infrared spectroscopy makes the cryogenic technique a very valuable tool for such elucidation of reaction pathways.

There are limitations of the matrix technique that must be recognized. There is little known about the molecular orientations in the "sustained collisional complex" nor is it yet clear how much this can be controlled either by matrix-cage size or elevated temperatures. Furthermore, there is no certainty that all intermediate species can be stabilized – experience shows that for exothermic reactions, this is a sensitive function of the competition between deactivation and continued reaction. Some bimolecular systems of interest cannot be studied because of pre-reaction during deposition (even with dual jet deposition). Careful attention must be paid to the degree of isolation and to reactant cluster size. Finally, there always remains the need for judgment and circumspection as we relate matrix reaction pathways to those of the gas phase.

This chapter surveys recent work in the field of selective photo-induced reactions in cryogenic solids, with focus on contributions from our laboratory. We will start with a brief review of vibrationally stimulated reactions in matrices, with special focus on evidence for mode selective chemistry (section 2). This will be followed by a discussion of selective vibronically induced bimolecular reactions with red and near-infrared photons (section 3). A few examples of chemistry induced by photo-excited group IIB metal atoms, which permit initiation of reactions on excited hypersurfaces not accessible by direct excitation of the reactants, will be presented in section 4. Then, we will discuss the study of a UV wavelength-controlled photo-isomerization, of fundamental importance both for molecular energy and information storage (section 5).

2. Infrared laser-induced chemistry

A cryogenic matrix is a unique environment for vibrational stimulation of reactions with low electronic ground-state barriers (a few kcal/mol), since under such conditions the kT energy available to the reactants is very small. As a consequence, thermal reactions are suppressed, hence high-energy conformers may be stabilized and accumulated for convenient spectroscopic study. The first such process, the *cis–trans* isomerization of nitrous acid induced by excitation of the OH stretching mode with broad band infrared light, was discovered in this laboratory more than 25 years ago (Baldeschwieler and Pimentel 1960, Hall and Pimentel 1963). However, new IR-induced processes in matrices have been discovered only in the past ten years, reflecting the growing interest in vibrational photochemistry and the increasing availability of tunable infrared laser sources. We have reviewed progress in this field very recently (Frei and Pimentel 1985), so only a brief summary will be given below.

Beyond the well-established IR-laser-induced interconversions of $Fe(CO)_4$ (Poliakoff and Turner 1980) and prototype cases with one conformational degree of freedom, like nitrous acid (Baldeschwieler and Pimentel 1960, Hall and Pimentel 1963, McDonald and Shirk 1982, Shirk and Shirk 1983) and 1,2-difluoroethane (Felder and Günthard 1984), an increasing number of vibrationally stimulated isomerizations of more complex molecules with two or more internal rotations are reported. Among these, the 2-haloethanols have been studied most extensively, 2-fluoroethanol under selective laser excitation (Pourcin et al. 1980, Hoffmann and Shirk 1983, 1985), and all of them under globar irradiation (Perttila et al. 1978a,b, Davidovics et al. 1983, 1984, Takeuchi and Tasumi 1982, Homanen and Murto 1982, Barnes and Bentwood 1984, Rasanen et al. 1985). An exciting feature of the vibrationally induced isomerization common to all 2-haloethanols is the one order of magnitude higher efficiency of the Gg′ → Tt interconversion (see scheme 1) upon OH stretch excitation when

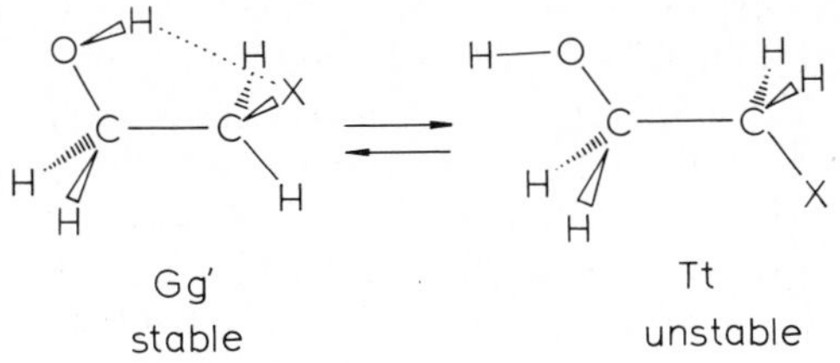

Gg′
stable

Tt
unstable

Scheme 1. Gg′ ⇌ Tt interconversion.

compared with the v(OH) induced, reverse Tt → Gg′ process (the capital letter in the designation Gg′ determines the conformation about the C–C bond, the lower case letter the conformation about the C–O bond). This is opposite to what is expected from energy considerations, since the OH stretching level of the less stable *trans* isomer, measured relative to the *gauche* ground vibrational state, lies several hundred cm^{-1} above the v(OH) level of the *gauche* form in the case of all 2-haloethanols. This suggests that the OH stretching mode couples with the delocalized CC, CO torsional overtones in the *gauche* form much more strongly than in the *trans* form, most probably because of the weak intramolecular hydrogen-bonding interaction in the former. Such conformer specificity in vibrationally induced interconversion processes has also been found in systems with three internal rotational degrees of freedom, like ethylene glycol (Frei et al. 1977, Takeuchi and Tasumi 1983), 2-nitroethanol (Rasanen et al. 1983), and 2-aminoethanol (Rasanen et al. 1982). This effect constitutes a very sensitive probe for conformer dependence of vibrational mode mixing.

A striking illustration of the use of IR-induced conformational interconversion in matrices for mapping of high-energy valleys on the electronic ground-state surface of flexible molecules is the very recent first synthesis of the *trans* form of a carboxylic acid (see scheme 2) by Günthard and co-workers (Hollenstein et al. 1986). Key to the elucidation of this complex conformational interconversion of glycolic acid was a vibrational analysis based on specific ^{18}O and ^{13}C substitution. Although the isotopic shifts are small, they can be measured very precisely due to the sharpness of the vibrational absorptions of matrix-isolated molecules. This example demonstrates that IR-induced isomerization of matrix-isolated molecules is a unique tool for conformational analysis.

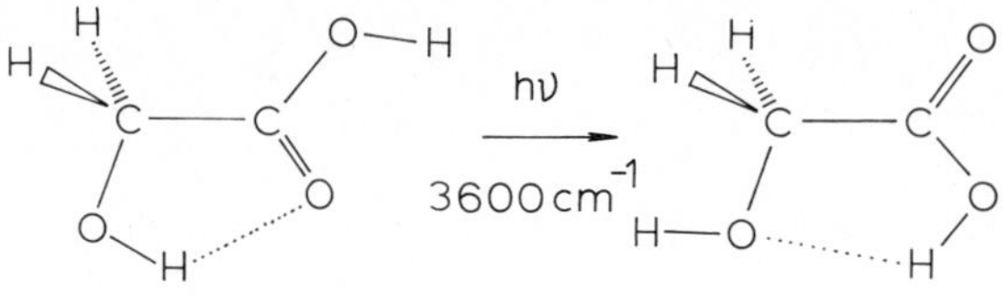

cis

stable

trans

unstable

Scheme 2. Synthesis of the *trans* form of a carboxylic acid

   The recent discovery of vibrationally stimulated rotamerization of an unsubstituted alkane shows that IR-induced isomerizations in matrices are not limited
to any group of molecules with particular substituents. Rasanen and Bondybey
(1984, 1985) have found that excitation of the CH stretching modes of a mixture
of *gauche* and *trans* n-butane in solid neon results in efficient conformational
interconversion. Broad-band globar source irradiation leads to a photostationary state consisting of about 90% *trans* n-butane, indicating that the quantum
efficiency to *gauche* → *trans* interconversion exceeds that of the reverse *trans*
→ *gauche* process by a factor of eight. The effect of selective deuteration on the
isomerization rates suggests a strong mode-selective effect: excitation of the
terminal methyl groups appears to be substantially more effective in promoting
internal rotation than irradiation of the methylene CH stretching modes,
although the energies of these two modes are similar.

   Examples of vibrationally induced bimolecular reactions in cryogenic
matrices are far more sparse than isomerizations. We would like to discuss here
briefly our most recent case of an IR-laser-induced chemical reaction, that of
allene with molecular fluorine (Knudsen and Pimentel 1983),

$$CH_2 = C = CH_2 + F_2 \rightarrow CH_2 = CF-CH_2F \quad (cis, \, gauche),$$

$$\rightarrow CH_2 = C = CHF + HF,$$

$$\rightarrow HC \equiv C-CH_2F + HF.$$

In a search for vibrational transitions that might stimulate the reaction, we
irradiated suspensions of allene and fluorine in solid nitrogen and rare-gas
matrices at 12 K with light from grating controlled cw CO_2, cw CO, or
continuously tunable cw F-center lasers. Concentrations of reactants were
selected so that about 10–20% of all allene molecules would have an F_2
molecule as a nearest neighbor. The laser frequency was tuned to an infrared
absorption attributable to one of these allene–fluorine nearest neighbor pairs,
whose position was located in advance by a concentration study and then later
confirmed by the observed loss of absorption upon chemical reaction. Reactant
loss and product growth upon infrared irradiation was monitored by FT–IR

spectroscopy, and assignment of bands to the *cis* and *trans* forms of 2,3-difluoropropene product was made on the basis of IR laser-induced interconversion of the two rotamers.

$2v_{10}$, the overtone of the CH_2 wagging motion at 1696 cm^{-1}, is the lowest infrared absorption frequency of allene, selective excitation of which led to reaction with fluorine. From the product absorbance growth curve obtained through prolonged CO laser irradiation at 1696 cm^{-1}, a rate constant was derived by fitting the measured curve to a first-order rate law. That allowed calculation of a quantum yield ϕ to reaction of 8×10^{-5}. As in the case of our earlier study of the vibrationally induced ethylene–fluoride reaction (Frei et al. 1981, Frei and Pimentel 1983a, Frei 1983), we encountered a general increase of ϕ with the energy per exciting photon, as shown in fig. 1. The increase is large, covering four orders of magnitude for transitions between 1700 and 3100 cm^{-1}, but it is not smooth. For example, the quantum yield increases by a factor of 30 from v_6 at 1955 cm^{-1} to $2v_9$ at 2002 cm^{-1}, while excitation at much lower frequency, 1696 cm^{-1}, induces reaction with an efficiency only nine times below

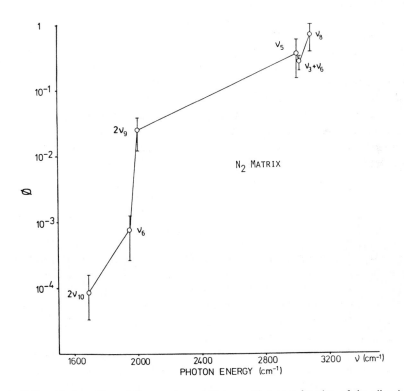

Fig. 1. $C_3H_4 + F_2$: logarithm of the quantum yield to reaction as a function of the vibrational frequency. Error bars denote uncertainties of relative quantum efficiencies.

that at 1955 cm^{-1}. A qualitatively similar, albeit less spectacular, observation was made in the case of an Ar matrix: $\phi(2v_9)/\phi(v_6) = 2.4$ (Knudsen and Pimentel 1983). While the general increase of the quantum yield with the energy of the exciting photon may originate for the most part from vibrational cascading during relaxation, which results in access to an increasing number of (reactive) vibrational states as the molecule is excited at higher and higher energies, the steep increase of ϕ from v_6 to $2v_9$ must reflect some distinct advantage of $2v_9$ over v_6 in promoting reaction. First, there is no vibrational state in the 47 cm^{-1} interval between the two levels, hence the entire change of ϕ from v_6 to $2v_9$ can be attributed to a higher quantum yield to reaction of the $2v_9$ vibrational state. Second, the known Fermi-resonance interaction between $2v_9$ and v_6 suggests an extra short residence time of the molecule in the upper state, implying that the high quantum efficiency originates from a high chemical reactivity of $2v_9$, i.e., a mode-specific effect. This can easily be interpreted in terms of the nuclear distortions involved: the CH_2 rocking overtone, $2v_9$, distorts the sp^2 hybridized methylene groups of allene toward the CH_2F–$CF = CH_2$ product tetrahedral structure, while it is hard to visualize how the asymmetric CCC stretching mode v_6 can contribute to the reaction coordinate. This is a very interesting result in view of the prevailing experimental evidence indicating rapid randomization of vibrational energy on the time scale of chemical reaction, and shows that selective infrared excitation of chemical reactions in a cryogenic matrix environment opens up a way to learn about types of vibrations that facilitate reaction.

The products of the allene–fluorine reaction, propargyl fluoride, fluoroallene, and 2,3-difluoropropene, are consistent with concerted addition of F_2 to the allene $C = C$ bond, forming a highly vibrationally excited, electronic ground-state $CH_2 = CF$–CH_2F intermediate. The branching ratio is determined by the competition between stabilization of this hot intermediate through transfer of the excess vibrational energy to the matrix environment, or elimination of HF. The postulated, vibrationally hot, difluoropropene intermediate is also consistent with the lack of observation of a mode-selective behavior with respect to product branching. Randomization of the vibrational energy of the intermediate is expected to be much faster than elimination of HF, hence the nature of the excited allene mode is not expected to influence the product ratio. However, the branching between elimination and stabilization changed dramatically when conducting the allene–fluorine reaction in different matrices, indicating substantial differences in the effectiveness of matrices to deactivate the hot intermediate. Interestingly, the branching between the two elimination products, fluoroallene and propargyl fluoride, also varied strongly with the matrix material. Although the origin of this matrix effect is not fully understood, the fact that product branchings can be influenced by the choice of the matrix bears considerable interest in the search for means to control chemical reaction pathways.

3. Selective vibronic excitation of bimolecular reactions with near infrared light

Control of an electronically induced bimolecular reaction is most likely achieved when it can be induced by excitation of a reactant into the bound region of an excited electronic state, especially in cases in which the excited vibronic level lies below the lowest dissociation limit. This tends to favor direct, molecular mechanisms over multi-step, atom or free-radical reactions induced by dissociation or predissociation of a reactant, with their propensity to cause secondary chemistry. A prototype reactant for chemistry of a small molecule in excited, bound states is molecular oxygen in the $^1\Delta_g$ ($\tilde{\nu}_{00} = 8000$ cm^{-1}) or $^1\Sigma_g^+$ state ($\tilde{\nu}_{00} = 13\,100$ cm^{-1}), both situated well below the dissociation limit to ground-state atoms ($D_0^0 = 41\,000$ cm^{-1}) (Herzberg 1950). $O_2(^1\Delta_g)$ is a particularly interesting reactant due to its long lifetime and strong electrophilic character, which makes its chemistry distinctly different from that of the remarkably unreactive, although radical-like $^3\Sigma_g^-$ ground-state molecule (Wasserman and Murray 1979). We have probed the chemical reactivity of selective vibronically excited $^1\Delta_g$ and $^1\Sigma_g^+$ oxygen with substituted furans and molecular fluorine in order to explore low-energy reaction pathways of these systems.

The first two members of the O_2 $^1\Delta_g, v' \leftarrow {}^3\Sigma_g^-, 0$ vibronic progression could be observed by near infrared Fourier transform spectroscopy of solid O_2 at 12 K. Figure 2 shows the 0–0 and the 1–0 transition of the α-phase at 7990 and 9480 cm^{-1}, respectively (Frei and Pimentel 1983b). These transitions, both spin and electric-dipole forbidden for the isolated molecule, become enhanced in the solid α-phase by about two orders of magnitude (Landau et al. 1962). This enhancement has been interpreted by Bhandari and Falikov (1973), in terms of

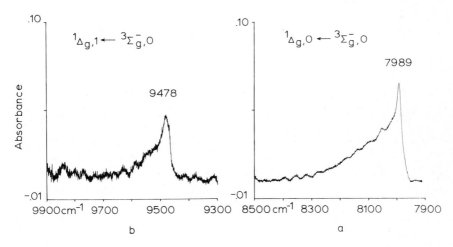

Fig. 2. Near-infrared spectra of the transitions (a) $^1\Delta_g, 0 \leftarrow {}^3\Sigma_g^-, 0$, and (b) $^1\Delta_g, 1 \leftarrow {}^3\Sigma_g^-, 0$ of solid α-oxygen at 12 K.

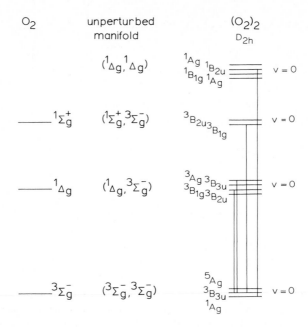

Fig. 3. Electric-dipole allowed transitions of interacting O_2 molecules in solid α-oxygen (see Bhandari and Falikov 1973).

factor group selection rules for interacting O_2 molecules in the α-O_2 unit cell, which lead to electric-dipole allowed transitions between substrates classified according to the D_{2h} symmetry of O_2 nearest neighbors. Figure 3 shows the two electric-dipole allowed transitions of $(O_2)_2$ to the $(^1\Delta_g, ^3\Sigma_g^-)$ manifold. These factor group selection rules also predict electric-dipole allowed transitions for the $^1\Sigma_g^+, v' \leftarrow ^3\Sigma_g^-, 0$ progression in solid α-O_2, but we were not able to detect any of its members by FT–IR absorption spectroscopy. However, we did observe this progression in Ar and O_2 matrices by means of the near-infrared laser-induced reaction of O_2 with 2,5-dimethylfuran (Frei and Pimentel 1983b). Figure 4 shows a plot of the endoperoxide product yield, measured by the growth of the CO stretch absorption at 1120 cm^{-1} over equal time intervals, as a function of the cw dye laser irradiation frequency in a matrix dimethylfuran:O_2:Ar = 1:15:85. This "laser reaction excitation spectrum" reveals the first two members of the progression, $^1\Sigma_g^+, 0 \leftarrow ^3\Sigma_g^-, 0$ at 13 100 and $^1\Sigma_g^+, 1 \leftarrow ^3\Sigma_g^-, 0$ at 14 500 cm^{-1}. Relative absorbances of the 0–0 and 1–0 transitions can be determined (except for a factor determined by the relative quantum efficiencies to reaction upon excitation of the two states) from measured product absorbance growths, irradiation times, and laser intensities, which puts the peak absorbance of the 1–0 transition 25 times below that of the 0–0 transition. Plainly, direct excitation of $^1\Sigma_g^+$ vibronic levels does induce

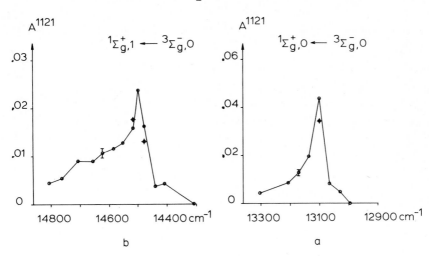

Fig. 4. Laser reaction excitation spectra of (a) $O_2 : {}^1\Sigma_g^+, 0 \leftarrow {}^3\Sigma_g^-, 0$ and (b) $O_2 : {}^1\Sigma_g^+, 1 \leftarrow {}^3\Sigma_g^-, 0$ of matrices DMF : O_2 : Ar = 1 : 15 : 85 at 12 K. DMF denotes dimethylfuran.

addition of O_2 to 2,5-dimethylfuran, but no conclusion can be drawn from this experiment with respect to the reactive state, i.e., whether O_2 reacts in the ${}^1\Sigma_g^+$ state or, after relaxation, in a lower lying ${}^1\Delta_g$ vibronic level.

In order to determine the reacting O_2 state(s), we measured the relative quantum yields to chemical reaction of dimethylfuran with O_2 upon selective excitation to ${}^1\Delta_g$ and ${}^1\Sigma_g^+$ vibronic levels. Excitation of O_2 to the ${}^1\Delta_g$, $v' = 0$ and ${}^1\Delta_g$, $v' = 1$ levels gave the same quantum yields to reaction, and an estimate of the absolute quantum efficiency led to a value around one. Assuming the physical quenching time of $O_2({}^1\Delta_g, 0)$ with a nearest-neighbor dimethylfuran molecule in an Ar matrix to be the same, or shorter than, as that measured by Crone and Kugler (1985) for $O_2({}^1\Delta_g)$ isolated in solid Ar, 85 s, the high quantum yield to reaction would imply a half-life to $O_2({}^1\Delta_g)$ + dimethylfuran reaction of 10 s or less. With this half-life, and a frequency factor equal to 10^{12} s^{-1}, a typical phonon frequency, classical Arrhenius behavior would require an activation energy no higher than 720 cal mol^{-1}. This is two thirds of the gas phase value of 1070 cal mol^{-1}, close enough to permit interpretation of the reaction of dimethylfuran with vibrational ground-state $O_2({}^1\Delta_g)$ in terms of thermal barrier crossing even in the cryogenic environment. On the other hand, in pure solid O_2, Akimoto and Pitts (1970) estimated a non-radiative lifetime of $O_2({}^1\Delta_g, 0)$ around 1µs or less, orders of magnitude shorter than in Ar. With this lifetime, the high quantum yield to reaction in solid O_2 requires a reaction time of 0.1 µs or less which, if explained in terms of thermal reaction, would imply a barrier no higher than 300 cal. This would represent an unusually large discrepancy between gas-phase and matrix barrier, and because of that we prefer interpretation in terms of O_2 molecular tunneling on the singlet hypersurface through a barrier with the gas-phase value (Frei and Pimentel 1983b). In any event, the fact that all observed quantum yields to reaction upon selective $^1\Delta_g$ and $^1\Sigma_g^+$ vibronic excitation are around one prevents determination of the reacting vibronic state. The high reaction efficiency would be observed even if relaxation among the singlet states down to the $^1\Delta_g$, $v' = 0$ level were faster than reaction, i.e., even if all reaction would occur from the $^1\Delta_g$ ground vibrational level.

Laser reaction excitation spectroscopy also unveiled a vibronic progression of oxygen in solid Ar in the red spectral range with bands at 15 830 and 17 300 cm^{-1}. These are assigned to the 0,0 and 1,0 members of the $(O_2)_2(^1\Delta_g, {}^1\Delta_g) \leftarrow (^3\Sigma_g^-, {}^3\Sigma_g^-)$ one-photon, two-molecule transition first observed by Ellis and Kneser (1933) in the absorption spectrum of liquid oxygen. The fact that these transitions were observed in a matrix of the composition dimethylfuran:O_2:Ar $= 1:2:100$ with its low concentration of reactive dimethylfuran–$(O_2)_2$ clusters demonstrates the high sensitivity of the laser reaction excitation technique. The factor group selection rules indicate that this transition, too, is electric-dipole allowed (fig. 3).

The frequencies of the first two members of the $(O_2)_2\, (^1\Delta_g, {}^1\Delta_g) \leftarrow (^3\Sigma_g^-, {}^3\Sigma_g^-)$ progression coincide with the corresponding absorptions of $(O_2)_2$ isolated in solid neon measured earlier by Goodman and Brus (1977) with the laser-fluorescence excitation technique. Both laser fluorescence and laser reaction excitation spectroscopy are very sensitive, complimentary techniques to monitor weak vibronic absorptions in condensed phase. We expect bimolecular reaction excitation spectroscopy to be a particularly useful tool when searching for weak absorptions in condensed-phase reactive samples where rapid non-radiative relaxation would quench laser-induced fluorescence.

FT–IR analysis of product spectrum in the range 4000–400 cm^{-1}, including analysis of the peroxidic ^{18}O counterpart, allowed empirical vibrational mode assignment and confirmation of the endoperoxide structure of the product. Thanks to the long wavelength of the exciting red and near-infrared photons, no excitation of the product endoperoxide occurred, hence the product spectrum and the growth kinetics could be studied in complete absence of secondary photolysis.

In contrast to the singlet oxygen–furan system, our cryogenic laser technique allowed us to uncover a strong vibronic dependence of the quantum yield to reaction of singlet O_2 with molecular fluorine (Frei 1984). We found the singlet O_2 induced reaction pathway when searching for new, low-energy paths of the reaction

$$F_2 + O_2 \rightarrow O_2F + F, \quad \Delta H°(298) = +31 \text{ kcal mol}^{-1};$$

$$\rightarrow O_2F_2, \quad \Delta H°(298) = +8 \text{ kcal mol}^{-1}.$$

The first of these reactions, induced in a cryogenic matrix by mercury-arc radiation, was used in this laboratory (Spratley et al. 1966), and independently by Arkell (1965), about two decades ago to generate and study spectroscopically the elusive O_2F radical. In those experiments, the near-UV and blue light emitted by the Hg arc source initiated reaction by dissociating F_2 into ground-state atoms through excitation of the molecule into the $^1\Pi_u$ repulsive state. Now we were interested in finding the lowest energy photons capable of promoting the reaction, with the goal of achieving control over the O_2F/O_2F_2 product branching. Given the $\Delta H°$ of 31 kcal mol^{-1} for the more endothermic $O_2F + F$ channel, we expected, e.g., $O_2(^1\Delta_g, v' \geq 2)$ or $O_2(^1\Sigma_g^+)$ to carry enough energy to initiate reaction with F_2.

The $F_2 + O_2$ laser reaction excitation spectrum in fig. 5a demonstrates that it is indeed possible to induce chemical reaction with photons in the red spectral

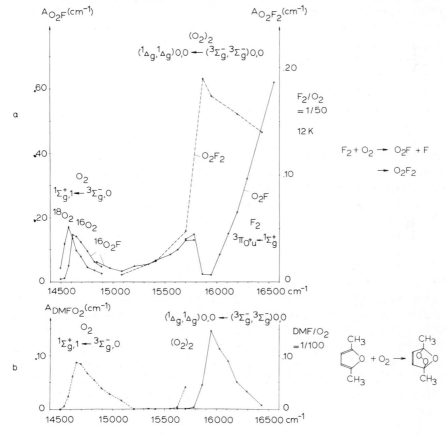

Fig. 5. (a) Laser excitation spectrum of the reaction $F_2 + O_2$ in the red spectral range. Matrix $F_2 : O_2 = 1 : 50$. (b) Laser excitation spectrum of the reaction 2,5-dimethylfuran + O_2. Ordinate: growth of the integrated absorbance of the 1330 cm^{-1} band of dimethylfuran endoperoxide (DMFO$_2$). Matrix DMF : $O_2 = 1 : 100$.

range with energies as low as 14 500 cm^{-1}. Each point of the solid curve represents the growth of the integrated absorbance of the O–O stretching mode of O$_2$F in solid oxygen at 12 K during an irradiation period of 15 min at 280 mW cm^{-2} cw dye laser intensity. The 70 cm^{-1} ^{18}O shift of the peak at 14 600 cm^{-1} (determined by laser photolysis of a suspension of F$_2$ in solid ^{18}O$_2$) is in agreement with that calculated for the O$_2$ $^1\Sigma_g^+$, 1 ← $^3\Sigma_g^-$, 0 transition, and the coincidence with the corresponding band in the O$_2$ + dimethylfuran laser reaction excitation spectrum (fig. 5b) supports this assignment. On the other hand, we were not able to induce F$_2$ + O$_2$ reaction by excitation of any of the lower lying O$_2$ vibronic transitions, the ground vibrational level of the $^1\Sigma_g^+$ state at 13 250 cm^{-1}, or the $v' = 0$ and $v' = 1$ levels of the lower lying $^1\Delta_g$ state. We estimate that excitation of O$_2$ $^1\Sigma_g^+$, $v' = 0$ is at least 180 times less effective in promoting reaction with F$_2$ than O$_2$ $^1\Sigma_g^+$, $v' = 1$. Goodman and Brus' laser fluorescence study of the non-radiative relaxation pathway of selective vibronically prepared oxygen dimers in a neon matrix permits a conclusion about the reactive O$_2$ vibronic state. These authors found that an O$_2$($^1\Sigma_g^+$, $v' = 1$) molecule with an O$_2$($^3\Sigma_g^-$, 0) ground-state nearest-neighbor relaxes directly to the $v' = 0$ level of the $^1\Sigma_g^+$ state, without cascading through intermediate $^1\Delta_g$ or $^3\Sigma_g^-$ vibronic levels. This, combined with our observation that O$_2$($^1\Sigma_g^+$, 0) did not promote F$_2$ + O$_2$ reaction, establishes the optically excited O$_2$ $^1\Sigma_g^+$, $v = 1$ level as the reactive state. Together with our recent direct observation of O$_2$($^1\Sigma_g^+$) in room-temperature solution (Chou and Frei 1985), this reopens the search for the elusive O$_2$($^1\Sigma_g^+$) chemistry in condensed phase. More generally, it shows that laser-induced cryogenic photochemistry is a valuable tool for the search of low-energy reaction pathways of bimolecular systems.

Quite to our surprise, the O$_2$F reaction excitation profile showed a minimum in the spectral range around 15 900 cm^{-1} where, according to the dimethylfuran + O$_2$ spectrum (fig. 5b), the (O$_2$)$_2$ ($^1\Delta_g$, $^1\Delta_g$), 0 ← ($^3\Sigma_g^-$, $^3\Sigma_g^-$), 0 transition absorbs. The O$_2$F$_2$ excitation profile, on the other hand, exhibits a peak at this frequency. Quantitative analysis of absorbance growth (decrease) upon irradiation at 15 900 cm^{-1} revealed that excitation of the (O$_2$)$_2$($^1\Delta_g$, $^1\Delta_g$) transition induces reaction of O$_2$F with F atoms to form O$_2$F$_2$, but very little or no F$_2$ + O$_2$ reaction. Excitation of the ($^1\Delta_g$, $^1\Delta_g$) transition turns out to be at least two orders of magnitude less efficient in inducing F$_2$ + O$_2$ reaction than excitation of the lower energy $^1\Sigma_g^+$, $v' = 1$ state. This surprising finding is most probably a manifestation of an extremely efficient relaxation process of (O$_2$)$_2$($^1\Delta_g$, $^1\Delta_g$) which is not accessible to O$_2$($^1\Sigma_g^+$, 1). There is spectroscopic evidence for picosecond exchange of $^1\Delta_g$ quanta between two nearest-neighbor O$_2$ molecules isolated in solid neon from the work of Goodman and Brus (1977), and the exchange of $^1\Delta_g$ excitation between oxygen molecules colliding in the gas phase is known to be close to gas kinetic (Jones and Bayes 1972),

$$O_2(^1\Delta_g) + O_2(^3\Sigma_g^-) \rightarrow O_2(^3\Sigma_g^-) + O_2(^1\Delta_g).$$

Hence, in solid oxygen, we can expect the $16\,000\ \mathrm{cm}^{-1}$ $(O_2)_2$ $(^1\Delta_g, {}^1\Delta_g)$ quanta to decay very fast into separate $8000\ \mathrm{cm}^{-1}$ $O_2(^1\Delta_g)$ quanta, a process that is endothermic possibly by a few cm^{-1}, but certainly not more than $17\ \mathrm{cm}^{-1}$

$$(O_2)_2(^1\Delta_g, {}^1\Delta_g) + O_2(^3\Sigma_g^-) \rightarrow (O_2)_2(^1\Delta_g, {}^3\Sigma_g^-) + O_2(^1\Delta_g).$$

As mentioned above, the energy of $^1\Delta_g$ $8000\ \mathrm{cm}^{-1}$ quanta does not suffice to induce reaction with F_2. An analogous break-up of the $O_2(^1\Sigma_g^+, v'=1)$ quantum through energy transfer to a nearest neighbor, ground-state O_2 molecule, however, is inhibited by an endothermicity of $150\ \mathrm{cm}^{-1}$ which exceeds kT at 12 K by far,

$$O_2(^1\Sigma_g^+, 1) + O_2(^3\Sigma_g^-, 0) \rightarrow O_2(^1\Sigma_g^+, 0) + O_2(^3\Sigma_g^-, 1),$$

$$\Delta E = +150\ \mathrm{cm}^{-1}.$$

We conclude that the inhibition of this relaxation process gives vibrationally excited $O_2(^1\Sigma_g^+)$ sufficient lifetime to react with F_2.

The most striking difference between the reaction excitation spectra of F_2 $+ O_2$ and dimethylfuran $+ O_2$ (fig. 5) is a continuous absorption clearly visible in the $F_2 + O_2$ profile which is absent in the furan $+ O_2$ case, hence cannot be due to absorption of molecular oxygen. The onset of the absorption is around $15\,000\ \mathrm{cm}^{-1}$, and it continuously increases through the red, yellow, and green

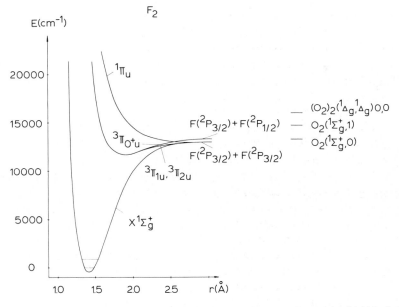

Fig. 6. Potential energy curves of low lying electronic states of F_2 according to ab initio calculations by Cartwright and Hay (1979).

spectral range. All experimental evidence points to F_2 as the absorber despite the fact that the absorption extends more than 200 nm to longer wavelengths than in the electronic spectra of F_2 reported in the literature [$\lambda_{max} = 285$ nm, with tail extending to 450 nm (Steunenberg and Vogel 1956)]. According to ab initio potential energy curves for the lowest excited electronic states of F_2 (Cartwright and Hay 1979) shown in fig. 6, the new, red fluorine absorption is either due to excitation to the $^1\Pi_u$ state, or to the repulsive limb of the weakly bound, hitherto unobserved $^3\Pi_{0u}^+$ state. Two arguments point to $^3\Pi_{0u}^+$ as the absorbing state. First, the $^3\Pi_{0u}^+$ transition has the more favorable Franck–Condon factors (fig. 6), and, second, there is an enhancement of the absorption by nearest-neighbor ground-state O_2 molecules (Frei 1984). The latter can easily be explained by removal of the spin forbiddance of the $F_2(^3\Pi_{0u}^+ \leftarrow {}^1\Sigma_g^+)$ transition by interaction with O_2,

$$F_2 \cdot O_2\text{:}\ {}^3(^3\Pi_{0u}^+, {}^3\Sigma_g^-) \leftarrow {}^3(^1\Sigma_g^+, {}^3\Sigma_g^-),$$

but no such enhancement is expected in the case of the spin-allowed excitation to the $^1\Pi_u$ state. The unveiling of the lowest triplet state of molecular fluorine again shows how laser reaction excitation spectroscopy in matrices can be used to locate very weak electronic transitions.

4.   New chemistry on excited haloethene hypersurfaces

Unique chemistry may occur on excited electronic hypersurfaces not accessible by direct optical excitation, e.g., on surfaces originating from reactant states with triplet spin multiplicity. Such surfaces may be reached, however, by use of triplet excited sensitizers like $Hg(^3P)$, and mercury-atom sensitization of gas-phase reactions has been used to this end quite extensively. Key to the understanding and control of the chemical reaction pathways made available through $Hg(^3P)$ initiation is the elucidation of the reaction mechanism, and in particular the role played by the excited metal atom. Does $Hg(^3P)$ initiate reaction by transient formation of chemical bonds, or does it merely transfer its electronic energy to a reactant? In the former case, there might be reaction products unique to $Hg(^3P)$ inducement, and not attainable by other triplet sensitizers. Matrix isolation, uniquely suited for stabilization of transient intermediates, offers a direct, convenient probe for $Hg(^3P)$ atom participation through chemical bonding. In addition, well-established metal atom–matrix deposition techniques allow comparison with ^3P-induced chemistry of the other group IIB metals Cd and Zn. Since the latter have their higher energy, $^1P_1 \leftarrow {}^1S_0$ absorptions at wavelengths longer than 200 nm, these metals permit convenient study and elucidation of differences between singlet and triplet metal-atom induced chemistry.

   An interesting example is provided by the ^3P mercury, cadmium, and zinc induced reactions of 2-chloro-1,1-difluoroethene in krypton matrices (Cartland and Pimentel 1986a). Krypton is most convenient for the study of $Hg(^3P)$ induced chemistry since the matrix shift brings the $^3P_1 \leftarrow {}^1S_0$ optical absorption into coincidence with the output maximum of our 1000 W Hg–Xe arc lamp at 250 nm. While gas-phase $Hg(^3P)$ sensitized decomposition of fluoroethenes leads to HF elimination (Norstrom et al. 1976), no evidence was found for hydrogen halide elimination in the matrix study. Instead, the FT–IR spectral record of reactions of $CF_2 = CHCl \cdot Hg(^3P)$ and $CF_2 = CDCl \cdot Hg(^3P)$ pairs (fig. 7) revealed products with vibrational spectra resembling that of the ethene parent, but with the bands shifted 30–70 cm^{-1} to lower frequencies. These shifts, in particular that of the $C = C$ stretching mode, are substantially smaller than those observed for metal–ethene Π complexes like $C_2H_4 \cdot Cu$, $C_2H_4 \cdot Ag$, and $C_2H_4 \cdot Au$ (Huber et al. 1976a,b), but close to shifts of group IIB metal insertion products reported in the literature (Mink et al. 1971, Nesmeyanov 1980). This points to $Hg(^3P)$ involvement through chemical bonding rather than through mere energy transfer, and comparison of product and reactant spectra showed that the excited mercury atom inserts into the C–Cl bond, constituting the first example of a photochemically prepared mercury–haloethene (see scheme 3).

While Zn $^3P_1 \leftarrow {}^1S_0$ (300 nm) and Cd $^3P_1 \leftarrow {}^1S_0$ (316 nm) induced chemistry of $CF_2 = CHCl$ gave analogous insertion products, excitation of the easily accessible $^1P_1$ states of these metals (Zn, 212 nm; Cd, 226 nm) did not lead to any reaction despite their higher energies relative to the ^3P state. This indicates that

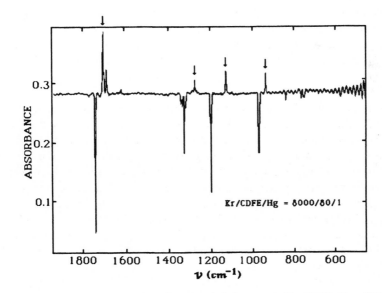

Fig. 7. FT–IR difference spectrum: 7 h Hg $^3P_1$ photolysis in a matrix Kr:CDFE:Hg = 8000:80:1 at 12 K. Product features are marked by arrows. CDFE denotes 2-chloro-1,1-difluoroethene.

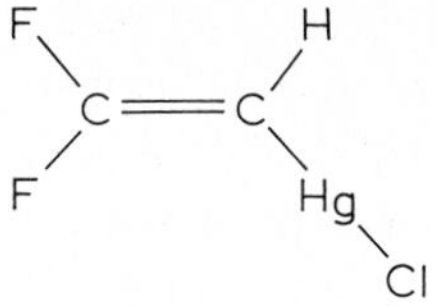

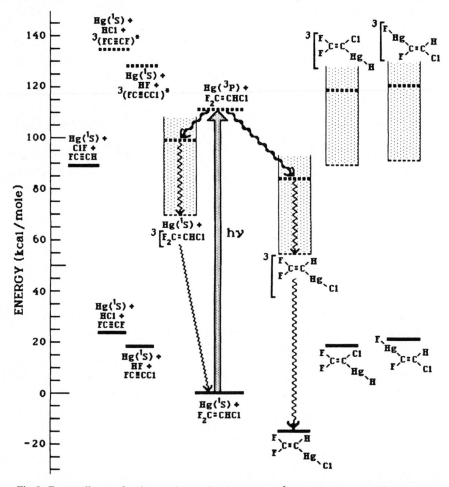

Scheme 3. Photochemically prepared mercury-haloethene.

the observed $M(^3P)$ induced chemistry is associated with a triplet reaction surface not accessible through $^1P_1$ excitation of the metals.

An interesting question opened by the observed product specificity concerns the factors that control the reaction pathway. Figure 8 shows an energy diagram

Fig. 8. Energy diagram for the matrix reaction between Hg ^{3}P and 2-chloro-1,1-difluoroethene.

for the matrix reaction of $CF_2 = CHCl \cdot Hg(^3P)$, which is based on thermochemical and UV spectral data available for the related ethene and a few haloethene systems (Cartland and Pimentel 1986a). Assuming the equilibrium geometry of chlorodifluoroethene in its lowest triplet state to be twisted like in the case of T_1 ethylene (Lee and Pimentel 1981), reaction of $Hg(^3P)$ with the planar haloethene on a triplet surface requires surpassing of a barrier associated with internal rotation around the CC bond. Taking again T_1 ethylene as a guide, we estimate triplet-state barriers to internal rotation to be around 30 kcal mol^{-1} (Lee and Pimentel 1981) (dotted lines in fig. 8). With these reaction heat, triplet-state energy, and barrier estimates, the observed product selectivity can be explained by restriction to a triplet reaction pathway. Under spin conservation, insertion of $Hg(^3P)$ into the C–Cl bond can proceed with the available energy, whereas insertion into the C–H or C–F bond would require substantial activation. Similarly, the energy diagram (fig. 8) shows that spin-conserving hydrogen halide elimination to produce either triplet HX or triplet excited acetylenic products would be endothermic. Hence our observation of only one reaction channel, C–Cl bond insertion, is consistent with a triplet surface pathway.

The three isomers of dichloroethene, *cis*-1,2, *trans*-1,2, and 1,1-dichloroethene have electronic absorptions at longer wavelengths than chlorodifluoroethene, with tails extending to approximately 235 nm. Hence we were able to compare $Hg(^3P)$ induced chemistry of these halo-olefins with reaction promoted by direct photo-excitation at $\lambda > 200$ nm. We found substantial differences in product distribution, indicating effective control of the reaction pathway by the method used for reactant excitation (direct versus Hg-atom mediated initiation) (Cartland and Pimentel 1986b).

Direct photolysis of *cis*- and *trans*-1,2-dichloroethene with unfiltered Hg–Xe arc radiation at $\lambda > 200$ nm in solid Kr led to *cis–trans* isomerization and, in the case of *cis*-1,2-dichloroethene, to HCl elimination (to form hydrogen chloride–chloroacetylene hydrogen-bonded complexes). Optical excitation of 1,1-dichloroethene at $\lambda > 200$ nm resulted in isomerization to the *cis* and *trans* form, and again elimination of HCl. Energy requirements strongly suggest that the observed photochemistry (isomerization plus HCl elimination, but no expulsion of Cl_2) is the result of a restriction of the reaction path to excited singlet states (Cartland and Pimentel 1986b).

In contrast, $Hg(^3P)$ induced chemistry of the dichloroethenes in Kr gives products consistent with reaction on a triplet surface not accessible by direct optical excitation of the energetically higher olefin S_1 state. The *cis* isomer and 1,1-dichloroethene undergo isomerization, but also Cl_2 elimination, a reaction channel not observed in the case of direct photo-excitation. In contrast to photolysis in the absence of Hg, only little HCl elimination could be detected. The energy diagram (fig. 9) shows that among the triplet elimination pathways only one is exothermic, namely the one leading to $HC \equiv CH + Cl_2$, while the H_2 and HCl elimination pathways are endothermic. Therefore, the observed elimination product branching can be rationalized on the bases of spin conservation coupled with energy requirements. This also gives an explanation for Hg atom insertion into the C–Cl bond without insertion into the C–H bond in the case of 1,1-dichloroethene and the *cis* isomer; only the carbon–chlorine bond reaction gives a product that can be reached on a triplet pathway with a barrier below the energy of the $C_2H_2Cl_2 \cdot Hg(^3P)$ reactant pair (fig. 9).

It is remarkable that the photoexcitation of *trans*-CHCl = CHCl·Hg pairs did not result in any elimination or metal-atom insertion at all, in contrast to the behavior of the *cis* isomer. This indicates that simple energy transfer from CHCl = CHCl·Hg(^3P) to ^{3}(CHCl = CHCl)·Hg(^1S) is not involved in Cl_2

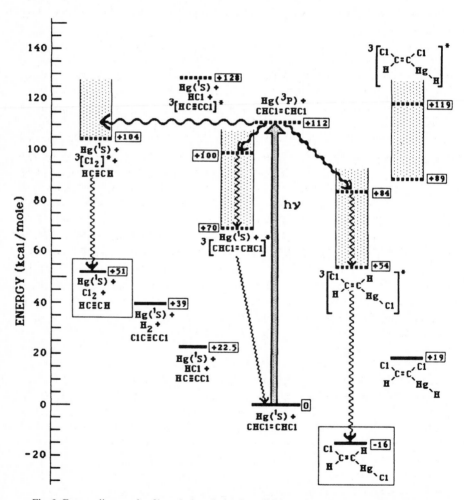

Fig. 9. Energy diagram for filtered photolysis of *cis*-dichloroethene in Kr with Hg present.

elimination and the insertion reaction, since if it were, the *cis* and the *trans* isomer would give identical products. Rather, the excited Hg atom appears to interact with the *cis* isomer in a way not possible in the case of the *trans* isomer, opening up a reaction pathway unique to the *cis* form.

A change of the product distribution in the presence of Hg atoms was also observed in the case of the UV photolysis of trichloroethene (Cartland and Pimentel 1989). Direct irradiation of the olefin at $\lambda > 200$ nm gave HCl elimination, while chemical reaction initiated by $Hg(^3P)$ resulted in elimination and Hg-atom insertion into the C–Cl bond to produce *trans*-CHCl = CCl–HgCl, similar to the case of the dichloroethenes. Again, these differences between direct and Hg atom mediated initiation of the reaction can most easily be explained by differences between excited singlet- and triplet-surface chemistry (Cartland and Pimentel 1989).

The participatory role of the excited metal atom found in the dichloro, trichloro, and chlorodifluoroethene–group IIB metal initiated photochemistry motivated us to explore other systems, like the vinyl and ethyl halides (Cartland and Pimentel 1989, Cartland 1985). $Hg(^3P)$ induced reaction of $CH_2 = CHF$, $CH_2 = CHCl$, and $CH_2 = CHBr$, and Zn and Cd initiated chemistry of $CH_2 = CHCl$, mostly in krypton, gave HX elimination in all cases. Infrared spectra revealed that the expelled hydrogen halide is engaged with acetylene, the other photo-fragment, in a hydrogen-bonded complex identical to the $C_2H_2 \cdot HX$ complexes prepared previously by Andrews and co-workers (McDonald et al. 1980, Andrews et al. 1982) through direct VUV photolysis of the vinyl halide precursor. In addition, metal-atom insertion into the C–Cl and C–Br bond occurred, but not into the C–F bond. Assuming our estimates of reactant and product triplet energies to be correct, this outcome in terms of insertion products is again consistent with a spin-conserving, triplet reaction pathway; $CH_2 = CHCl \cdot Hg(^3P)$ and $CH_2 = CHBr \cdot Hg(^3P)$, with the vinyl halide in its planar configuration, has sufficient energy to undergo metal-atom insertion, while $CH_2 = CHF \cdot Hg(^3P)$ does not carry enough energy to undergo C–F bond insertion (fig. 8). However, reactant and product excited-state energies give no immediate clue as to what pathway may lead to HX elimination. These systems are under further study.

Reaction of $CH_3–CH_2X \cdot Hg(^3P)$ pairs gave HX elimination (ethyl chloride and bromide) and, to less extent, H_2 elimination in the case of the chloride and fluoride. Excited Hg atom insertion was only observed into the C–Cl and C–Br bond, but not into the C–F bond, no surprise in the light of the haloethene results. However, in contrast to the haloethenes, chemical reactivity in the case of ethyl chloride appeared not to be limited to $CH_3–CH_2Cl \cdot Hg$ nearest-neighbor pairs. The asymptotic growth of the $C_2H_4 \cdot HCl$ product alone indicates that 16% of all ethyl chloride molecules react, a far higher percentage than represented by the statistical fraction of $CH_3–CH_2Cl \cdot Hg$ nearest neighbors. A possible reason is energy transfer between $Hg(^3P)$ and non-nearest-neighbor ethyl chloride molecules, leading to excitation of the chloride to its lowest triplet state.

5. Azirine–nitrile ylide photo-isomerization

Azirine–nitrile ylide systems (see scheme 4) constitute prototypes of reversible photo-induced isomerizations which can be controlled by the wavelength of the exciting photon. Although systems of this type have been studied by visible/UV

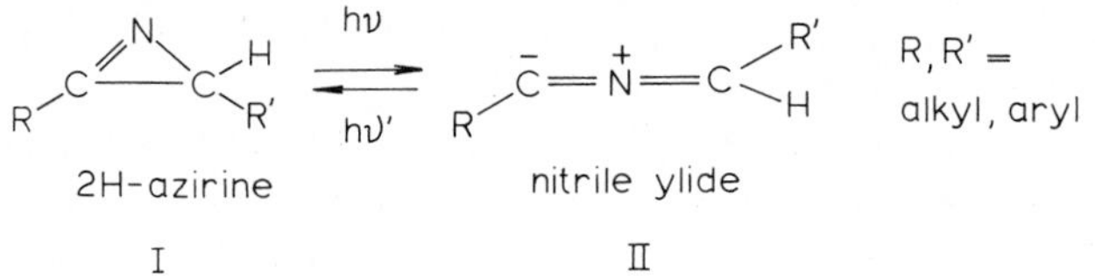

Scheme 4. Azirine–nitrile ylide systems.

absorption techniques (Sieber et al. 1973, Orahovats et al. 1975), there is little definitively known about the structure of the high-energy ylide form, mainly due to the lack of structurally informative infrared data. Moreover, the energy pathway of these photo-initiated interconversions has not yet been elucidated. This is despite the intriguing possibility that these systems exhibit adiabatic, upper-state surface chemistry of interest to photon conversion.

Favored systems for the determination of the structure of the high-energy ylide form and the wavelength study of the interconversion are those with small substituents R, R', like H or CH_3, since these substituents would interfere the least with arizine and ylide vibrational modes and electronic chromophores. However, such small azirines and ylides tend to be reactive and unstable at room temperature. A cryogenic matrix, then, offers an ideal environment for the study of their IR spectra and monitoring of wavelength-dependent photochemistry initiated with tuned UV radiation.

The simplest system, R = R' = H, might be attainable through the cryogenic reaction of NH radicals with acetylene. In fact, Jacox and Milligan (1963) and Jacox (1979) have initiated this reaction in solid Ar by photolyzing HN_3 (to form $NH + N_2$) in the presence of C_2H_2. The only product observed was ketenimine, $CH_2 = C = NH$, but these authors postulated the isomeric 1H–azirine (III; see scheme 5) as an undetected intermediate.

Assuming the initial reaction step to be addition of NH to the triplet bond, the observed ketenimine implies H-atom migration between the two carbons. So, if undesired isomerization to ketenimine is to be suppressed, H has to be replaced by a substituent that is less amenable to migration and/or acts as a sink for the excess energy of the hot azirine as it is generated upon NH addition, e.g., a

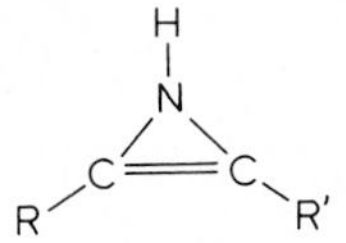

H
|
N
C══C
R   R'

1H–azirine

III

Scheme 5. 1H–azirine with R = R′ = H.

methyl group. Therefore, we attempted an azirine synthesis by photolyzing hydrazoic acid in the presence of dimethylacetylene (DMA), with the hope of being able to trap the 1H–azirine (III) or the isomeric 2H–azirine (I) (R = R′ = CH$_3$).

Mercury-arc photolysis of dimethylacetylene/HN$_3$ suspensions in Ar and Xe gave more than one product in both matrices as inferred from absorbance growth behavior (Collins and Pimentel 1984). In Ar, two bands were observed in the region around 2000 cm^{-1}, which is characteristic for multiple bond-stretching motions of possible product structures shown in scheme 6.

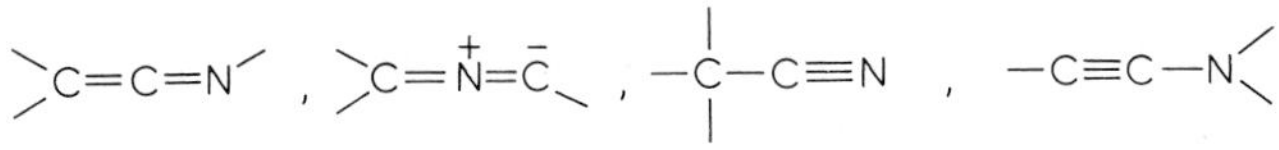

$$\text{C}=\text{C}=\text{N} \quad , \quad \overset{+}{\text{C}}=\overset{-}{\text{N}}=\overset{-}{\text{C}} \quad , \quad -\text{C}-\text{C}\equiv\text{N} \quad , \quad -\text{C}\equiv\text{C}-\text{N}$$

Scheme 6. Possible structures for dimethylacetylene + HN$_3$ products in Ar and Xe matrices.

The growth behaviors of the two absorptions at 2031.5 and 2007.9 cm^{-1} were identical, indicating that they probably originate from one and the same species. Comparison of the observed ^{15}N and D isotope shifts with those predicted by the diatomic approximation for the stretching motions of the structures above led to assignment to a Fermi doublet of 3,3-dimethylketenimine (see scheme 7).

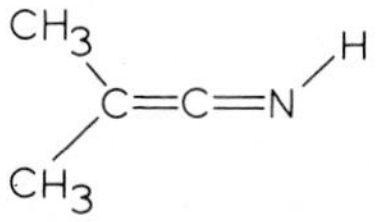

CH$_3$
       C══C══N
CH$_3$ H

Scheme 7. 3,3-dimethylketenimine.

This product also appeared in Xe, but only as a minor by-product relative to the principal product N-methyl-1-amino-propyne (see scheme 8). The structure of this main product was established by a band at 2038.6 cm^{-1}, which did not exhibit an ^{15}N shift and hence was assigned to $v(\text{C}\equiv\text{C})$, and observation of a strong absorption at 870 cm^{-1} with an ^{15}N isotope shift of 14 cm^{-1} characteristic for the C–N stretching motion. There is an additional set of product bands both in Ar and Xe, so close that they most probably originate from one and the

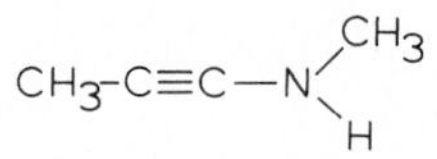

$$CH_3\text{-}C\equiv C\text{---}N\overset{CH_3}{\underset{H}{\diagup}}$$

Scheme 8. N-methyl-1-amino-propyne.

same product. Since it does not possess an absorption in the 2000 cm^{-1} region, this molecule cannot have a cumulene or triple bond, which suggests a cyclic azirine structure (I or III). The lack of a C = N stretching absorption around 1700 cm^{-1}, combined with observation of a strong band around 890 cm^{-1} featuring ^{15}N and D-shifts characteristic for an N–H bending motion points to the 1H–azirine form III. We believe that it is formed in the initial reaction step, with its yield determined by the competition between deactivation of the hot 1H–azirine and rearrangement to ketenimine (Ar) and/or N-methyl-1-amino-1-propyne (Xe). If this identification is correct, it is the first spectral detection of a 1-azirine.

Although the cryogenic reaction of NH with dimethylacetylene apparently did not lead to the desired 2H–azirine–nitrile ylide pair, the strong dependency of the rearrangement path on the matrix material is a most interesting finding. We believe that it is the heavy atom induced spin–orbit coupling by Xe which is responsible for the observed difference in product branching between the two matrices. In the gas phase, UV photolysis of HN$_3$ produces 95% of NH in its excited $^1\varDelta$ state, and it is reasonable to assume that in an inert-gas matrix, too, a substantial fraction of photodecomposing HN$_3$ produces NH in the $^1\varDelta$ excited state. Spin–orbit coupling induced by Xe can be expected to increase the rate of crossing from the initial singlet reaction surface to a triplet surface somewhere along the reaction coordinate. This might be early in the reaction through Xe catalysis of the NH $^1\varDelta \rightarrow {}^3\varSigma^-$ relaxation, which would enhance the triplet reaction pathway connected with vibrationally excited NH($^3\varSigma^-$) in Xe relative to the NH($^1\varDelta$) singlet reaction path favored in Ar. At the other extreme, NH($^1\varDelta$) may react rapidly with DMA to form singlet excited 1H–azirine followed, in Ar, by rearrangement on the singlet surface to form ketenimine, but in Xe, through matrix-induced intersystem crossing, to give triplet 1H–azirine. The triplet 1H–azirine would then isomerize to N-methyl-1-amino-1-propyne. While the present data do not discriminate between these two extremes (or the possible singlet–triplet crossings between), the reaction is a fascinating prototype that shows the influence and control that can be exerted by the matrix environment on excited-state reaction paths.

In our next attempt to synthesize a nitrile ylide in a cryogenic matrix, we took 3-phenyl-2H–azirine (see scheme 9) as starting material. This azirine has the advantage that it can be stored indefinitely at room temperature, and then be suspended in a cryogenic solid by standard matrix-deposition techniques. Irradiation in solid nitrogen with mercury arc light around 250 nm resulted in interconversion to a single new product (see scheme 9). Subsequent photolysis of

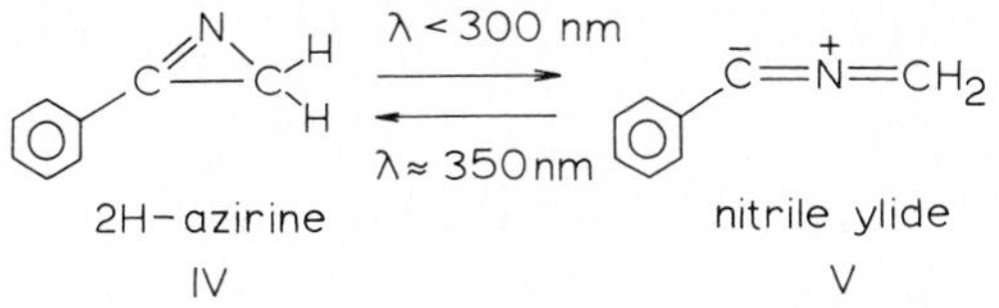

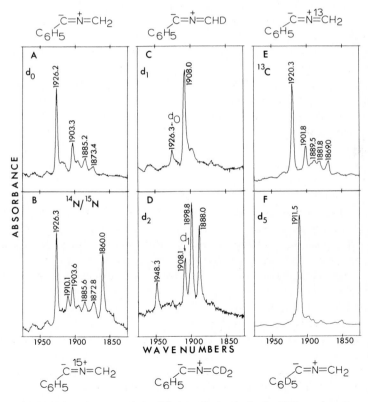

Scheme 9. Interconversion of 3-phenyl-2H-azirine and 3-phenyl-nitrile ylide at the wavelengths indicated.

the product at longer wavelengths, around 350 nm, allowed nearly complete recovery of the 2H–azirine form (Orton et al. 1986).

Specific isotopic labeling of this photochromic system with ^{15}N, D, and ^{13}C turned out to be crucial for elucidation of the nitrile ylide structure. We obtained the isotopically modified species by synthesis of azidostyrene from labeled styrene or labeled sodium azide, followed by UV-induced elimination of N_2 in the matrix to give the nitrile ylide product in high yield.

The triple bond/cumulene stretching region around 1900 cm^{-1}, shown in fig. 10 for six isotopic species, was found to contain the most decisive information

Fig. 10. Infrared spectra of six ylide isotopic species in the 1900 cm^{-1} region.

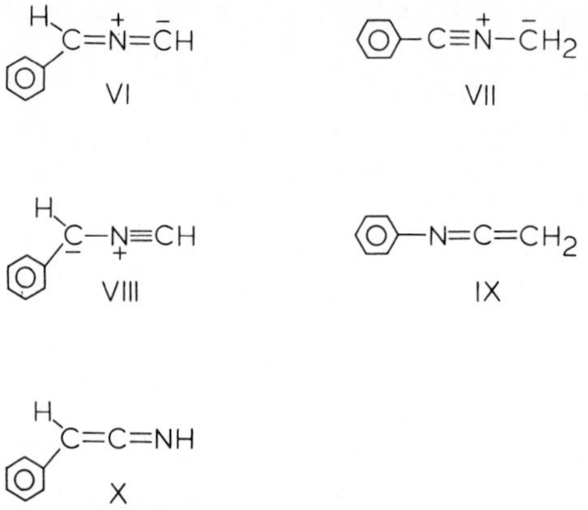

Scheme 10. Possible cumulene and triple-bond structures for the ylide.

about the structure of the ylide form. This spectral region, which is free of vibrational absorptions of the aromatic ring, exhibits only one band in the case of two of the six isotopes, $C_6H_5-C_2HDN$ (fig. 10C) and $C_6D_5-C_2H_2N$ (fig. 10F). This shows that there is a single ylide structure, and the monodeutero isotope result, in addition, suggests that the nitrile ylide possesses two equivalent CH bonds (otherwise one would expect two skeletal stretching frequencies in the 1900 cm^{-1} region). This rules out possible cumulene and triple-bond structures shown in scheme 10, namely compounds VI and VIII. While N-phenylketen-imine, compound IX, has been identified as a photostable side product of the matrix photolysis of azidostyrene (it does not absorb around 1900 cm^{-1}), the ketenimine structure, compound X, could be ruled out on the basis of other isotope data, in particular the lack of NH modes with their characteristic ^{15}N and D frequency shifts.

Key to the distinction between the remaining ylide structures V and VII are the isotope shifts of the 1900 cm^{-1} stretching absorption. Determination of these shifts from the spectra shown in fig. 10 is rendered difficult by the appearance of doublets (multiplets) in the case of $C_6H_5-C_2H_2N$ (fig. 10A), $C_6H_5-C_2H_2{}^{15}N$ (fig. 10B), $C_6H_5-C_2D_2N$ (fig. 10D) and $C_6H_5-C^{13}CH_2N$ (fig. 10E). Since the d_1 and d_5 species have only a single absorption in that spectral range, and all components of the multiplets of the other isotopic species showed uniform photolysis behavior, we are confident in attributing the splittings to Fermi resonances between combinations of skeletal vibrations and the multiple bond-stretching mode. This interpretation is strongly supported by the fact that the calculated discrepancy between the unperturbed stretching frequency and the closest combination is on the order of 1–2 cm^{-1} for the isotopes which exhibit multiplets around 1900 cm^{-1} (d_0, ^{15}N, d_2 and ^{13}C), but larger, 8–9 cm^{-1}, in the case of d_1 and d_5 species which do not show splittings. Taking the location halfway between the two most prominent features of d_0, ^{15}N, d_2, and ^{13}C as the unperturbed stretching frequency, we have determined the isotope shifts and compared them with those calculated for structures V and VII in terms of the diatomic approximation. Substantially better agreement was obtained for the nitrile ylide structure V than for structure VII. Hence we conclude that the isomer of 3-phenyl-2H–azirine obtained upon irradiation at 250 nm is the nitrile ylide with the linear heterocumulene skeleton, structure V, which has double bonds of about equal force constant and the nitrogen in the center position.

With the structures of this azirine–nitrile ylide photochromic system now established, we are interested in learning about the energy pathway of the photo-induced isomerization. We have begun measuring the yields of the azirine → ylide and the reverse, ylide → azirine interconversion in solid nitrogen as a function of irradiation wavelength, using a continuously tunable pulsed laser source. From these yields and the UV spectra of the two isomers, wavelength-dependent quantum yields to isomerization will be determined, and from them we hope to be able to deduce the shape of the excited reactive surface of this system. If successful, this would be a prototype study that shows how to map reaction surfaces of electronically excited molecules.

Acknowledgement

The authors gratefully acknowledge support by the Office of Basic Energy Sciences, Chemical Sciences Division, US Department of Energy, under contract No. DE-AC03-76SF00098.

References

Akimoto, H., and J.N. Pitts Jr, 1970, J. Chem. Phys. **53**, 1312.
Andrews, L., G.L. Johnson and B.J. Kelsall, 1982, J. Phys. Chem. **86**, 3374.
Arkell, A., 1965, J. Am. Chem. Soc. **87**, 4057.
Baldeschwieler, J.D., and G.C. Pimentel, 1960, J. Chem. Phys. **33**, 1008.
Barnes, A.J., and R.M. Bentwood, 1984, J. Mol. Struct. **112**, 31.
Bhandari, R., and L.M. Falikov, 1973, J. Phys. C **6**, 479.
Cartland, H.E., 1985, Ph.D. Dissertation (University of California, Berkeley, CA).
Cartland, H.E., and G.C. Pimentel, 1986a, J. Phys. Chem. **90**, 1822.
Cartland, H.E., and G.C. Pimentel, 1986b, J. Phys. Chem. **90**, 5485.
Cartland, H.E., and G.C. Pimentel, 1989, in preparation.
Cartwright, D.C., and P.J. Hay, 1979, J. Chem. Phys. **70**, 3191.
Chou, P.-T., and H. Frei, 1985, Chem. Phys. Lett. **122**, 87.
Collins, S.T., and G.C. Pimentel, 1984, J. Phys. Chem. **88**, 4258.
Crone, K.P., and K.J. Kugler, 1985, Chem. Phys. **99**, 293.
Davidovics, G., J. Pourcin, M. Carles, L. Pizzala, H. Bodot, L. Abouaf-Marguin and B. Gauthier-Roy, 1983, J. Mol. Struct. **99**, 165.

Davidovics, G., J. Pourcin, M. Monnier, P. Verlaque, H. Bodot, L. Abouaf-Marguin and B. Gauthier-Roy, 1984, J. Mol. Struct. **116**, 39.

Ellis, J.W., and D. Kneser, 1933, Z. Phys. **86**, 583.

Felder, P., and Hs.H. Günthard, 1984, Chem. Phys. **85**, 1.

Frei, H., 1983, J. Chem. Phys. **79**, 748.

Frei, H., 1984, J. Chem. Phys. **80**, 5616.

Frei, H., and G.C. Pimentel, 1983a, J. Chem. Phys. **78**, 3698.

Frei, H., and G.C. Pimentel, 1983b, J. Chem. Phys. **79**, 3307.

Frei, H., and G.C. Pimentel, 1985, Ann. Rev. Phys. Chem. **36**, 491.

Frei, H., T.-K. Ha, R. Meyer and Hs.H. Günthard, 1977, Chem. Phys. **25**, 271.

Frei, H., L. Fredin and G.C. Pimentel, 1981, J. Chem. Phys. **74**, 397.

Goodman, J., and L.E. Brus, 1977, J. Chem. Phys. **67**, 4398.

Hall, R.T., and G.C. Pimentel, 1963, J. Chem. Phys. **38**, 1889.

Herzberg, G., 1950, Spectra of Diatomic Molecules (Van Nostrand, New York).

Hoffman III, W.F., and J.S. Shirk, 1983, Chem. Phys. **78**, 331.

Hoffman III, W.F., and J.S. Shirk, 1985, J. Phys. Chem. **89**, 1715.

Hollenstein, H., T.-K. Ha and Hs.H. Günthard, 1986, J. Mol. Struct. **146**, 289.

Homanen, L., and J. Murto, 1982, Chem. Phys. Lett. **85**, 322.

Huber, H., D. McIntosh and G.A. Ozin, 1976a, J. Organomet. Chem. **112**, C50.

Huber, H., D. McIntosh and G.A. Ozin, 1976b, J. Am. Chem. Soc. **98**, 6508.

Jacox, M.E., 1979, J. Chem. Phys. **43**, 157.

Jacox, M.E., and D.E. Milligan, 1963, J. Am. Chem. Soc. **85**, 278.

Jones, I.T.N., and K.D. Bayes, 1972, J. Chem. Phys. **57**, 1003.

Knudsen, A.K., and G.C. Pimentel, 1983, J. Chem. Phys. **78**, 6780.

Landau, A., E.J. Allin and H.L. Welsh, 1962, Spectrochim. Acta **18**, 1.

Lee, Y.-P., and G.C. Pimentel, 1981, J. Chem. Phys. **75**, 4241.

McDonald, P.A., and J.S. Shirk, 1982, J. Chem. Phys. **77**, 2355.

McDonald, S.A., G.L. Johnson, B.W. Keelan and L. Andrews, 1980, J. Am. Chem. Soc. **102**, 2892.

Mink, J., Y.A. Pentin, 1971, Acta Chim. Acad. Sci. Hung. **67**, 435.

Nesmeyanov, A.N., V.T. Aleksanyan, L.A. Leites and A.K. Prokof'ev, 1980, Izv. Akad. Nauk. SSSR, Ser. Khim. **10**, 2297.

Norstrom, R.J., H.E. Gunning and O.P. Strausz, 1976, J. Am. Chem. Soc. **98**, 1454.

Orahovats, A., H. Heimgartner, H. Schmid and W. Heinzelmann, 1975, Helv. Chim. Acta **58**, 2662.

Orton, E., S.T. Collins and G.C. Pimentel, 1986, J. Phys. Chem. **90**, 6139.

Perttila, M., J. Murto, A. Kivenen and K. Turunen, 1978a, Spectrochim. Acta A **34**, 9.

Perttila, M., J. Murto and L. Halonen, 1978b, Spectrochim. Acta A **34**, 469.

Poliakoff, M., and J.J. Turner, 1980, Infrared Laser Photochemistry in Matrices, in: Chemical and Biochemical Applications of Lasers, Vol. 5, ed. C.B. Moore (Academic Press, New York) p. 175.

Pourcin, J., G. Davidovics, H. Bodot, L. Abouaf-Marguin and B. Gauthier-Roy, 1980, Chem. Phys. Lett. **74**, 147.

Rasanen, M., and V.E. Bondybey, 1984, Chem. Phys. Lett. **111**, 515.

Rasanen, M., and V.E. Bondybey, 1985, J. Chem. Phys. **82**, 4718.

Rasanen, M., A. Aspiala, L. Homanen and J. Murto, 1982, J. Mol. Struct. **96**, 81.

Rasanen, M., A. Aspiala and J. Murto, 1983, J. Chem. Phys. **79**, 107.

Rasanen, M., J. Murto and V.E. Bondybey, 1985, J. Phys. Chem. **89**, 3967.

Shirk, A.E., and J.S. Shirk, 1983, Chem. Phys. Lett. **97**, 549.

Sieber, W., P. Gilgen, S. Chaloupka, H.-J. Hansen and H. Schmid, 1973, Helv. Chim. Acta **56**, 1679.

Spratley, R.D., J.J. Turner and G.C. Pimentel, 1966, J. Chem. Phys. **44**, 2063.

Steunenberg, R.K., and R.C. Vogel, 1956, J. Am. Chem. Soc. **78**, 901.

Takeuchi, H., and M. Tasumi, 1982, Chem. Phys. **70**, 275.

Takeuchi, H., and M. Tasumi, 1983, Chem. Phys. **77**, 21.

Wasserman, H.H., and R.W. Murray, 1979, Singlet Oxygen (Academic Press, New York).

CHAPTER 7

Production Methods for ESR studies of Neutral and Charged Radicals

Lon B. KNIGHT Jr

Department of Chemistry
Furman University
Greenville, South Carolina 29613
USA

Chemistry and Physics of
Matrix-Isolated Species
L. Andrews and M. Moskovits, eds

Contents

1. Introduction

Generation techniques for neutral and charged radicals with a variety of illustrative examples taken primarily from studies in the author's laboratory will be the major emphasis of this chapter on matrix-isolation electron spin resonance spectroscopy (MI-ESR). This will not be a comprehensive review of the field, and space constraints require that most radical examples be restricted to rare-gas matrix studies even though important advances in condensed-phase radical experiments have been achieved in other inert media such as solid hydrocarbons, freons, TMS, N_2, SF_6 and CO, just to name a few.

Developments in radical production techniques for rare-gas matrices since Weltner's monograph (1983) on MI-ESR will be presented in more detail than earlier methods. This excellent source should be consulted for the historical development of the field, interpretation of powder ESR spectra including high-spin radicals, literature coverage and magnetic parameters of practically all radicals studied in rare-gas matrices up to 1983. Exciting applications of laser vaporization applied alone or in combination with ion-generation schemes are providing the means to study numerous classes of radicals that would probably not be possible by the more conventional approaches. The coupling of MI-ESR with various high-energy experimental conditions should broaden the usefulness of the method and could make it an important spectroscopic monitoring technique for surface bombardment processes, ion–neutral reactions and laser-induced reactions of solid mixtures. Laser-induced fluorescence (LIF) studies of metal clusters formed by laser vaporization and trapped in rare gases have recently been reported (Rasanen et al. 1987). The recent commercial availability of small closed-cycle refrigerators capable of 4 K operation will help make MI-ESR a more practical approach for applications involving charged radicals which have to be trapped in neon matrices if the electron affinity (EA) of the cation is greater than ~ 10–11 eV.

Recent progress has been made in several important areas of matrix ESR which are not directly related to radical production techniques. One of these is the adaptation of electron nuclear double resonance (ENDOR) for rare-gas matrix studies of volatile and non-volatile radicals including C_2H_5 (Toriyama et al. 1981a), CH_3 and vanadium dibenzene (Cirelli et al. 1982), SiH_3 (Van Zee et al. 1985), and MnH (Van Zee et al. 1986a). Important design features for a matrix

169

apparatus have been presented which should make ENDOR measurements more common in future studies (Weltner and Van Zee 1986). The observation of magnetophotoselectivity for the formyl radical in a CO matrix and earlier work on metastable triplets in glassy media suggest that the high optical quality of neon matrices would be an appropriate medium for such experiments (Adrian et al. 1984). Magnetophotoselectivity can provide information on photochemical pathways, temperature-dependent reorientation phenomena and the optical transition moment. The considerable progress made in the ESR matrix study of naked metal clusters has been recently reviewed and will not be included in this account (Ozin and Mitchell 1983, Weltner and Van Zee 1984, Morse 1986). A recent achievement in this area is the ESR characterization of Cu_3 generated by Cu atom deposition (Lindsay et al. 1987, Howard et al. 1983). A review of ESR metal-cluster results utilizing the rotating-cryostat method with adamantane matrices has been presented which concludes that the larger cavities in this host matrix provide trapping sites of comparable inertness to the smaller cavities of rare-gas matrices (Howard and Mile 1987). An alternative matrix-preparation method involving pulsed deposition into a tube closed on one end has been described (Adrian et al. 1985). This type of matrix sample apparently shows less aggregation and can be raised to higher temperatures without severe evaporative losses. The ability to achieve higher temperatures increases the chances for obtaining an isotropic spectrum – information that can be quite useful in interpreting the complex powder spectra of the more rigid matrix.

1.1. Matrix ESR of charged radicals

The first reported use of MI-ESR applied to free or isolated molecular cation radicals was as recent as 1982 (Knight and Steadman 1982b). This important area has been described in a recent review article (Knight 1986) which covers the earlier literature and gives the developmental relationship between ESR matrix ion studies and other spectroscopic areas such as infrared and ultraviolet-visible absorption that are covered in the chapter by Andrews (ch. 2). Since most cations of the undissociated neutral parents are radicals, the number of new molecular systems amenable to MI-ESR is quite large! Recent studies have shown that NH_3^+ can be trapped in argon (McKinley et al. 1986) and that transition-metal carbonyl cations can be generated in the heavier rare gases by γ-irradiation (Fairhurst et al. 1984a, Morton and Preston 1985, Fairhurst et al. 1984b). The only doublet radicals, neutral or charged, that cannot be studied are those with orbitally degenerate ground electronic states. In practice, this restriction prevents the ESR matrix detection of Π and Δ radicals because of g-tensor anisotrophy which produces extreme broadening of the absorption envelope. In principle, all non-linear doublet radicals can be studied since the orbital angular momentum is quenched in such non-degenerate cases. The detectability and ESR spectral characteristics of high-spin ($S > \frac{1}{2}$) radicals in rare-gas matrices have been thoroughly analyzed and only a few of these types will be treated in this account (Weltner 1983).

The ESR study of relatively large organic cation radicals in freon matrices has yielded much valuable information (Kato and Shida 1979, Symons and Smith 1979, Wang and Williams 1980, Toriyama et al. 1981b, Symons 1984; see also Faraday Discuss. Chem. Soc. 1984, Vol. 78, which is an issue entirely devoted to radicals in condensed phases). A summary of a detailed comparison between neon and freon matrix ESR results for the acetaldehyde cation (CH_3CHO^+) will be presented in this chapter (Knight et al. 1988a). Apparently this is the first such comparison, and others are needed to determine the full extent of matrix perturbations that might be present in the more reactive freon type environment. A comparison of ESR results for the isolated free anion radical, $F_2^-(X^2\Sigma)$, in neon and various ionic lattices has also been reported (Knight et al. 1986a). An analysis of how the electronic distribution in F_2^-, as revealed by the A and g-tensors, varies as a function of the lattice environment was conducted.

1.2. Radical identification

In most matrix studies of highly reactive species or intermediates, especially those trapped under high-energy conditions, confirming the spectral assignment is often the most difficult and time consuming phase. ESR matrix experiments have several advantages in this regard relative to most other spectroscopic methods: only radicals are detected; the detection of $\sim 10^{12}$ radicals is feasible under favorable conditions; magnetic isotopic substitution for small radicals usually allows a definitive assignment to be made; and information contained in the lineshape of powder spectra is often helpful in the analysis process provided rigorous comparisons with simulated spectra are conducted. Photobleaching or photolysis of the anion traps with low-intensity visible light allows a distinction to be made between neutral and charged radicals. Unless the chemical origins of a new radical produced under such extreme conditions is straightforward, it is often necessary to produce the species by an entirely independent method in order to confirm the spectral assignment. The impetus for developing several of the generation methods described below has been this confirmation objective. Comparisons with state-of-the-art theoretical calculations for the nuclear hfs can also be used as supporting evidence for a particular assignment.

A good example of an assignment problem arose in our investigation of the $AlF^+(X^2\Sigma)$ cation radical which should produce a sextet of doublets hyperfine pattern since Al has $I = 5/2$ and ^{19}F has $I = 1/2$ (Knight et al. 1986b). The original generation attempt involved the high-temperature vaporization reduction reaction of $AlF_{3(s)} + Al_{(s)}$ in a tantalum effusion oven. Previous mass spectrometric studies had shown that $AlF_{(g)}$ was a dominant vapor species under such conditions (Snelson 1967). The $AlF_{(g)}$ was deposited into a neon matrix during photo-ionization at 16.8 eV and for every matrix formed under these conditions, a clear sextet of doublets centered near g_e was observed. The Al hfs was large (~ 340 G) as expected and the ^{19}F A tensor also seemed quite reasonable. Evidence that the ESR signals belonged to an ion radical was provided by photobleaching. However, ab initio CI calculations yielded an Al hfs of approximately 980 G for AlF$^+$ which is consistent with the calculated charge distribution of Al^{+2}F$^-$. This large discrepancy between observed and calculated Al hfs prompted attempts to generate AlF$^+$ by an alternate method, namely reactive laser sputtering with photo-ionization during deposition. By this procedure, AlF$^+$ was produced which showed excellent agreement with theory. The impurity species originally seen in the high-temperature vaporization experiments is most probably AlOF$^+$ where the ^{16}O with $I = 0$ would not contribute to the hyperfine pattern. Subsequent high-temperature experiments showed that carefully dried samples of AlF$_{3\,(s)}$ did not produce the impurity radical ion but did yield AlF$^+$ in agreement with the reactive laser sputtering results.

1.3. Experiment versus theory

Our matrix ESR studies have concentrated on small radicals which are difficult or impossible to manage in the gas phase, yet sufficiently "simple" as to allow comparisons with ab initio calculations of the nuclear hyperfine properties (A tensors). The A- and g-tensors obtained from the ESR spectrum provide the most direct experimental information available for describing the electronic structure of the singly occupied MO (SOMO). In some cases, such detailed knowledge of the SOMO also provides insight into the nature of the lower lying filled MOs. An approximate description of the excited orbitals can sometimes be obtained from matrix data provided g-tensor components are shifted sufficiently from the free-spin value. Neon matrix effects usually cause small g shifts in the range of 0.0005 or approximately 0.03% for doublet radicals.

One strategy employed jointly with Davidson and Feller to probe the electronic structure of small radicals has involved calculating ab initio CI wavefunctions which yield A-tensors in good agreement with experiment (Knight et al. 1985). Such "tested" wavefunctions can then be used to obtain electronic structure information which is more comprehensive than ESR results alone can provide. Examples utilizing this approach include the following groups of isoelectronic radicals: CO$^+$, BO, BeF, BF$^+$; AlO, MgF, SiO$^+$, AlF$^+$; and C$_2$O$_2^+$, N$_4^+$ with work on N$_2$CO$^+$ in progress.

These calibrated wavefunctions can also be used to gauge the accuracy of the commonly applied procedure in ESR studies which compares the isotropic and dipolar nuclear hyperfine parameters (A_{iso} and A_{dip}) with a set of free-atom parameters for estimating the spin populations in the valence "s" and "p" orbitals. The reasonableness of this free-atom comparison method (FACM) has really not been tested in a systematic and independent manner. We have attempted to evaluate FACM results by comparing them to a Mulliken spin population analysis (MSPA) conducted on ab initio wavefunctions which yield calculated A-tensors in close agreement with the directly measured nuclear hyperfine interactions. This FACM versus MSPA comparison has been made for only a few diatomic cases, but it has been able to identify an electronic structure type that is not appropriate for a simple FACM analysis. This topic is discussed in greater detail in the comparison of CO^+ and SiO^+ in section 4.2.4.

2. ESR matrix apparatus

The ESR matrix apparatus (fig. 1) used by our group is designed for a high degree of flexibility in combining several different generation methods in order to produce and trap the radical of interest. Specific procedures for neutral and charged radicals are described in the next section. A synchronized hydraulic system is used to raise and lower the liquid-helium cryostat (Air Products' Heli-Tran) with the attached matrix-deposition surface over a travel distance of 6 cm. This has proven to be a very convenient method of moving the 4 K matrix target between the molecular-beam deposition position and its location inside the microwave cavity. The ability to lower the matrix sample and observe the ESR spectra in a short time period is important for adjusting the various generation conditions, which is very much a trial-and-error procedure.

The commercial microwave cavity operates in the TE_{102} mode and employs 100 kHz modulation. The cavity remains at ambient temperature and is not

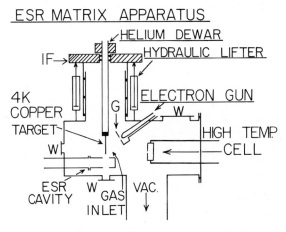

Fig. 1. ESR matrix apparatus which combines high-temperature vaporization and electron bombardment for production of radicals, especially molecular cations. Details of the electron source are shown in fig. 11. The labels denote: G, electron control grid; W, windows; IF, electrically insulated flange. (From Knight et al. 1983a.)

cooled to 77 K as done in earlier versions of the ESR matrix apparatus (Jen et al. 1958). Both single-crystal sapphire and copper-matrix targets have been employed. The copper target ($4 \times 0.6 \times 0.15$ cm) is carefully positioned in the node of the electric field component of the microwave radiation so that the Q of the cavity is not significantly reduced. Weak pitch samples mounted on the copper target indicate that the sensitivity reduction is less than a factor of two. During deposition the matrix temperature is maintained at 5 ± 1 K as indicated by a thermocouple embedded in the copper target and no warming is noted when the matrix sample is lowered into the 1 mm opening of the microwave cavity. Having the cryostat and generation sections combined into one system saves considerable time compared to the alternative of connecting and disconnecting these two units in order to record spectra. The ESR magnet with a 9 cm air gap and mounted on tracks can be rolled into position for recording spectra. A major design characteristic of the apparatus is that all vital components are located quite close to the deposition target. These distances are given in the descriptions of the various generation methods.

During deposition with a neon-matrix gas flow of 5 sccm, the pressure in the deposition area is approximately 5×10^{-5} Torr compared to background pressures of 3×10^{-7} Torr. Experience has shown that it is important to keep this pressure as low as possible to prevent the rare-gas flow from scattering radicals in the molecular beam which is also directed at the deposition target. In addition to the cryopumping that occurs on the target and heat shield, approximately $400 \ell s^{-1}$ of pumping capacity is provided by diffusion pumps which remain permanently trapped with a $-100°C$ cooling system. The actual matrix gas flow and that through the open tube photo-ionization resonance lamp are continuously monitored by thermal conductivity flowmeters. These gases also pass through 77 K traps which contain a molecular sieve in order to remove ppm levels of contaminants such as N_2 and O_2 in the research grade rare gases. To further reduce the background pressure, the copper walls of the entire matrix apparatus are equipped with coils used for cooling and heating. The apparatus is cooled to 10°C just before and during deposition, otherwise it is continuously maintained at 50°C to facilitate outgassing. A quadrupole mass spectrometer (0–300 amu) is permanently mounted in the 10 cm pump-out line. This has proven to be an extremely valuable source of information for establishing matrix-deposition conditions, regulating gas flows, confirming isotopic purities and reproducing various ratios of flows involved in ion–neutral co-deposition reaction schemes. Unusual contaminants outgassed from materials undergoing laser vaporization or contained in commercial samples can also be detected.

The matrix apparatus described above is used with a Varian Century Series ESR spectrometer interfaced with an HP 9836A computer system and a HP 9872C plotter. Another version of an apparatus for MI-ESR that is also used in our laboratory, utilizes a Bruker ER-200D spectrometer with a 10 cm air-gap magnet. This system makes greater use of commercially available components although some rather extensive modifications have been made. Only the major features of this system which differ substantially from the one already described will be discussed. The cryostat is a closed-cycle refrigerator (Air Products' Heliplex CS 304) with a capacity of approximately 1 Watt at 5 K. We have found it to be as effective in forming delicate neon matrices as the more conventional liquid-helium Dewars or Heli-Tran cryostats. A major advantage is the adjustable and stable temperature control in the useful neon range of 5–12 K. An Air Products' vacuum shroud (DMX-1) has been modified by installing a section which allows the 4 K refrigerator to move 6 cm on o-ring seals in order to lower and raise the matrix sample in and out of the microwave cavity by means of a hydraulic system. An 11 mm quartz tube extends below the vacuum shroud and into the Bruker ESR X-band cavity. The 3 mm matrix-deposition target travels inside this quartz tube and into the cavity which is not contained in the vacuum system. This arrangement prevents cavity contamination from laser-sputtered materials and is more compact than the first apparatus described. The magnet does not move and the entire system is mounted on top of the polecaps by contoured aluminium plates and padded side plates inside the magnet coils.

3. Generation methods for neutral radicals

Tables 1 and 2 list generation methods and recent radical examples for neutral and charged species, respectively. This non-comprehensive listing of predominantly small radicals illustrates the wide variety of chemical systems that can be studied by rare-gas MI-ESR with examples ranging from metal hydride cations to high-spin carbon clusters to several types of metal carbonyls. A complete listing of radicals studied by ESR under all conditions can be found in the appropriate Landolt–Börnstein volumes.

Methods for generating neutral radicals for MI-ESR studies will be described first since several of these are used in combination with the ion techniques for producing and trapping charged radicals. Methods which have been in general use for some time will not be treated in detail but are included to indicate the variety of available procedures. As illustrated above in the case of AlF^+, it can be essential to utilize independent methods in order to confirm the ESR assignment of a new radical. Some of the radical examples used to illustrate the capabilities of the various generation techniques are discussed in greater detail in subsequent sections of this chapter.

3.1. Laser vaporization

The relative positions of components for laser-vaporization MI-ESR experiments are given in fig. 2. Typical depositions utilize a Nd:YAG laser doubled to 532 nm operating in the range of 0.5–10 mJ/pulse at 10 Hz. Under these

Table 1

An illustrative, non-comprehensive review of recent examples of neutral-radical production methods for rare-gas matrix ESR studies. The number in brackets refers to the corresponding reference.

Methods	Neutral radicals
Laser vaporization	
One material	B_2 [1]; C_6 [2]; Cu, B, P and other atoms [3]
Reactive laser vaporization	
Two reactants; gas–solid or solid–solid	GaO [3]; CP [3]; PO_2 [3]; AsO_2 [3]; CH_3CuF [3]
Thermal vaporizations	
High-temperature effusion ovens	VO [4]; GdO [5]; HSiO [6]; ScH_2 [7]
Co-deposition reactions	
Atom with reactant gas; metal-cluster formation	MnH [8]; BeH [9]; $Cu(O_2)$ [10]; KO_2 [11]; $Ga(CO)_2$ [12]; $V(CO)_n$ [13]; $(C_6H_6)V$[14]; Na^+HCl^- [15]; Si_2 [6]; AgMg [16]; CrCu [17]; Cu_3 [18]; Li_7 [19]; Sc_{13} [20]
Photolysis	
Vis, UV or γ-irradiation of isolated reactants	$HNi(CO)_3$ [21]; XeCl [22]; $HC^{17}O$ [23]; $HAlCH_3$ [24]; HBBH [3]; $KrMn(CO)_5$ [25]; $V(CO)_4$ [26]; $V(CO)_5$ [26]
Discharge sources	
Including collisions with metastables	C_6 [2]; CH_3CH_2N [27]
Fast atom bombardment	
1–10 keV neutral rare-gas atoms	CdOH [28]; MgH [28]; CdH [28]; Cu [28]; Ag and H atoms [28]
Cathode sputtering	
Gas passed over reactant metal cathode	MoN [29]; Mo and Al atoms [29]

References:
[1] Knight et al. (1987b).
[2] Van Zee et al. (1987).
[3] Knight (1989).
[4] Kasai (1968).
[5] Van Zee et al. (1981).
[6] Van Zee et al. (1985).
[7] Knight et al. (1981).
[8] Van Zee et al. (1986a).
[9] Knight et al. (1972).
[10] Kasai and Jones (1986).
[11] Lindsay et al. (1974).
[12] Kasai and Jones (1985).
[13] Van Zee et al. (1986b).
[14] Andrews et al. (1986).
[15] Lindsay et al. (1982).
[16] Kasai and McLeod (1975).
[17] Baumann et al. (1983).
[18] Lindsay et al (1987);
     Howard et al. (1983).
[19] Garland and Lindsay (1984).
[20] Knight et al. (1983b).
[21] Morton and Preston (1984a).
[22] Adrian and Bowers (1976).
[23] Adrian et al. (1985).
[24] Parnis and Ozin (1986).
[25] Fairhurst et al. (1984a).
[26] Morton and Preston (1984b).
[27] Ferrante et al. (1987).
[28] Knight et al. (1984a).
[29] Knight and Steadman (1982a).

Table 2

A non-comprehensive review of recent examples of ion radical production methods for rare-gas matrix ESR studies. Only free-ion and isolated-ion data are included (chemically bonded and highly ionic molecules are excluded). The number in brackets refers to the corresponding reference.

Methods	Ion radicals
Photo-ionization (16.8 eV) of neutrals from	
Volatile gases	CO^+ [1]; F_2^- [2]; CH_4^+ [3]; H_2O^+ [4]; $C_8H_8^+$ [5]
Laser vaporization	$GaAs^+$ [6]; C_2^+ [19]
Reactive laser vaporization	AlF^+ [7]; AlH^+ [8]; BF^+ [9]
Thermal vaporization	SiO^+ [10]; AlF^+ [7]; BF^+ [9]
Laser MPI	Mg^+ [6]; H_2O^+ [4]; CH_3F^+ [6]
Electron bombardment	
Volatile gas	N_2^+ [11]; CO^+ [2]; H_2O^+ [4]; H_2CO^+ [12]; F_2CO^+ [6]; F_2^- [2]
Reactive laser vaporization	AlF^+ [7]; BF^+ [9]
Thermal vaporization	SiO^+ [10]; AlF^+ [7]
Ion–neutral co-deposition reactions	
Photo-ionization or electron bombardment during deposition	N_4^+ [13]; O_4^+ [6]; $C_2O_2^+$ [14]; N_2CO^+ [6]; Mg_N^+ ($N = 1$–6) [6]
γ-irradiation	
After deposition of neutral precursor	$Cr(CO)_4^+$ [15]; $KrFe(CO)_5^+$ [16]; $Ni_2(CO)_6^+$ [17]
Photolysis	
In-situ electron transfer from metal atom	Cr^+ [18]; Cd^+ [18]; Mn^+ [18]; $B_2H_6^-$ [18]
Mass-selected cation beam	CO^+ [6]; H_2O^+ [6]

References:
 [1] Knight and Steadman (1982b).
 [2] Knight et al. (1986a).
 [3] Knight et al. (1984c).
 [4] Knight and Steadman (1983).
 [5] Knight et al. (1987d).
 [6] Knight (1989).
 [7] Knight et al. (1986b).
 [8] Knight et al. (1987a).
 [9] Knight et al. (1986c).
[10] Knight et al. (1985).
[11] Knight et al. (1983a).
[12] Knight and Steadman (1984).
[13] Knight et al. (1987c).
[14] Knight et al. (1984b).
[15] Fairhurst et al. (1984b).
[16] Fairhurst et al. (1984a).
[17] Morton and Preston (1985).
[18] Kasai (1971).
[19] Knight et al. (1988b).

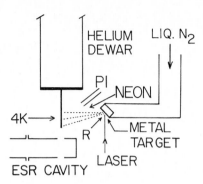

Fig. 2. ESR matrix apparatus for trapping radicals at 4 K generated by the reactive laser vaporization of a volatile gas (R) deposited onto a metal target at 77 K. The photo-ionization (PI) of these products can be accomplished with the open-tube neon resonance lamp (16.8 eV) powered by an Evenson microwave cavity (not shown).

conditions and with a continuous neon flow of 5 sccm, 20 min deposits usually give intense ESR signals for favourable systems. We have achieved the best results with the laser target quite close (~ 1 cm) to the matrix-deposition surface. For such close distances, the angular relationships between the laser-target surface, the laser beam, and the matrix do not seem to be especially critical, provided a small fraction of the sputtered material is intercepted by the deposition surface. With the ESR magnet rolled away from the matrix apparatus, the laser vaporization process can be monitored through side windows adjacent to the deposition surface. Observation of the shadow pattern caused by the matrix target blocking part of the sputtered beam is quite useful in establishing initial conditions of laser power, time of deposit, angles and relative distances. Contamination on the vacuum side of the window used to transmit the laser beam is reduced by locating it 40 cm from the target and using an external focusing lens to produce a spot size of about 0.5 mm. The external lens, mounted on an X–Y translator, is periodically moved every 2–4 min to prevent the formation of deep holes in the target. However, for most applications it is not necessary to continuously move or rotate the target as is commonly done in gas-phase molecular-beam experiments.

3.1.1. The B_2 radical

Only a few MI-ESR studies have utilized laser vaporization but the potential for new radical chemistry is considerable, as illustrated by the following examples. Boron vapor is notoriously difficult to produce by conventional methods given the extreme temperatures required and the corrosive nature of molten boron. Laser vaporization of the solid is quite straightforward and has been used to trap B atoms in argon and the B_2 molecule in neon and argon matrices (Knight et al. 1987b). The first direct experimental evidence that the electronic ground

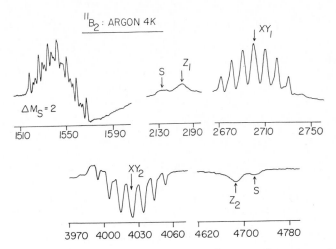

Fig. 3. The overall ESR spectrum for the $^{11}B_2$ radical in its $^3\Sigma_g^-$ ground state trapped in an argon matrix at 4 K. The XY_1 and XY_2 features are the $\Delta M_S = 1$ transitions for the direction perpendicular to the molecular axis; Z_1 and Z_2 are the parallel transitions. The weaker parallel features labeled "S" result from an alternate trapping site in the argon lattice. The $\Delta M_S = 2$ and the perpendicular transitions show septet nuclear hyperfine patterns resulting from two equivalent $^{11}B(I = \frac{3}{2})$ atoms.
    Note the phase consistency between Z_1 and XY_2; XY_1 and Z_2. (From Knight et al. 1987b.)

state of B_2 is $^3\Sigma_g$ was obtained from these ESR matrix results. Detailed electronic-structure information has been obtained from a comparison of the observed and calculated ^{11}B hyperfine structure shown in the neon matrix ESR spectrum of fig. 3.

3.1.2. The C_6 radical

Laser equilibrium vaporization of graphite has produced the C_6 radical whose electronic ground state has also been assigned as $^3\Sigma_g$ based on MI-ESR results (Van Zee et al. 1987). The matrix-deposited samples produced by laser vaporization seem to contain a higher percentage of the larger carbon molecules compared to conventional high-temperature methods – more C_6 than C_4. The C_6 radical was also produced by independent procedures involving the thermal vaporization of graphite from thin-walled tantalum cells and by the microwave discharge of a dilute mixture of triacetylene (C_6H_2) in argon gas.

3.1.3. Atomic radicals

The equilibrium vapor composition is clearly altered in some cases as shown by the intense ESR spectrum of phosphorus atoms (^4S) shown in fig. 4 obtained by the laser vaporization of red phosphorus. The P_4 molecule is the dominant vapor species above thermally heated phosphorus. These experiments were conducted as part of a study designed to generate P_4^+ for comparison with the

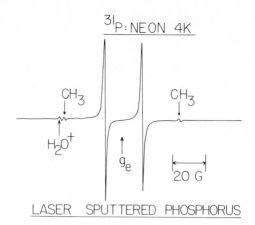

Fig. 4. The ^{31}P(^{4}S$_{3/2}$) ESR spectrum in a neon matrix at 4 K which consists of a nuclear hyperfine doublet centered near g_e [^{31}P($I = \frac{1}{2}$)]. The lowest field line of the hydrogen triplet in H$_2$O$^+$, and the first and fourth lines of the CH$_3$ quartet are also indicated. This matrix sample was produced by the laser vaporization of red phosphorus.

previously reported N$_4^+$ cation radical (Knight et al. 1987c). We have compared the ESR spectral quality of such atomic radicals as Cu, Ag, Au, Al, B and Mo, produced by laser vaporization versus the conventional high-temperature effusion cell method. In all cases, the laser-deposited samples exhibited narrower ESR lines and occupied a smaller number of multiple trapping sites in the rare-gas lattice. The P atom ESR spectrum has apparently not been previously observed in neon matrices but its A value has been measured in argon (29 G), krypton (30 G) and xenon (31 G) and in the gas phase (20 G) (Jackel et al. 1968). The A-value observed in this neon study of 23 G shows much closer agreement with the gas-phase parameter. We have observed a similar improvement in agreement with the gas phase for N and As atoms in neon matrices, both of which have the ^{4}S ground state. The ability to generate phosphorus, boron and other atoms in such an efficient manner by laser vaporization suggests numerous co-deposition experiments for producing new radicals of interest which would be difficult to achieve by conventional high-temperature methods.

3.2. Reactive laser vaporization

We will use the term reactive laser vaporization (RLV) to refer to laser-initiated reactions between different materials during the vaporization process. Two versions of RLV used in our group for MI-ESR experiments will be illustrated, namely gas–solid and solid–solid reactions. The passage of a reactant gas over the material undergoing laser vaporization can generate new radicals that are not formed by the more standard approach of co-depositing metal atoms from thermal sources with the reactant molecule. The potential of this method has been demonstrated in a series of experiments by the laser vaporization of Cu, Al, Ga, and Sc, in the presence of $CH_3F_{(g)}$ using both neon and argon matrices at 4 K (Knight et al. 1987b). The unusual insertion product radicals, CH_3MF, have been produced for the first time by this procedure; these types are especially interesting since they provide an opportunity to study metal–carbon bonding by ESR in radicals sufficiently small for detailed theoretical calculations.

An alternate approach to passing the reactive gas over the metal target during laser vaporization is to condense a thin film of the gas on the metal surface by cooling it to liquid-nitrogen temperatures. Our scheme for conducting this type of condensed phase RLV matrix experiments is presented in fig. 2. The small liquid nitrogen Dewar is constructed so that its cold end can be moved under high vacuum very near (~ 1 cm) the matrix-deposition surface. Following the laser vaporization process, the N_2 Dewar is withdrawn several inches so the matrix sample can be lowered into the microwave cavity. This condensed-phase approach changes the reaction conditions and increases the contact between the reactant and the metal. The objective of this approach is to determine if certain metal products can be generated that would not form by passing the gas over the surface. The CH_3CuF product described below is readily generated by laser sputtering a copper target cooled to 77 K with $CH_3F_{(g)}$ condensed on its surface. Numerous other combinations of reactant gases and metals are currently being investigated as well as the optimization of various parameters such as the film thickness, laser power per unit area, pulse duration, pulse rate and frequency dependence.

3.2.1. CH_3CuF

The neon matrix ESR spectrum of $^{12}CH_3CuF$ (fig. 5) exhibits large quartet hfs from the two copper isotopes [$^{63}Cu(I = \frac{3}{2})$, 69% abundance and $^{65}Cu(I = \frac{3}{2})$, 31% abundance] for both the parallel and perpendicular directions of this axially symmetric radical. Each quartet component is split into $^{19}F(I = \frac{1}{2})$ doublets which are further split by the methyl hydrogens. The methyl splitting of 2.8 G is too small to be recognized in the wide-scale sweep of fig. 5. The substitution of $^{13}CH_3F$ confirms the assignment by producing additional $^{13}C(I = \frac{1}{2})$ doublet hfs. These complete A- and g-tensor results which show large shifts from g_e allow a rigorous comparison to be made with the CuF_2 radical which was one of the first high-temperature radicals studied by MI-ESR (Kasai et al. 1966). The extreme ionicity of $CuF_2(F^- Cu^{++} F^-)$ is moderated by replacing one F with CH_3. The resulting changes in the magnetic parameters (Knight et al. 1988c) provide an interesting probe into the electronic changes that accompany such a substitution. The similar molecule CH_3CuH assigned by Parnis et al. (1985) in a methane matrix has magnetic parameters substantially different from those we have observed for CH_3CuF in neon which are $g_{\parallel} = 1.966(2)$; $g_{\perp} = 2.362(1)$; $A_{\parallel}(^{63}Cu) = 3081(2)$ MHz and $A_{\perp}(^{63}Cu) = 2987(2)$

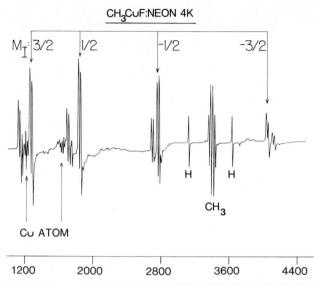

Fig. 5. The matrix apparatus shown in fig. 2 was used to generate the CH_3CuF radical whose overall ESR spectrum is shown. The large quartet hfs resulting from $^{63}Cu(I = \frac{3}{2};$ 69% abundance) is indicated; the weaker $^{65}Cu(I = \frac{3}{2};$ 31% abundance) hfs is also clearly recognizable. Each copper line is split into an $^{19}F(I = \frac{1}{2})$ doublet which, on expanded sweeps, shows clearly resolved CH_3 quartet hfs (see text).

MHz compared with the following reported values for CH_3CuH; $g_{\parallel} \sim 2.10$; $g_{\perp} \sim 1.98$; $A_{\parallel}(^{63}Cu) = 514$ MHz and $A_{\perp}(^{63}Cu) = 235$ MHz.

3.2.2. Metal oxide radicals

Another class of RLV matrix ESR studies involves passing $O_{2(g)}$ over metals and trapping the diatomic metal oxide radicals. A major advantage of this approach is the ability to generate $^{17}O(I = \frac{5}{2})$ substituted molecules which are extremely difficult to do by the high-temperature thermal vaporization of the refractory metal oxides. The ^{17}O A-tensor has not been determined for any metal oxide type radical thus preventing a complete determination of the electronic distribution for the SOMO. The original matrix ESR investigation of the group III oxides (ScO, YO and LaO) employed thermal generation and was not able to determine the metal dipolar hfs because of multiple trapping site complications and relatively broad lines (Weltner et al. 1967). Matrix RLV experiments recently conducted in our laboratory for $Al^{17}O$, $Ga^{17}O$, $In^{17}O$, $Sc^{17}O$ and $Y^{17}O$ have been able to resolve the isotropic and dipolar constants for both the metal and oxygen atoms which provide a complete description of the electronic distribution of the unpaired electron except in assigning relative amounts of "p" and "d" orbital character. An example of the spectral quality observed for these MO radicals is shown in fig. 6 for the ^{45}Sc: $M_I = -\frac{5}{2}$ transition of $Sc^{16}O$ and $Sc^{17}O$ in a neon matrix at 4 K ($I = \frac{7}{2}$ for ^{45}Sc). The parallel ($\parallel$) and per-

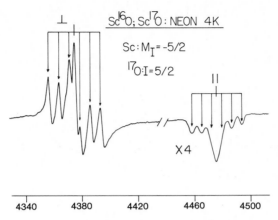

Fig. 6. Reactive laser vaporization was used to generate ScO ($X^2\Sigma$) whose overall ESR spectrum shows octet nuclear hfs [^{45}Sc($I = \frac{7}{2}$)]. The parallel and perpendicular components for one of these transcriptions are shown which also yield clearly resolvable ^{17}O($I = \frac{5}{2}$) hfs. Scandium metal was laser vaporized while passing a mixture of ^{16}O$_2$ and ^{17}O$_2$ over its surface.

pendicular ($\perp$) transitions labeled in fig. 6 exhibit a classical lineshape for a randomly oriented powder sample with each component split into an ^{17}O sextet with a small A value. The ^{45}Sc $A_\parallel$ and $A_\perp$ parameters have values of 2067(2) and 1993(2) MHz, respectively, in neon. The observed argon parameters are within 1% of these neon results indicating that only small matrix shifts are involved.

3.2.3. PO$_2$

Oxygen passed over red phosphorus during laser vaporization produces intense ESR lines of the bent ^{31}PO$_2$ radical whose spectrum is shown in fig. 7. A comparison with the simulated spectra (also shown) yields the anisotropic g and ^{31}P A-tensors which have been compared with the isovalent NO$_2$ radical and recent gas-phase LMR (laser magnetic resonance) results for PO$_2$ (Kawaguchi et al. 1985). The ^{31}P A_{iso} parameter observed in neon is 1683(1) MHz which is shifted only +0.8% from the gas phase. The sharp well-defined matrix spectra allow a complete determination of the ^{31}P dipolar tensor which could not be fully resolved in the gas-phase LMR experiment. The unpaired electron on phosphorus in PO$_2$ has orbital occupancies of 0.13 (3s), 0.31 (3p$_z$) and 0.091 (3p$_x$) compared to the distribution on nitrogen orbitals in NO$_2$ of 0.085 (2s), 0.36 (2p$_z$) and 0.045 (2p$_x$). See fig. 7 for assignment of directions. Similar MI-ESR experiments are being conducted for the AsO$_2$ radical by the RLV procedure.

3.2.4. CP; laser-initiated solid/solid reactions

An example of a laser-initiated reaction between solids during vaporization is C(s)/P(s) to generate the C^{31}P radical whose nuclear hyperfine interactions have not been measured previously by any spectroscopic method. A search for this

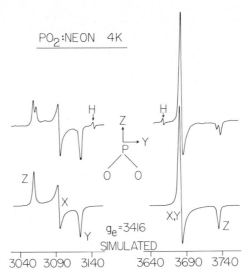

Fig. 7. The experimental (upper trace) and simulated ESR spectra of the PO_2 radical are shown. The large $^{31}P(I = \frac{1}{2})$ doublet hfs is resolved into three g-components for this non-linear case. Background H atom ESR signals are indicated.

phosphorus analog of the thoroughly studied CN radical (Weltner 1983) is being made in interstellar space and the availability of the ^{31}P hfs should allow a definitive assignment to be made.

The technique we have employed to produce CP for MI-ESR is quite straightforward and should be applicable to numerous types of condensed-phase reactions for radical generation purposes. Carbon powder and red phosphorus were thoroughly mixed and pressed under 8000 psi into a pellet about 1 cm in diameter to form the laser target for RLV. Surprisingly, the $CP(X^2\Sigma)$ ESR signals were more intense on some experiments than the ^{31}P atom (^4S) lines (see fig. 4). The powder pattern ESR spectrum assigned to CP was confirmed by repeating the above procedure with 40 mg of 99% ^{13}C powder mixed with 100 mg of red phosphorus. The convenience and feasibility of utilizing such small sample sizes for several deposition experiments is a distinct advantage of this generation procedure especially when expensive isotopes are involved.

A complete determination of the ^{13}C and ^{31}P A-tensors and the molecular g-tensor was possible in this study which will be compared to ab initio calculations in a forthcoming report. Based on a simple FACM analysis, the A-tensors reveal SOMO compositions on ^{13}C of 0.13 (2s); 0.49 ($2p_z$) and for ^{31}P, -0.01 (3s) and 0.38 ($3p_z$). These results compare with CN in the following manner: 0.19 (2s); 0.48 ($2p_z$) for ^{13}C and -0.008 (2s); 0.32 ($2p_z$) for ^{14}N (Weltner 1983).

3.3. Thermal vaporization

For nearly 25 years, conventional high-temperature (~ 1000 K) effusion ovens for vaporizing materials have been used in numerous rare-gas matrix studies. The vapor species in equilibrium with the condensed phase is trapped and high-temperature mass spectral data is usually available to assist in the ESR assignment if decomposition is thought to occur. Molecular radicals utilizing this technique include HSiO (Van Zee et al. 1985), ScF_2 (Knight and Wise 1979) and the high-spin oxides VO (Kasai 1968) and GdO (Van Zee et al. 1981).

A non-equilibrium high-temperature method was employed in the generation of the dihydride metal radicals, ScH_2 and YH_2 (Knight et al. 1981). In this arrangement, solid-metal hydride powder was slowly "dripped" into an open ended tantalum tube heated to 2000 K. A small effusion hole in the side of the tantalum tube was directed towards the matrix-deposition target. The small particles of the metal hydride powder were rapidly heated and vaporized in a non-equilibrium manner upon contacting the heated cell. This process resulted in the formation of MH_2 radicals which were trapped in argon matrices for ESR characterization. The conventional equilibrium heating of the metal hydride solid produces only metal-atom vapor and hydrogen, not the MH_2 molecules. Evidence was presented in this investigation to show that the MH_2 molecules formed under the non-equilibrium ("dripper experiments") conditions did not result simply from the co-deposition of metal and hydrogen atoms during the matrix-condensation process.

3.4. Co-deposition techniques

There are many versions of the co-deposition approach which date back to the very beginning of the matrix technique in several spectroscopic areas (Harvey and Brown 1959, Andrews and Pimentel 1966). The procedure is particularly appropriate for highly reactive radicals and in those cases where reaction needs to be abruptly stopped at some intermediate stage. The thermal deposition of metal atoms often produces metal clusters and many studies have been conducted to control the distribution of the various cluster sizes generated. Recent examples include Li_7 (Garland and Lindsay 1984), Si_2 (Van Zee et al. 1985), Sc_3 and Sc_{13} (Knight et al. 1983b) and the mixed-metal radicals AgMg (Kasai and McLeod 1975) and CrCu (Baumann et al. 1983). Many other ESR rare-gas matrix studies of paramagnetic metal molecules can be found in a recent review (Weltner and Van Zee 1984).

Another widely applied version is the matrix co-deposition of metal atoms with a reactant gas to form adducts or metal insertion products. If reaction does not occur spontaneously during deposition, photolysis is often employed. Examples include Cu, Ag, Au complexes with O_2 (Kasai and Jones 1986), $Ga(CO)_2$ (Kasai and Jones 1985), several metal hydride radicals such as MnH

(Van Zee et al. 1986a) and BeH (Knight et al. 1972), metal insertion products HAlCH$_3$ (Parnis and Ozin 1986), HAlOH, HScOH and HYOH (Knight et al. 1987b) and the charge-transfer type radicals M$^+$HX$^-$ (Davies and Lindsay 1985, Lindsay et al. 1982).

3.5. Photolysis and γ-irradiation

γ-irradiation from a ^{60}Co source has been used in numerous MI-ESR studies particularly with the heavier rare gases. The irradiation process is usually conducted after the deposition of some volatile precursor into the solid matrix and apparently no applications have been reported with non-volatile materials which would have to be vaporized by one of the generation methods discussed in this summary. Examples of radicals produced by γ-irradiation in rare gases include KrMn(CO)$_5$ (Fairhurst et al. 1984a), V(CO)$_4$ (Morton and Preston 1984b), BF$_2$ (Nelson and Gordy 1969) and several atomic radicals (Jackel et al. 1968).

Photolysis with visible or ultraviolet sources is commonly used in matrix studies and three ESR samples are cited as illustrations: the reaction between HI and Ni(CO)$_4$ to produce HNi(CO)$_3$ (Morton and Preston 1984a), the generation of HC^{17}O from HI and C^{17}O (Adrian et al. 1985) and XeCl (Adrian and Bowers 1976). The HNi(CO)$_3$ radical was found to have C$_{3v}$ symmetry in its ^{2}A ground state.

3.6. Cathode sputtering; FAB and discharge sources

Hollow cathode sputtering has been used to generate MoN(X$^4\Sigma$) radicals for MI-ESR by passing N$_2$(g) over a Mo cathode in a DC arc environment (Knight and Steadman 1982a). This method should be appropriate for producing other metal nitrides although reactive laser sputtering would probably be more convenient. The sputter method was first developed for electronic and vibrational studies of matrix-isolated species by others (Green and Gruen 1972, Carstens et al. 1972, Schock and Kay 1973). Fast atom neutral beams (FAB) in the 1–10 keV energy range have been adapted for use with MI-ESR experiments for vaporizing metal atoms and for initiating metal–gas reactions. The CdOH radical was characterized by ESR for the first time by this method, and other metal atoms were also generated and trapped with a FAB source (Knight et al. 1984a). Microwave and RF discharge sources were used in some of the first ESR matrix experiments and in more recent investigations, the C$_6$ triplet molecule has been generated by introducing C$_6$H$_2$ into a microwave discharge (Van Zee et al. 1987), although these energetic conditions usually cause the dissociation of most molecules.

4. Generation methods for ion radicals

The various methods used to generate charged radicals for MI-ESR along with representative examples are listed in table 2. The examples and following discussion include only "free" or isolated ions in rare-gas matrices and not ionic molecules or chemically bonded ion pairs in the same matrix cage. The highly reactive nature and short lifetimes of ion radicals coupled with generation and isolation difficulties have presented major obstacles to overcome in applying MI-ESR to the study of cation radicals. Open-shell cations are especially difficult to study in the gas phase by high-resolution microwave spectroscopy or LMR, and nuclear hyperfine interactions which can provide much detailed electronic structure information have been determined for only about ten cation radicals, although rapid progress is being made in this important area. It has been demonstrated for $^{13}CO^+$ and H_2O^+ (Knight 1986) that neon-matrix perturbation of the nuclear hyperfine interaction is quite small (2–3%). The contraction of the electron distribution that occurs for cations relative to anion radicals probably makes them less susceptible to matrix perturbations. Neon-matrix ESR spectra of ions are simpler to analyze than comparable gas-phase spectra since at 4 K all transitions originate from ground-state levels. Obtaining sufficient numbers of ions for detection in matrices is also more readily achieved than in the gas phase since deposition times can be extended if necessary.

ESR rare-gas matrix experiments were originally conducted on several atomic ions and a few molecular anions which were produced by in-situ visible light photolysis of neutral precursors (Kasai 1971). Our emphasis has been the development of new generation methods which will enable the direct deposit of any type of molecular cation radical for detailed ESR study and comparison with theoretical calculations.

The methods discussed will emphasize non-volatile inorganic cation radicals since so few have been studied in rare-gas media, however, the ability to trap small organic cations in neon has also been demonstrated. Unsuccessful attempts to trap such small ions as CH_4^+ and H_2CO^+ in freon matrices have been reported (Symons et al. 1983, Shida et al. 1984). Even argon matrices can perturb ions having EAs above 10–11 eV. Neon's extreme inertness, low polarizability and large conduction-band gap make it an ideal host for isolating highly reactive cation radicals. The narrower linewidths and lack of matrix hyperfine interaction in neon usually allow a complete determination of the A- and g-tensors even for powder spectra with the help of isotopic substitution and computer simulated spectra. The limited temperature range (2–12 K) for neon is a disadvantage since motional averaging and other temperature-dependent effects cannot usually be observed. The necessity of 4 K operation and high-vacuum conditions cause neon-matrix studies to be more time consuming and expensive than cation radical studies in other condensed phases such as freons or the heavier rare gases.

4.1. Ion trapping process

The detailed mechanism for the ion generation and trapping process is not well understood for the methods described above. However, all of the evidence obtained with these techniques strongly supports the model of well-isolated cations located in the rigid neon lattice some distance away from the counter anion which must be present to maintain electroneutrality. We have found that the intensity of the cation ESR signal is not significantly affected by the intentional introduction of an electron acceptor such as Cl_2 or HI. Ionization events which are successful in producing matrix-stabilized ions are most likely to occur right at the matrix surface and the liberated electrons probably are thermalized by neon collisions before being captured by a background molecule such as O_2, N_2 or OH. Practically all anions that would be formed from background impurities are either not radicals or do not have the proper electronic ground state to be observed by matrix ESR. To demonstrate that cation and anion radicals are both present, we have doped the matrix gas with trace amounts of an electron acceptor species such as $F_{2(g)}$ whose anion can be detected (Knight et al. 1986a). In such neon matrices, the cation radicals (H_2O^+, CO^+, N_2^+, etc.) and F_2^- ($X^2\Sigma$) produce intense and narrow ESR absorptions with linewidths in the range of 0.3–1.0 G. Furthermore, the magnetic parameters and all spectral characteristics for cations in these matrices are identical to those obtained in matrices not containing $F_{2(g)}$. This observation plus the total lack of ^{19}F hyperfine structure on the cation absorptions strengthen the arguments for classifying such radicals as free or isolated ions with virtually no perturbation by the counter anion.

The photobleaching effect is extremely important in discriminating between ESR lines belonging to neutral and ion radicals especially in making assignments for new radicals produced under such extremely energetic conditions where numerous impurity species are likely to be present. In optimizing conditions and methods for ion generation, the photobleaching effect provides extremely useful information even before the ion's identity can be established. In the F_2^- investigation, the ESR signal intensity of H_2O^+, N_2^+, CH_3, H atoms and F_2^- was monitored as a function of time that the matrix was photolyzed with visible light. The ESR signals of all charged radicals decreased at approximately the same rate while no intensity changes were observed for the neutral radicals. Typically, 30 min of visible light photolysis of a deposited matrix sample is sufficient to reduce the ESR ion intensity by ∼75%. Visible photons are not sufficiently energetic to break most bonds but they can photo-ionize the anion traps whose ionization energies are in the 1–3 eV range. The liberated electrons migrate throughout the solid rare-gas lattice and neutralize cation radicals in a nonselective manner. The only neutral radical we have observed that is susceptible to photobleaching is the formyl radical (HCO).

4.2. Photo-ionization

The matrix apparatus we have used to produce isolated ions by photo-ionization is similar to the one shown in fig. 1 except the electron gun is replaced with a neon resonance lamp whose output at 16.8 eV is sufficient to photo-ionize practically all neutral molecules during the neon matrix deposition process (Knight and Steadman 1982b). The resonance lamp is a 9 mm o.d. quartz tube whose open end is located 5 cm from the deposition target. The lamp is carefully tuned to the Evenson microwave cavity so that the active part of the discharge extends to the tube's open end; the lamp is typically operated at 50 W forward and 2 W reflected power with a continuous neon flow of 3 sccm. The lamp is mounted above the matrix-deposition surface (off axis at 45°) so that the laser target and high-temperature oven or other components used to produce the neutral precursors can be located on-line and close to the matrix-deposition target. In addition to the gas flow through the resonance lamp, neon-matrix gas input of 5 sccm is used for the standard 30 min deposit. The ions are generated during the deposition process and using the crude estimate that only about 0.1% of the neutrals are ionized and trapped, we estimate that the ratio of cation radicals to neon is approximately $1:10^6$. Neon resonance lamp irradiation of a matrix sample after deposition of the neutral precursor is completed does not usually produce cation radicals.

4.2.1. The acetaldehyde radical cation, CH_3CHO^+

The neutrals to be photo-ionized can be produced by a variety of methods as shown by the examples in table 2. The simplest type is the introduction of a volatile gas through a separate inlet tube directed at the same spot on the matrix target as that of the neon resonance lamp output. The neon matrix ESR spectrum of the acetaldehyde (CH_3CHO^+) cation produced by this procedure is shown in fig. 8. This radical cation has been the subject of a detailed comparison between freon ($CFCl_3$) and neon-matrix ESR results (Knight et al. 1988a). Several combinations of magnetic isotopes have been analyzed including CD_3CHO^+, CH_3CDO^+, $^{13}CH_3CHO^+$ and $CH_3^{13}CHO^+$ and the neon absorptions are sufficiently well resolved to allow a complete determination of the various A- and g-tensors. As seen in the spectrum in fig. 8, two strong g-components are detected for field directions in the matrix surface plane (YZ) while another set of weaker line components becomes strongly enhanced when the external magnetic field is applied normal to this plane – the "X" lines at $\theta = 0°$ position. The cations are therefore aligned with their CCHO planes parallel to the deposition surface. This is the first charged radical to exhibit strong preferential orientation effects which have been observed for several neutral molecules in rare-gas hosts (Weltner 1983). The largest principal value of the g-tensor is found to be perpendicular to this CCHO plane as previously determined for the formaldehyde cation (H_2CO^+) (Knight and Steadman

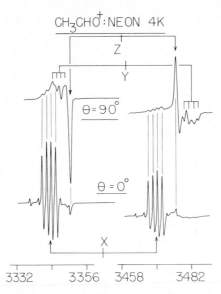

Fig. 8. The preferential orientation characteristics for CH_3CHO^+ are shown in these neon matrix ESR spectra. The large aldehydic H doublet hfs (~ 120 G) is resolved into g-components which show additional CH_3 splitting. θ is the angle between a line normal to the flat trapping surface and the external magnetic field. (From Knight et al. 1987d.)

1984). The large doublet hfs (~ 360 MHz) results from the aldehydic hydrogen with additional hyperfine interaction of 4.3(1) MHz being resolved for the CH_3 protons. The deviation from a $1:3:3:1$ intensity ratio for the CH_3 quartet hfs results from some form of restricted rotational motion, possibly via tunneling. The measured values of the g- and A-tensors are consistent with the assignment of the unpaired electron to the nonbonding oxygen-centered $10a'$ molecular orbital which is about 50% delocalized onto the aldehydic hydrogen (1s) and methyl carbon (2p) orbitals. These neon results for CH_3CHO^+ are simpler to interpret than earlier freon studies which showed hyperfine coupling to a chlorine atom of the matrix host (Symons and Boon 1982, Snow and Williams 1983). Also, the neon linewidths are ~ 20 times narrower thus allowing a complete resolution of the g- and A-tensors which was not possible in freon.

4.2.2. The gallium arsenide cation, $GaAs^+$

Laser vaporization of a gallium arsenide crystal in combination with photo-ionization has recently been conducted in our laboratory to produce the $GaAs^+$ radical in a neon matrix. Theoretical calculations predict a $^4\Sigma$ ground state for $GaAs^+$, although no previous spectroscopic measurements have been reported (Balasubramanian 1987). Mass spectrometric monitoring of the vapor species generated from laser sputtering $GaAs_{(s)}$ does indicate the presence of a small amount of $GaAs_{(g)}$ although the predominant species are Ga^+, Ga and As_2

(Mohlmann and Kuzmin 1986). Laser vaporization of GaAs has also been used in molecular-beam cluster studies (Liu et al. 1986). A $^4\Sigma$ state can be readily confirmed by searching in the g_e magnetic field region for cases with relatively small D-tensors (zero-field parameters) or for the perpendicular transitions in the $g = 4$ region for those radicals with large D-values. We have observed the $\Delta M_s = 1$ transition ($\frac{1}{2} \leftarrow -\frac{1}{2}$) near $g = 4$ which shows quite clearly the two quartet hyperfine patterns expected for the two gallium isotopes, each with $I = \frac{3}{2}$. The ESR results are consistent with an electronic structure description approximating $Ga^+\cdots As$, with the 3 unpaired electrons primarily occupying the 3p orbitals on As split by the axial field of Ga^+. The bond energy for $GaAs^+$ is estimated to be only 0.25 eV.

4.2.3. AlF^+; AlH^+; BF^+

These $^2\Sigma$ cation radicals were generated by combining the reactive laser vaporization technique previously described with photo-ionization during matrix deposition. For example, in the case of AlH^+, $H_{2(g)}$ was passed over aluminum metal undergoing laser vaporization (Knight et al. 1987a) (see fig. 2). The laser sputtering process alone produced weak AlH^+ ESR signals which increased five-fold when an argon resonance lamp was used with an argon matrix. Two of the six $Al(I = \frac{5}{2})$ hyperfine transitions in the ESR spectrum of AlD^+ are shown in fig. 9 where $D(I = 1)$ triplet hfs is clearly evident for both the parallel and perpendicular directions in the radical ion. The ESR spectrum previously assigned to AlH^+ is actually that of the divalent neutral aluminum radical AlHOH whose charge distribution resembles AlH^+OH^- with 90% of the

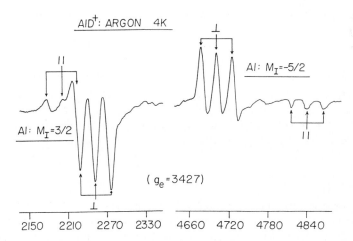

Fig. 9. Reactive laser vaporization combined with photo-ionization was employed to generate this argon-matrix ESR spectrum of AlD^+ which consists of a widely spaced $Al(I = \frac{5}{2})$ sextet, each member of which is further split into $D(I = 1)$ triplets. Two of these Al hf components are shown for both the parallel and perpendicular directions. Analogous ESR spectra for AlH^+ were also obtained.

unpaired spin on the AlH⁺ part (Knight et al. 1979). This revised assignment has been confirmed by ^{17}O and D isotopic substitution utilizing both high-temperature and laser-sputtering co-deposition reactions between $Al_{(g)}$ and $H_2O_{(g)}$ in neon and argon matrices.

These latest ESR results for AlH⁺ and AlD⁺ show excellent agreement with ab initio CI calculations of the aluminum and hydrogen A-tensors. The molecular orbital of the unpaired electron contains about 0.35 (3s) and 0.35 ($3p_z$) on Al and 0.31 (1s) on H which are distinctly different from the odd-electron distribution in AlHOH. These results can be compared to the isoelectronic neutral radical, MgH, which has 0.45 (3s) and 0.30 ($3p_z$) on Mg and 0.21 (1s) H. An interesting isotope effect was observed in that the Al A_{iso} parameter is about 1% larger in AlD⁺ relative to AlH⁺. This was accounted for by the electronic structure of the cation and the slightly shorter bond distance in AlD⁺.

The matrix ESR results for AlF⁺ and BF⁺ (Knight et al. 1986b,c) have been presented previously and compared to the isoelectronic neutral radicals AlO and BO, respectively. Figure 10 shows two of the four transitions for ^{11}B($I = \frac{3}{2}$) in ^{11}BF⁺. Each is split into anisotropic ^{19}F($I = \frac{1}{2}$) doublets with parallel and

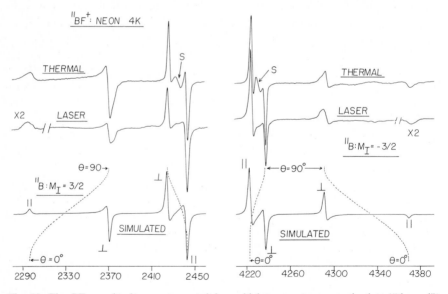

Fig. 10. The $BF_{(g)}$ molecule was generated by a high-temperature vaporization ("thermal") technique and by reactive laser vaporization ("laser") involving the passage of $F_{2(g)}$ over $B_{(s)}$. The cation radical, ^{11}BF⁺, was produced by photo-ionization during deposition into neon at 4 K. The highly anisotropic ^{19}F hfs for two of the four ^{11}B($I = \frac{3}{2}$) lines is shown which is compared to simulated ESR spectra in the bottom trace. The dashed lines give the H_{res} versus θ dependence for the various transitions. The weaker features in the thermal spectra labeled "S" result from BF⁺ being trapped in an alternate site in neon.

perpendicular features indicated on the spectra. The $^{11}BF^+$ spectra marked "thermal" was produced by the photo-ionization of $BF_{(g)}$ obtained from the high-temperature vaporization reactions $CaF_{2(g)} + B_{(s)}$ at 1600 K; the $BF_{(g)}$ source for the spectra marked "laser" was from reactive laser vaporization. The well-resolved powder spectra are nearly identical, except for the multiple trapping site labeled "S" in the thermal spectrum. The simulated spectrum (bottom trace) and θ versus H_{res} plots (dashed lines) obtained from an exact diagonalization program show excellent agreement with the two independent experimental results. Note the increased peak height intensity when the associated parallel and perpendicular transitions occur in the same magnetic field vicinity and the much weaker intensity when these components are well separated. Detection of weak parallel lines can be the most difficult aspect of MI-ESR experiments, especially when significant anisotropy is present.

4.2.4. CO^+ and SiO^+

Another example of high-temperature vaporization/photo-ionization for producing a non-volatile cation is SiO^+ (Knight et al. 1985). $SiO_{(g)}$ was produced by vaporizing both $SiO_{(s)}$ and $SiO_{2(s)}$ in the 1500–1900 K range from a tantalum Knudsen cell. This radical is especially interesting since magnetic parameters can be compared with isovalent CO^+ to probe the electronic changes that occur as carbon is replaced by silicon in a "simple" diatomic molecule. The results of a detailed experimental and theoretical study of the $^{29}Si(I = \frac{1}{2})$ and $^{13}C(I = \frac{1}{2})$ A-tensors in these two cation radicals have been presented. Striking differences were noted in the composition of the 7σ MO containing the unpaired electron when free-atom comparison method (FACM) populations were compared with a Mulliken spin population analysis (MSPA) conducted on CI wavefunctions which yielded A_{iso} and A_{dip} values in good agreement with experiment. The FACM results for the carbon populations in $^{13}CO^+$ of 0.42:0.43 for $2s/2p_z$ show reasonable agreement with the MSPA values of 0.51:0.41. However, for $^{29}SiO^+$, the Si $3s:3p_z$ results are 0.17:0.56 and 0.30:0.27 for FACM and MSPA, respectively.

These large differences in the two methods, especially the predicted $s:p$ ratios, have been accounted for by monitoring the effects of oxygen valence orbital spin density overlap with the core orbitals of silicon (referred to simply as core–other-valence overlap). Going from CO^+ to SiO^+, the MSPA results show a large spin density increase in the O $2p_z$. Unfortunately, this has not been confirmed experimentally given the difficulty of producing $Si^{17}O^+$ by high-temperature methods, although the reactive laser vaporization approach might allow this isotope to be studied. The core–other-valence overlap effect makes a large negative contribution to A_{iso} and a large positive contribution to A_{dip}, thus the uncorrected FACM procedure would underestimate the 3s and overestimate the $3p_z$ spin density as the above data indicate. In other words, all of the anisotropic hfs on ^{29}Si should not be attributed solely to Si $3p_z$ spin density.

These findings indicate that FACM results should be interpreted very cautiously when an adjacent atom has significant spin density in an orbital that overlaps strongly with the magnetic nucleus whose hfs is being analyzed, especially if that atom has core p levels. In the case of CO^+, no core p levels are present and the spin density in the O $2p_z$ is probably much less relative to SiO^+. As discussed in this theoretical study, the core–other-valence overlap effect is not a spin polarization phenomenon.

4.3. Electron bombardment

The electron bombardment (EB) method for generating ion radicals serves as a convenient and independent confirmation technique for photo-ionization studies. Most of the molecular cation and anion radicals generated by photo-ionization in our group have also been trapped in rare-gas matrices with the EB method. It also provides more experimental control than discharge resonance lamps since the electron flux and energy can be varied independently over wide ranges. Figure 1 shows the location of the electron gun in the matrix apparatus and fig. 11 gives a more detailed description of the simple circuitry involved. The Heli-Tran cryostat is completely isolated from ground in order to maintain the copper matrix-deposition target at some positive potential relative to the electron gun filament, a tungsten wire (0.005 in.) which requires about 2.4 A for sufficient thermionic emission. The distance between the control grid used to accelerate electrons to the desired energy and the matrix target is about 5 cm. The electron flow to the target is maintained at 50 μA for a typical 30 min deposit with the matrix and other gas flows similar to those used in the photo-ionization experiment. Neutrals to be ionized by EB can be generated from the variety of methods previously discussed.

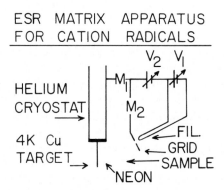

Fig. 11. Electrical circuitry associated with the ESR matrix apparatus shown in fig. 1 for conducting electron bombardment matrix-isolation experiments. M_1 and M_2 are current meters. (From Knight et al. 1983a.)

4.3.1. N_2^+

For a given sample, the radicals actually trapped in neon show an interesting dependence on the electron energy that is similar to energy-dependent mass spectrometric cracking patterns. This has been demonstrated with several molecules including the monitoring of the ESR intensity ratio of neutral N atoms to N_2^+ ions as a function of the voltage applied to the control grid (Knight et al. 1983a). An apparent negative surface charge has also been observed in a series of depositions where the ESR intensity of $^{13}CO^+$ signals is plotted against the applied electron voltage. The ionization energy of CO^+ is 14 eV, however, its ESR intensity extrapolates to zero for applied potentials in the 40–50 V range. Apparently, the condensing gases on the copper target form a partially insulating film which causes a surface-charge accumulation. This surface charge must be overcome for electrons to produce cations in a region sufficiently close to the deposition surface. It is possible that the only cations trapped in a stable lattice site are those formed from the ionization of neutrals trapped near the surface.

The first application of EB for MI-ESR experiments was the generation of the nitrogen cation radicals, $^{14}N_2^+$ and $^{15}N_2^+$, whose ESR spectra and hyperfine interactions had not been previously determined. The magnetic parameters for N_2^+ and its electronic structure have been compared with several other isoelectric 13 electron radicals such as BeF, CN, BO and CO^+. In contrast to these other radicals, the unpaired electron in N_2^+ occupies predominantly the $2p_z$ orbital and has a surprisingly low s/p ratio of $\sim 1/7$ compared to a 1/1 ratio for CO^+.

4.3.2. The HBBH triplet radical

Matrix ESR experiments designed to produce BH^+ by the electron bombardment of $B_2H_{6(g)}$ have thus far been unsuccessful. However, the previously unreported HBBH radical was accidentally generated which can be assigned as $^3\Sigma$ based on a very clear ESR spectrum shown in fig. 12. Since the zero-field parameter is small ($D \sim 0.1$ cm^{-1}), all of the expected features for a powder-pattern axially symmetric triplet state could be observed. Each of the two perpendicular fine structure lines labeled XY_1 and XY_2 are split into 1:2:1 triplets by the two equivalent hydrogens. Each component of each triplet is split into a septet by two equivalent $^{11}B(I = \frac{3}{2})$ atoms in HBBH. The lower field (XY_1) triplet of septets is shown in fig. 13. The ^{11}B hfs is quite small since the unpaired electrons occupy 2p orbitals with virtually no 2s character. This HBBH radical is also produced by photo-ionization and fast atom bombardment of diborane. In these experiments the BH_2 radical has also been observed for the first time. The bonding and electronic structure of these unusual radicals are especially interesting and will be presented in detail in a future report.

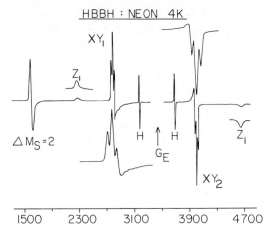

Fig. 12. The neon-matrix ESR spectrum of the HBBH ($^3\Sigma$) radical is presented which shows triplet H hfs for each of the perpendicular fine structure features labeled XY_1 and XY_2. The matrix sample was produced by the electron bombardment of $B_2H_{6(g)}$ during deposition. See fig. 13 for resolution of ^{11}B hfs.

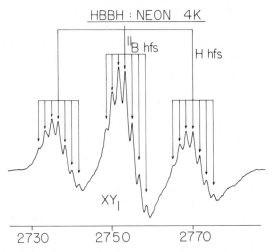

Fig. 13. The high-resolution ESR spectrum of the XY_1 transition (see fig. 12) for the HBBH triplet radical shows each component of the H triplet split further into septets resulting from two equivalent $^{11}B(I = \frac{3}{2})$ nuclei.

4.4. Ion–neutral-product radicals; $C_2O_2^+$, N_4^+, N_2CO^+ and O_4^+

One of the most interesting applications of these matrix-ion methods is the study of ion–neutral reactions which are extremely difficult to characterize spectroscopically in the gas phase. By carefully controlling the concentration of neutrals deposited into neon matrices under ionizing conditions, we have been able to trap several ion–neutral-product radicals not previously reported. For example, the $C_2O_2^+$ radical species was trapped by depositing a CO : neon ratio of 1 : 200 (Knight et al. 1984b). Apparently, deposition at these higher concentrations allows CO^+ and CO to be trapped in the same matrix site thus stabilizing the $C_2O_2^+$ molecule which has a C–C bond energy of 22 kcal mol^{-1} (Linn et al. 1981). An alternative mechanism might be the formation of the $(CO)_2$ Van der Waals molecule which is ionized during the deposition process, although it would seem that pressures are too low for this to occur. For more dilute matrices (1 : 800), the only radical detected is CO^+.

Six different isotopic combinations of $C_2O_2^+$ involving ^{13}C and ^{17}O magnetic isotopes were used to confirm the spectral assignment. The electronic ground state was determined by experiment and CI calculations to be the planar trans-configuration with a CCO bond angle of 141°. The unpaired electron is delocalized over the entire radical with an occupancy of 0.19 (2s) and 0.11 ($2p_y$) on each carbon atom and 0.17 on the $2p_y$ of each oxygen, where Y is the direction perpendicular to the C–O axis. The matrix ESR data allowed the principal g- and A-tensors to be determined even though these tensors are not coincident.

The trapping of $^{14}N_4^+$ and $^{15}N_4^+$ from the ion–neutral co-deposition reaction $N_2^+ + N_2 \rightarrow N_4^+$ has been studied by MI-ESR and compared with CI theoretical calculations. The only previous spectroscopic information for N_4^+, which is potentially important in upper atmospheric chemistry, is from photodissociation (PD) spectra which are broad and structureless in both the visible and ultraviolet (Smith and Lee 1978, Ostrander and Weisshaar 1986). The isotropic and dipolar nuclear hyperfine interactions for both the inner and outer nitrogen atoms were determined. These data indicate that the unpaired electron in this linear $^2\Sigma$ radical resides primarily on the inner N atoms with significant $2p_z$ and 2s character. Theoretical calculations show that the SOMO is anti-bonding and the bond length between the inner nitrogens is approximately 1.99 Å. The N_4^+ radical is a good example of a dimer cation whose bond energy lies between that of a Van der Waals complex and a normal single bond.

The cation radicals N_2CO^+ (Knight et al. 1988d) and O_4^+ have also been generated by this ion–neutral co-deposition approach. Based on MI-ESR results both are non-linear and the ground state of O_4^+ was found to be a quartet in agreement with earlier theoretical calculations (Conway 1969). Preliminary evidence obtained utilizing ^{17}O substitution in these O_4^+ experiments indicates the presence of O_4^- which appears to have a doublet ground electronic state. A full account of these systems will be presented in a future report. The neon-matrix ESR spectra of $^{13}C_2O_2^+$ and $^{15}N_2CO^+$ are shown in fig. 14.

4.5. γ-irradiation ion production

Several metal carbonyl neutral molecules have been dissolved in krypton matrices and subjected to γ-irradiation from a ^{60}Co source. Several types of ions

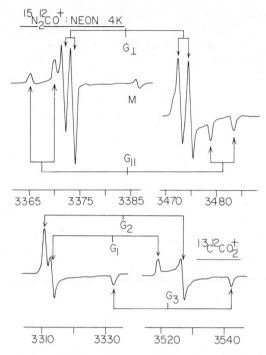

Fig. 14. ESR spectra for the molecular cation radicals $^{15}N_2CO^+$ and $^{12,13}C_2O_2^+$. These unusual molecules were generated by ion–neutral reactions ($CO^+ + N_2$ and $CO^+ + CO$) during the neon-matrix deposition process. Large $^{15}N(I = \frac{1}{2})$ and $^{13}C(I = \frac{1}{2})$ nuclei hfi is readily apparent as well as the inequivalency of the ^{15}N atoms in $^{15}N_2CO^+$. The lowest field methyl quartet line is denoted "M" in the upper spectrum. The magnetic field position of g_e is 3422 G. (From Knight 1986.)

have been produced in this manner and studied by spin resonance, including: $Cr(CO)_4^+$ (Fairhurst et al. 1984b), $KrFe(CO)_5^+$ (Fairhurst et al. 1984a), and the cluster ions $Co_2(CO)_8^-$ (Fairhurst et al. 1983) and several binuclear nickel carbonyls (Morton and Preston 1985). The ground state of $Cr(CO)_4^+$ is $^6A_1$ and ^{13}C hfs was used to assign the geometry as T_d and determine the spin density in each carbon 2p orbital (0.08). More experimental work is required to determine if γ-irradiation of samples deposited into rare-gas matrices could be used as a general ion production method.

4.6. *Trapping ions from cation beams*

The matrix isolation of ion radicals trapped directly from ion beams would enable the mass selection of the trapped species and greatly simplify the ESR assignment procedure. We have found this to be a very difficult experiment, but limited success has been achieved using a quadrupole mass filter and a standard EI ion source. The radical cations H_2O^+ and CO^+ have been selectively trapped in neon by this approach, although the signals were quite weak. If this method could be made to function more efficiently, the number of more complicated radical ions that could be studied would be enormous! The numerous advances made in the generation of ions for mass spectrometric studies could then be combined with matrix-isolation ESR spectroscopy.

Acknowledgments

Financial support from the National Science Foundation (CHE–8508085), Dreyfus Foundation, Pew Memorial Trust, Research Corporation and the Amoco Foundation is gratefully acknowledged. Duke Endowment and General Electric grants to Furman University also provided valuable support. I am indebted to the following student colleagues for their contributions to these matrix ESR studies: Jim Bostick, Dwayne Bowman, Edward Baldwin, Cheryl Cleveland, John Cleveland, Daryl Cobranchi, Scott Cobranchi, Jim Derrick, Edward Earl, Thomas Fisher, Brian Gregory, Karl Johannessen, Glenn Jones, Andrew Ligon, Patricia Miller, Jeff Petty, Jhobe Steadman, Michael Winiski, Marcus Wise and Robert Woodward.

References

Adrian, F.J., and V.A. Bowers, 1976, J. Chem. Phys. **65**, 4316.
Adrian, F.J., J. Bohandy and B.F. Kim, 1984, J. Chem. Phys. **81**, 3805.
Adrian, F.J., B.F. Kim and J. Bohandy, 1985, J. Chem. Phys. **82**, 1804.
Andrews, L., and G.C. Pimentel, 1966, J. Chem. Phys. **44**, 2361.
Andrews, M.P., M.M. Saba and G.A. Ozin, 1986, J. Phys. Chem. **90**, 744.
Balasubramanian, K., 1987, J. Chem. Phys. **86**, 3410.
Baumann, C.A., R.J. Van Zee and W. Weltner Jr, 1983, J. Chem. Phys. **79**, 5272.
Carstens, D.H.W., J.F. Kozlowski and D.M. Gruen, 1972, High Temp. Sci. **4**, 301.
Cirelli, G., A. Russu, R. Wolf, M. Rudin, A. Schweiger and Hs.H. Günthard, 1982, Chem. Phys. Lett. **92**, 223.
Conway, D.C., 1969, J. Chem. Phys. **50**, 3864.
Davies, M.A., and D.M. Lindsay, 1985, Surf. Sci. **156**, 335.
Fairhurst, S.A., J.R. Morton and K.F. Preston, 1983, Organometallics **2**, 1869.
Fairhurst, S.A., J.R. Morton, R.N. Perutz and K.F. Preston, 1984a, Organometallics **3**, 1389.
Fairhurst, S.A., J.R. Morton and K.F. Preston, 1984b, Chem. Phys. Lett. **104**, 112.
Ferrante, R.F., S.L. Erickson and B.M. Peek, 1987, J. Chem. Phys. **87**, 2421.
Garland, D.A., and D.M. Lindsay, 1984, J. Chem. Phys. **80**, 4761.
Green, D.W., and D.M. Gruen, 1972, J. Chem. Phys. **57**, 4462.
Harvey, K.B., and H.W. Brown, 1959, J. Chim. Phys. **56**, 745.
Heimbrook, L.A., M. Rasanen and V.E. Bondybey, 1987, J. Phys. Chem. **91**, 2468.
Howard, J.A., and B. Mile, 1987, Acc. Chem. Res., in press.
Howard, J.A., K.F. Preston, R. Sutcliffe and B. Mile, 1983, J. Phys. Chem. **87**, 536.
Jackel, G.S., W.H. Nelson and W. Gordy, 1968, Phys. Rev. **176**, 453.
Jen, C.K., S.N. Foner, E.L. Cochran and V.A. Bowers, 1958, Phys. Rev. **112**, 1169.

Kasai, P.H., 1968, J. Chem. Phys. **49**, 4979.

Kasai, P.H., 1971, Acc. Chem. Res. **4**, 329.

Kasai, P.H., and P.M. Jones, 1985, J. Phys. Chem. **89**, 2019.

Kasai, P.H., and P.M. Jones, 1986, J. Phys. Chem. **90**, 4239.

Kasai, P.H., and D. McLeod Jr, 1975, J. Phys. Chem. **79**, 2324; 2325.

Kasai, P.H., E.B. Whipple and W. Weltner Jr, 1966, J. Chem. Phys. **44**, 2581.

Kato, T., and T. Shida, 1979, J. Am. Chem. Soc. **101**, 6869.

Kawaguchi, K., S. Saito and E. Hirota, 1985, J. Chem. Phys. **82**, 4893.

Knight Jr, L.B., 1986, Acc. Chem. Res. **19**, 313.

Knight Jr, L.B., 1989, to be published.

Knight Jr, L.B., and J. Steadman, 1982a, J. Chem. Phys. **76**, 3378.

Knight Jr, L.B., and J. Steadman, 1982b, J. Chem. Phys. **77**, 1750.

Knight Jr, L.B., and J. Steadman, 1983, J. Chem. Phys. **78**, 5940.

Knight Jr, L.B., and J. Steadman, 1984, J. Chem. Phys. **80**, 1018.

Knight Jr, L.B., and M.B. Wise, 1979, J. Chem. Phys. **71**, 1578.

Knight Jr, L.B., J.M. Brom and W. Weltner Jr, 1972, J. Chem. Phys. **56**, 1152.

Knight Jr, L.B., R.L. Martin and E.R. Davidson, 1979, J. Chem. Phys. **71**, 3391.

Knight Jr, L.B., M.B. Wise, T.A. Fisher and J. Steadman, 1981, J. Chem. Phys. **74**, 6636.

Knight Jr, L.B., J.M. Bostick, R.J. Woodward and J. Steadman, 1983a, J. Chem. Phys. **78**, 6415.

Knight Jr, L.B., R.J. Woodward, R.J. Van Zee and W. Weltner Jr, 1983b, J. Chem. Phys. **79**, 5820.

Knight Jr, L.B., P.K. Miller and J. Steadman, 1984a, J. Chem. Phys. **80**, 4587.

Knight Jr, L.B., J. Steadman, P.K. Miller, D.E. Bowman, E.R. Davidson and D. Feller, 1984b, J. Chem. Phys. **80**, 4593.

Knight Jr, L.B., J. Steadman, D. Feller and E.R. Davidson, 1984c, J. Am. Chem. Soc. **106**, 3700.

Knight Jr, L.B., A. Ligon, R.J. Woodward, D. Feller and E.R. Davidson, 1985, J. Am. Chem. Soc. **107**, 2857.

Knight Jr, L.B., E. Earl, A.R. Ligon and D.P. Cobranchi, 1986a, J. Chem. Phys. **85**, 1228.

Knight Jr, L.B., E. Earl, A.R. Ligon, D.P. Cobranchi, J.R. Woodward, J.M. Bostick, E.R. Davidson and D. Feller, 1986b, J. Am. Chem. Soc. **108**, 5065.

Knight Jr, L.B., A.R. Ligon, S.T. Cobranchi, D.P. Cobranchi, E. Earl, D. Feller and E.R. Davidson, 1986c, J. Chem. Phys. **85**, 5437.

Knight Jr, L.B., S.T. Cobranchi, B.W. Gregory and E. Earl, 1987a, J. Chem. Phys. **86**, 3143.

Knight Jr, L.B., B.W. Gregory, S.T. Cobranchi, D. Feller and E.R. Davidson, 1987b, J. Am. Chem. Soc. **109**, 3521.

Knight Jr, L.B., K.D. Johannessen, D.C. Cobranchi, E. Earl, D. Feller and E.R. Davidson, 1987c, J. Chem. Phys. **87**, 885.

Knight Jr, L.B., C.A. Arrington, B.W. Gregory, S.T. Cobranchi, S. Liang and L. Paquette, 1987d, J. Am. Chem. Soc. **109**, 5521.

Knight Jr, L.B., B.W. Gregory, S.T. Cobranchi, F. Williams and X.Z. Qin, 1988a, J. Am. Chem. Soc. **110**, 327.

Knight Jr, L.B., S.T. Cobranchi and E. Earl, 1988b, J. Chem. Phys. **88**, 7348.

Knight Jr, L.B., S.T. Cobranchi, G.C. Jones Jr and B.W. Gregory, 1988c, J. Chem. Phys. **88**, 524.

Knight Jr, L.B., J. Steadman, P.K. Miller and J.A. Cleveland Jr, 1988d, J. Chem. Phys. **88**, 2226.

Lindsay, D.M., D.R. Herschbach and A.L. Kwiram, 1974, Chem. Phys. Lett. **25**, 175.

Lindsay, D.M., M.C.R. Symons, D.R. Herschbach and A.L. Kwiram, 1982, J. Phys. Chem. **86**, 3789.

Lindsay, D.M., G.A. Thompson and Y. Wang, 1987, J. Phys. Chem. **91**, 2630.

Linn, S.H., Y. Ono and C.Y. Ng, 1981, J. Chem. Phys. **74**, 3342.

Liu, Y., Q.L. Zhang, F.K. Tittel, R.F. Curl and R.E. Smalley, 1986, J. Chem. Phys. **85**, 7434.

McKinley, A.J., R.F.C. Claridge and P.W. Harland, 1986, Chem. Phys. **102**, 283.

Mohlmann, G.R., and V.A. Kuzmin, 1986, Laser Chem. **6**, 349.

Morse, M.D., 1986, Chem. Rev. **86**, 1049.

Morton, J.R., and K.F. Preston, 1984a, J. Chem. Phys. **81**, 5775.

Morton, J.R., and K.F. Preston, 1984b, Organometallics **3**, 1385.

Morton, J.R., and K.F. Preston, 1985, Inorg. Chem. **24**, 3317.

Nelson, W.H., and W. Gordy, 1969, J. Chem. Phys. **51**, 4710.

Ostrander, S.C., and J.C. Weisshaar, 1986, Chem. Phys. Lett. **129**, 220.

Ozin, G.A., and S.A. Mitchell, 1983, Angew. Chem. Int. Ed. Engl. **22**, 674.

Parnis, J.M., and G.A. Ozin, 1986, J. Am. Chem. Soc. **108**, 1699.

Parnis, J.M., S.A. Mitchell, J. Garcia-Prieto and G.A. Ozin, 1985, J. Am. Chem. Soc. **107**, 8169.

Schock, F., and E. Kay, 1973, J. Chem. Phys. **59**, 718.

Shida, T., E. Haselbach and T. Bally, 1984, Acc. Chem. Res. **17**, 180.

Smith, G.P., and L.C. Lee, 1978, J. Chem. Phys. **69**, 5393.

Snelson, A., 1967, J. Phys. Chem. **71**, 3202.

Snow, L.D., and F. Williams, 1983, Chem. Phys. Lett. **100**, 198.

Symons, M.C.R., 1984, Chem. Soc. Rev. **13**, 393.

Symons, M.C.R., and P.J. Boon, 1982, Chem. Phys. Lett. **89**, 516.

Symons, M.C.R., and I.G. Smith, 1979, J. Chem. Res. Synop., p. 382.

Symons, M.C.R., T. Chen and C.J. Glidewell, 1983, J. Chem. Soc., Chem. Commun., p. 326.

Toriyama, K., M. Iwasaki, K. Nunome and H. Muto, 1981a, J. Chem. Phys. **75**, 1633.

Toriyama, K., K. Nunome and M. Iwasaki, 1981b, J. Phys. Chem. **85**, 2149.

Van Zee, R.J., R.F. Ferrante and W. Weltner Jr, 1981, J. Chem. Phys. **75**, 5297.

Van Zee, R.J., R.F. Ferrante, K.J. Zeringue and W. Weltner Jr, 1985, J. Chem. Phys. **83**, 6181.

Van Zee, R.J., D.A. Garland and W. Weltner Jr, 1986a, J. Chem. Phys. **85**, 3237.

Van Zee, R.J., S.B.H. Bach and W. Weltner Jr, 1986b, J. Phys. Chem. **90**, 583.

Van Zee, R.J., R.F. Ferrante, K.J. Zeringue and W. Weltner Jr, 1987, J. Chem. Phys. **86**, 5212.

Wang, J.T., and F. Williams, 1980, J. Phys. Chem. **84**, 3156.

Weltner Jr, W., 1983, Magnetic Atoms and Molecules (Van Nostrand-Reinhold, New York).

Weltner Jr, W., and R.J. Van Zee, 1984, Ann. Rev. Phys. Chem. **35**, 291.

Weltner Jr, W., and R.J. Van Zee, 1986, Rev. Sci. Instrum. **57**, 2763.

Weltner Jr, W., D. McLeod Jr and P.H. Kasai, 1967, J. Chem. Phys. **46**, 3172.

CHAPTER 8

Reactive Organic Species in Matrices

Ian R. DUNKIN

Department of Pure and Applied Chemistry
University of Strathclyde
295 Cathedral Street
Glasgow, G1 1XL
United Kingdom

Chemistry and Physics of
Matrix-Isolated Species
L. Andrews and M. Moskovits, eds

Contents

1. Introduction

There are several important areas in which matrix-isolation studies have contributed to the understanding of organic chemistry. One of these is the spectroscopy of complexes such as those formed by H-bonding or by interactions between metal atoms and organic species. A second is the trapping and interconversion of individual conformations of stable molecules, e.g., *trans* and *gauche* forms of substituted ethanes or *s-cis* and *s-trans* forms of butadienes and enones. A third is in the use of low-temperature matrices to stabilize highly reactive organic species such as radicals, carbenes and anti-aromatic ring-systems. Discussion of the applications of matrix spectroscopy to the first two areas of organic chemistry can be found elsewhere in this book (see chs. 6, 7 and 10). The present chapter is devoted almost entirely to the third application: the stabilization and characterization of reactive organic species, and the study of their structures and reactions.

Two early, though still useful, books dealing with low-temperature spectroscopy (Bass and Broida 1960, Meyer 1971) list a large number of organic molecules and radicals which have been trapped in frozen solvents and observed by optical spectroscopy or ESR. After Whittle et al. (1954) published their seminal paper on noble-gas matrices, thus adding IR spectroscopy to the collection of practical low-temperature techniques, reactive organic fragments, such as CH_3, CCl_2 and C_2H, were amongst the many new species characterized in the ensuing decade or so (Meyer 1971). Nevertheless, despite this long history, it would be fair to say that for most organic chemists matrix isolation made its first significant impact in the early and mid-1970s, when several research groups started to examine larger organic molecules by matrix IR spectroscopy. Most notable at this time were extensive studies of cyclobutadiene, which led eventually to a decisive answer to the long standing question of the geometry of this unstable molecule (Bally and Masamune 1980). Since then, the number of research groups using matrix isolation to investigate organic species has grown steadily, and the inclusion of matrix spectroscopy, along with other techniques, in studies of organic reaction mechanisms or in new syntheses is becoming more usual, contrary to the pessimistic claim of one writer (Horspool 1983).

The growing body of published work in the field of organic matrix spectroscopy has not been reviewed as thoroughly or as frequently as it deserves.

None of the books on matrix isolation published in the last fifteen years (Hallam 1973, Cradock and Hinchcliffe 1975, Moskovits and Ozin 1976, Barnes et al. 1981) has devoted very much space to this topic; and my own previous but far from comprehensive review (Dunkin 1980), now seriously out of date, seems to provide the last general coverage. Fortunately, this apparent gap in the more recent literature is partially filled by an overall review of matrix isolation (Perutz 1985a), which deals with some aspects of organic matrix spectroscopy, by a valuable tabulation of vibrational energy levels of polyatomic transient molecules with up to sixteen atoms (Jacox 1984b), and by reviews of specific topics, such as organic photochemistry (Horspool 1983), decompositions of azides (Platz 1984, Wentrup 1984), multiple bonding in organosilicon compounds (Raabe and Michl 1985, Maier 1986), and highly strained bonds (Michl et al. 1983, Maier 1986), all of which contain significant reference to matrix studies. My aim in this chapter, therefore, is to provide a general account of recent matrix work on reactive organic species (from about 1978), and within the available space to try to cover the field as fully as possible. Further restrictions on the contents of this chapter, however, are the omission of any reference to matrix experiments with organic ions and a deliberately reduced discussion of radical (doublet) species. These topics are currently of great interest to organic chemists, but both are dealt with in other chapters (see chs. 2 and 4) of this book.

2. Experimental techniques

In the majority of investigations, reactive organic species have been generated either by photolysis of matrix-isolated precursors or by flow pyrolysis followed by dilution with a suitable host gas and condensation at low temperature. Some interesting variations or extensions of these methods have been reported, however. For instance, Tseng and Michl (1977) and Otteson and Michl (1984), have described a procedure for the gas-phase dehalogenation of organic dihalides with alkali metal vapours, involving microwave or ultrasound excitation; details have been published (Vaccani et al. 1977, Kühne 1978) of a gas-phase linear reactor attached to a matrix-isolation cell, which has been used in the study of ozone–alkene reactions; experiments with haloethenes in Kr matrices containing Hg atoms have shown (Cartland and Pimentel 1986a,b) that photosensitization can, in favourable cases, lead to triplet-derived products which are not formed by higher energy singlet excitation; and the technique of synchronizing flash photolysis with pulse deposition has improved the yields of matrix-isolated primary photoproducts in certain reactions where cage recombination is very efficient (Allamandola et al. 1978). The use of IR radiation to induce matrix photochemistry has produced some interesting results (Frei and Pimentel 1985). Reactions between ozone and C_2H_2 or C_2H_4, apparently, cannot be induced by vibrational excitation (Frei et al. 1981), nor can reactions between butadiene and Cl_2, ClF, Br_2, I_2 or XeF_2, or between vinyl chloride and F_2. In contrast, reactions between F_2 and C_2H_4, vinyl bromide or allene have been brought about in this way (Frei et al. 1981, Cesaro et al. 1983, Frei 1983, Frei and Pimentel 1983a, Knudsen and Pimentel 1983). Quantum yields varied widely with the particular vibrational mode excited, but there appeared to be no mode selectively in the distribution of products. For example, the IR-induced reaction of F_2 with allene in N_2, Ar, Kr or Xe matrices yielded four products: *cis*- and *gauche*-2,3-difluoropropene and complexes of HF with propargyl fluoride and fluoroallene,

$$CH_2 = C = CH_2 + F_2 \;\rightarrow\; \underset{cis \text{ and } gauche}{CH_2 = CF–CH_2F}$$

$$\rightarrow\; HC \equiv CCH_2F·HF + CH_2 = C = CHF·HF.$$

The relative abundance of each of the four products depended markedly on the matrix host, but was, in a particular matrix, independent of the exciting frequency in the range 3076–1679 cm^{-1}. The relative quantum yield at 3076 cm^{-1}, however, was about 10^3 times greater than at lower frequencies (Knudsen and Pimentel 1983). Selective vibronic excitation of O_2, using near-IR light, induced reaction between O_2 and furans in Ar matrices at 12 K (Frei and Pimentel 1983b). In the case of 2,5-dimethylfuran, which has been most fully reported on, the reaction seemed to be a single-photon process with unit quantum yield, and the IR spectrum of the product suggested that it was the endoperoxide formed by addition of O_2 across the 2,5-positions.

Despite the continued development of new techniques for sample preparation and the elaboration of older ones, it is well to remember that many of the classes of reactive intermediates that have been of traditional importance in organic chemistry – e.g., carbenium ions, carbanions, and a wide variety of ions with charges on heteroatoms – have not so far proved amenable to study by matrix-isolation spectroscopy. In addition, large molecules, especially those of biological interest, such as oligopeptides or saccharides and their derivatives, have such low volatility that any sampling technique requiring vaporization is likely to be useless. Even the problem of transferring relatively volatile, uncharged intermediates from the liquid phase to solidified gas matrices has not yet been tackled in a concerted manner. Andrews and Kohlmiller (1982), however, have demonstrated that alkene ozonolysis can be carried out in CF_3Cl solution at 130–150 K and the secondary ozonides transferred to Ar matrices by solvent evaporation, volatilization of the residue, and co-deposition with Ar. It would be interesting to see this approach developed further.

Matrix preparation methods provide one set of severe limitations for the organic matrix chemist, and a second set results from the restricted availability of appropriate analytical techniques. Organic chemists who deal only with stable molecules are now accustomed to the study of quite complex reactions.

Efficient chromatographic methods can separate complicated product mixtures even down to very low percentage levels, while subsequent identification of the pure substances is relatively straightforward with a whole battery of analytical techniques, including mass-spectrometry, IR and UV–visible spectroscopy, and – most powerful of all – nuclear magnetic resonance. In contrast, matrix studies have to rely on IR spectroscopy as the most widely applicable means of identifying new species, a technique which is good for functional groups but comparatively poor for determining an overall molecular geometry. A sustained effort is in progress to develop low-temperature ^{13}C-NMR as a useful structural tool (Zilm et al. 1978, 1980, Zilm and Grant 1981, Strub et al. 1983, Beeler et al. 1984, Facelli et al. 1985, 1986, 1987, Orendt et al. 1985, Solum et al. 1986), but so far the most reactive species to be examined has been sodium cyclopentadienylide (Strub et al. 1983). A similar effort to develop secondary-ion mass-spectrometry (SIMS) could also provide an additional analytical tool for matrix chemists (Jonkman and Michl 1978, Jonkman et al. 1978, Michl 1983). In the meantime, most investigations have been pursued by means of IR spectroscopy (occasionally supplemented by Raman measurements), backed up by UV–visible absorption and, more rarely, emission spectroscopy. ESR spectroscopy has, of course, been widely used for studies of radicals and other paramagnetic species in matrices, and recently electron–nuclear double resonance (ENDOR) spectra have been reported for C_2H_5 matrix isolated in Xe (Toriyama et al. 1981) and CH_3 in Ar (Cirelli et al. 1982). Nevertheless, even with such increased information as can be provided by isotopic substitution, linear dichroism measurements (Dunkin 1986, Michl and Thulstrup 1986), or ESR combined with photoselection (Adrian et al. 1984, Dunkin et al. 1986a), identification of novel matrix-isolated organic species is seldom as positive or as rigorous as one would wish.

Limitations in the scope of the technique and the quality of the analytical methods which can be used, probably go a long way to explain why organic chemists have not taken up matrix isolation in greater numbers or with more enthusiasm. These difficulties, however, and the general problem of carrying out matrix reactions on a preparative scale present significant challenges for future work.

3. Hypovalent organic species

3.1. Carbenes and their reactions

Carbenes (methylene and its derivatives) are uncharged divalent carbon species. The simplest carbene, CH_2, as noted in my earlier review (Dunkin 1980), has been difficult to trap in matrices. The most obvious precursors for CH_2 are diazomethane and its cyclic isomer, diazirine; and the history of matrix studies of the photolysis of these molecules and various isotopomers has recently been summarized by Perutz (1985b). Because of its small size, CH_2 can diffuse relatively easily in matrices. It is also extremely reactive. From experiments with ^{15}N-labelled diazomethane in $^{14}N_2$ matrices (Moore and Pimentel 1964), it seems that CH_2 will react even with N_2, regenerating the diazocompound. The result is that it has been relatively easy to obtain the ESR spectrum of CH_2 – a non-linear triplet in its ground state – but concentrations high enough to permit observation of its IR absorption have been hard to achieve. In the most recent re-investigation of the matrix photolysis of diazomethane, Lee and Pimentel (1981b) identified CH_2NH, C_2H_4, C_2H_2 and CH_4 as stable photoproducts, and assigned IR bands at 1115 (Ar matrix) and 1109 cm^{-1} (Xe) to the v_2 (bending) mode of CH_2 and a band at 819 cm^{-1} (Ar) to the equivalent mode of CD_2. The assignments confirm earlier suggestions but still leave some room for doubt. When such matrices supposedly containing CH_2 were warmed, chemilumines-cence occurred, part of which (a broad emission centred at 599 nm) was attributed to the formation of ethene in an excited state. As the ultimate products indicated, however, the processes taking place in these matrices were complex, and more than one species was responsible for the emission. In an attempt to find an alternative precursor for CH_2, Maier and Reisenauer (1986) examined the UV photolysis of dihalomethanes (H_2CIX, $X = I$, Br, Cl) in Ar and N_2 matrices. They discovered instead a novel isomerization process, yielding coloured complexes ($H_2CX \cdots I$), which readily reverted to the starting materials when irradiated in their near-UV or visible absorptions. A normal coordinate analysis suggests that the complexes might best be described as contact ion pairs ($H_2C^+ - X\, I^- \leftrightarrow H_2C = X^+\, I^-$).

Simple acyclic alkyl or dialkyl carbenes are also unlikely to be easy to trap in matrices, owing to their facile rearrangements to the corresponding alkenes. Fluorine substitution is one way of inhibiting carbene rearrangement. Thus photolysis of bis(trifluoromethyl)diazirine ($\underline{1}$) in Ar at 12 K with a high-pressure Hg arc (Mal'tsev et al. 1985a) gave the isomer ($\underline{4}$) [v(CNN) 2135 cm^{-1}] and carbene ($\underline{2}$), with IR bands at 1344, 1197, 1157 and 965 cm^{-1} (scheme 1). It is

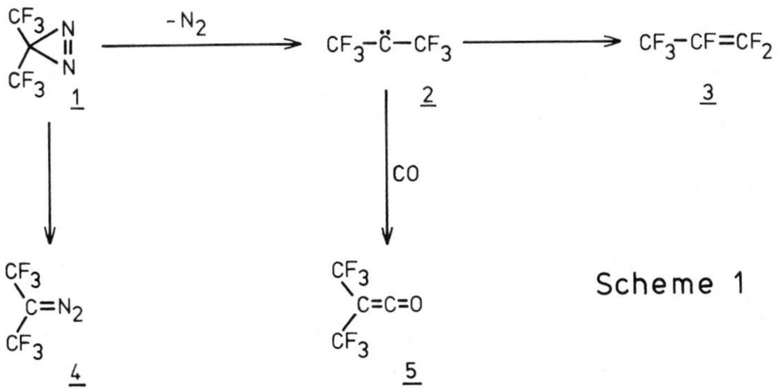

Scheme 1

not known whether the diazocompound (4) was formed by unimolecular isomerization of (1) or by cage recombination of carbene (2) and N_2, but once formed it was surprisingly stable towards further photolysis. In CO-doped matrices, ketene (5) was produced, giving reasonable confirmation of the identity of the carbene. IR spectra recorded after flow pyrolysis (0.001 torr, 350–500°C) of (1) followed by matrix isolation showed absorptions of the rearrangement product (3) rather than those of the carbene; at 350–500°C an upper limit of 10^{-5} s was thus estimated for the carbene lifetime.

Kesselmayer and Sheridan (1986b) generated phenoxychlorocarbene in *cis* and *trans* forms by photolysis of phenoxychlorodiazomethane in Ar and N_2 matrices at 10 K. The identity of the carbene was confirmed by its reaction with HCl in suitable doped N_2.

No evidence for the intermediacy of a diazocompound could be found in the IR spectra. Partial interconversion of the two carbene conformers could be brought about by selective monochromatic irradiation, but at the same time decomposition to benzoyl chloride, chlorobenzene and CO occurred. Analogous *cis* and *trans* isomers of methoxychlorocarbene have also been observed (Sheridan and Kesselmayer 1984, Kesselmayer and Sheridan 1986a). Carbenes of the type RO–C̈–Cl have broad UV absorptions with λ_{max} near 320 nm, and appear to have anomalously high antisymmetric v(C–O–C) frequencies at about 1300 cm^{-1} (^{18}O-shift approximately 28 cm^{-1}), indicating significant C–O double bond character. The much simpler analogue, hydroxymethylene, has been implicated as an intermediate in the matrix photochemistry of formaldehyde (Sodeau and Lee 1978), but has not so far been observed directly.

There have been a number of matrix studies of alicyclic carbenes. Cyclopropenylidene (7) along with benzene has been trapped in Ar at 10 K following flow pyrolysis of the quadricyclane derivative (6) (Reisenauer et al. 1984). Subsequent irradiation ($\lambda > 360$ nm) caused the disappearance of IR bands at 1279, 1063, 888 and 789 cm^{-1}, assigned to (7), and the appearance of the known IR absorptions of propynylidene (8). When the same process was monitored by ESR, the initial pyrolysate showed no triplet signal, because (7) has a singlet ground state. The subsequent formation of (8), however, was accompanied by the appearance of its triplet spectrum. The bicyclic cyclopropylidene (9) could not be trapped in Ar matrices following its generation at high temperatures, but rearranged readily to the highly strained cyclic allene, 1,2-cyclohexadiene (10) (Wentrup et al. 1983). The asymmetric $v(C = C = C)$ band of (10) at 1886 cm^{-1} is shifted approximately 70 cm^{-1} from those of normal allenes, reflecting the high degree of strain. Cyclopentadienylidenes (11, R = H, Cl) (Baird et al. 1981, Bell and Dunkin 1985), the benzo analogues of (11) indenylidene and fluorenylidene (Bell and Dunkin 1985), and cycloheptatrienylidene (12)

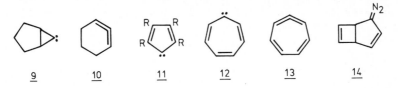

9	10	11	12	13	14

(McMahon and Chapman 1986) have all been matrix isolated by photolysis of diazoprecursors, and characterized by IR and UV absorptions. ESR showed that (12) has a triplet ground state, and there is apparently no tendency for (12) to be converted to the singlet allene (13), either thermally or on further photolysis. All these carbenes gave the corresponding ketenes in CO-doped matrices, although with varying degrees of reactivity. Interestingly, matrix photolysis of the diazocompound (14) (Chapman and Abelt 1987) did give allene (13), with characteristic $v(C = C = C)_{as}$ bands at 1818 and 1810 cm^{-1}, and so did matrix photolysis of phenyldiazomethane (West et al. 1982). The allene is unreactive towards CO. It is strange that (12) and (13) differ only in their distribution of electrons, that apparently both can be isolated separately in matrices, but that no interconversion between them seems to be possible.

The ring expansion of phenyl carbenes to derivatives of cyclic allene (13) is not a completely general reaction. Chapman and co-workers (West et al. 1982, Chapman et al. 1984, McMahon and Chapman 1987) have made matrix IR, UV–visible and ESR studies of the photolyses of phenyldiazomethane and its four monomethyl derivatives. In each case the primary photoproduct ($\lambda > 470$ nm) was the corresponding triplet carbene. On further irradiation or warming to 80 K (Xe matrix), PhC̈CH$_3$ rearranged to styrene but did not ring expand, while the three tolylmethylenes each underwent photorearrangement ($\lambda > 416$ nm) to methyl derivatives of (13), with IR bands near 1810 cm^{-1}. Interconversion of methyl derivatives of (13) could be brought about by irradiation at still shorter wavelengths ($\lambda > 216$ nm), presumably via the tolylmethylenes. In the case of *o*-tolylmethylene, however, a second pathway seemed to predominate: a thermal [1,4]-hydrogen shift producing *o*-xylylene (Chapman et al. 1984, McMahon and Chapman 1987). This reaction occurred in the dark even at 4.6 K, and a tunnelling mechanism is implicated. Phenylchlorocarbene also seems not to undergo facile ring expansion. It has been generated in Ar matrices by photolysis of phenylchlorodiazirine, characterized by IR spectroscopy, and found to react with both HCl and Cl_2, yielding the corresponding chlorinated toluenes (Mal'tsev et al. 1985b, Ganzer et al. 1986).

The route by which aryl carbenes ring expand is often supposed to involve initial formation of bicyclic singlet species, benzocyclopropenes. In none of the matrix studies of phenyl carbenes carried out so far has such a species been detected. Matrix photolysis of naphthyldiazomethanes, however, resulted in the IR detection of several of these primary rearrangement products, e.g. (15) (West et al. 1986). The cyclopropenes were characterized by ring vibrations at $1765-1750$ cm^{-1}, shifted by about 40 cm^{-1} on deuterium substitution. They were found, however, to be surprisingly photostable, and could not be induced to undergo ring opening to benzocycloheptatetraenes on further irradiation.

15   R = H, D

Before their work on aryl carbenes appeared in print, Chapman's group (Chapman et al. 1978, Chapman and Sheridan 1979) had already published studies of pyridylmethylenes. They found similar carbene reactions and ring expansions to those of the carbocyclic analogues, with the added complication of a ring nitrogen atom, which led to the formation of both allenes and ketenimines and also to the involvement of phenylnitrene. The reader is referred especially to two reviews of this work for discussion of the details (Chapman 1979, Chapman et al. 1979).

A characteristic matrix reaction of carbenes which is of great current interest is the addition of O_2. The resulting adducts, carbonyl oxides, sometimes known as Criegee intermediates, have for a long time been suspected of participation in alkene ozonolyses and photo-oxygenations of diazocompounds. An interest in matrix studies of carbene–oxygen reactions has been evident since 1979 (Chapman 1979, Chapman and Hess 1979, Lee and Pimentel 1981a), but the first report of the matrix IR spectrum of a carbonyl oxide was made some years later by Bell and Dunkin (1983). Carbonyl oxide (16) was formed by narrow-band UV-irradiation (296 ± 5 nm) of diazocyclopentadiene in N_2 matrices containing 10% O_2, or by annealing N_2 matrices containing the corresponding carbene and 1% O_2 (scheme 2). Fourteen IR bands of (16) were identified, and these all rapidly disappeared upon irradiation of the matrix with light of $\lambda > 330$ nm, the major secondary photoproducts being α-pyrone (18), its ring-opened isomer (19) and cyclopentadienone. Shortly afterwards, Chapman and

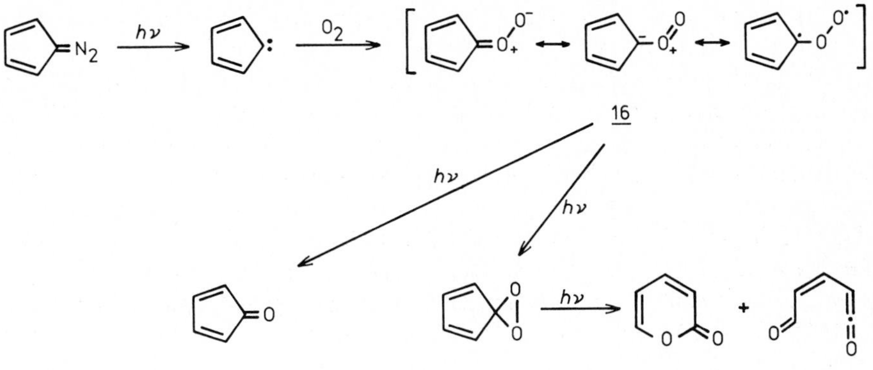

Scheme 2

Hess (1984) reported the formation of (16) by wide-band visible irradiation ($\lambda > 418$ nm) of diazocyclopentadiene in Ar matrices containing 20% O_2, but the intermediate that they discovered had a different IR spectrum from that of the first report. It has now been shown (Dunkin and Shields 1986) that Chapman and Hess's intermediate is a photo-isomer of (16), and a precursor to (18) and (19) but not apparently to cyclopentadienone. It is most probably the dioxirane (17). Work with isotopically labelled oxygen has been crucial to this conclusion. The strongest IR band of (16), which occurs at 1014 cm^{-1}, also has the largest $^{18}O_2$-shift (28 cm^{-1}). When (16) was generated from a mixture of $^{16}O_2$, $^{16}O^{18}O$ and $^{18}O_2$, this band showed a very well-defined doublet due to the $^{16}O^{18}O$-isotopomers, thus proving the presence of two non-equivalent O-atoms (fig. 1). A similar doublet due to $^{16}O^{18}O$-species could not be found for the photo-isomer of (16), consistent with a symmetrical disposition of the two O atoms, as in (17). Such negative evidence, however, obviously remains open to question. Related matrix studies, in which both carbonyl oxides and dioxiranes were detected, have been carried out for the oxygenation of diphenylcarbene (Sander 1986) and phenylchlorocarbene (Ganzer et al. 1986); while carbonyl oxides, but not dioxiranes, have been observed in the matrix reactions of O_2 with tetrachlorocyclopentadienylidene (11, R = Cl), indenylidene and fluorenylidene (Bell et al. 1985, Dunkin and Bell 1985, Dunkin et al. 1986b).

Also of special interest, because of their relevance to gas- and liquid-phase mechanistic studies, are various attempts to generate α-ketocarbenes and the isomeric oxirenes in matrices. Unless hampered by severe geometrical constraints, α-ketocarbenes (21) rearrange extremely readily to ketenes (24) (scheme 3). Scrambling of isotopic labels, however, has indicated that in some cases an equilibrium [(21) ⇌ (23)] exists prior to this Wolff rearrangement, and the intervention of an oxirene intermediate (22) has for a long time been considered possible. Oxirenes are amongst the most elusive members of the class of

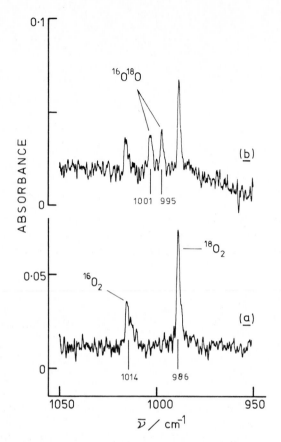

Fig. 1. IR spectra of (<u>16</u>) in the 1000 cm^{-1} region, recorded after photolysis (295 $\pm$ 5 nm) of diazocyclopentadiene at 20 K in N$_2$ matrices containing 4–5% O$_2$, labelled with 70 at% ^{18}O. (a) Only ^{16}O$_2$ and ^{18}O$_2$ present. (b) ^{16}O$_2$, ^{16}O^{18}O, and ^{18}O$_2$ present in roughly statistical proportions. (Reproduced with permission from Bell and Dunkin (1983).)

anti-aromatic three-membered heterocycles, and are therefore of considerable theoretical significance (Torres et al. 1980b). Although triplet ESR signals, assigned to (<u>21</u>, R^1 = R^2 = CF$_3$), were detected after low-temperature photolysis of polycrystalline (<u>20</u>, R^1 = R^2 = CF$_3$) (Murai et al. 1981), photolyses of a wide range of diazoketones in Ar matrices resulted in Wolff rearrangement in all cases and failed to produce any detectable yields of ketocarbenes or oxirenes (Maier et al. 1982c, Torres et al. 1983a). Similar negative results were obtained from attempts to trap examples of these species from cycloreversion reactions (Maier et al. 1982d) or elimination of COS from vinylene thioxocarbonates (Torres et al. 1980a). Recently, however, several strained cyclic ketocarbenes, e.g. (<u>25</u>), have been generated in matrices from diazo precursors and characterized very fully

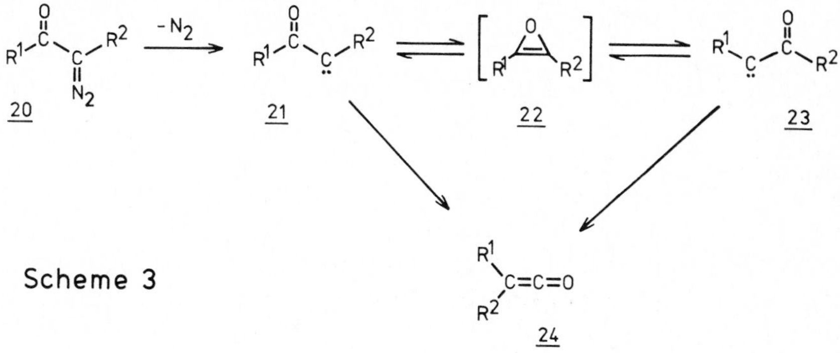

Scheme 3

by IR, UV–visible and ESR spectroscopy (Hayes et al. 1983, McMahon et al. 1985). These carbenes react normally with CO and O_2, and on further irradiation undergo Wolff rearrangement to very strained ketenes, e.g. ($\underline{26}$).

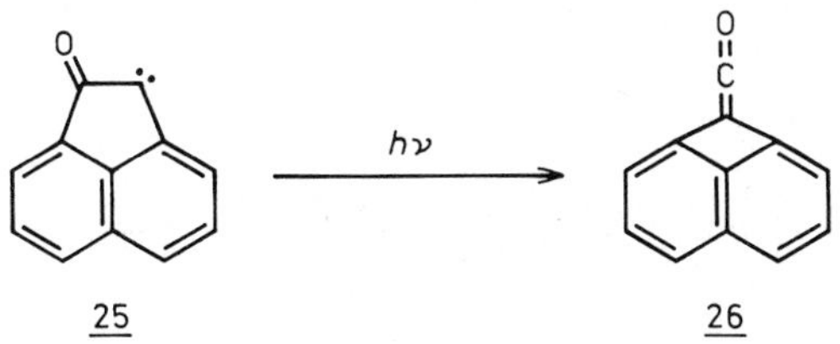

Oxirenes remain elusive, though. One report (Torres et al. 1983b) claimed that IR spectra of oxirenes ($\underline{22}$, $R^1 = R^2 = CF_3$) and ($\underline{22}$, $R^1 = CF_3$, $R^2 = C_2F_5$), as well as the corresponding ketocarbenes, were identified following matrix photolysis of perfluorinated diazoketones ($\underline{20}$, $R^1 = R^2 = CF_3$), ($\underline{20}$, $R^1 = CF_3$, $R^2 = C_2F_5$) and ($\underline{20}$, $R^1 = C_2F_5$, $R^2 = CF_3$). Serious doubts, however, were raised by Laganis et al. (1983), who showed that the spectrum originally assigned to ketocarbene ($\underline{21}$, $R^1 = R^2 = CF_3$) belonged instead to the diazirine isomer of diazoketone ($\underline{20}$, $R^1 = R^2 = CF_3$). At present, therefore, it seems that a definite identification of an oxirene is still awaited.

3.2. Nitrenes and their reactions

Nitrenes are uncharged monovalent nitrogen derivatives, the nitrogen analogues of carbenes. Unlike CH_2, the simplest nitrene, NH, was observed very early in the history of matrix spectroscopy through its IR and electronic spectra (Dows et al. 1955, McCarty and Robinson 1959, Milligan and Jacox 1964, Rosengren and Pimentel 1965). More recently, its magnetic circular dichroism spectra in Ar and Xe matrices have also been reported by Lund et al. (1982). The reaction of NH (from photolysis of HN_3) with but-2-yne gave the propynamine,

$MeC \equiv CNHMe$, as the major product in Xe matrices, but the ketenimine, $Me_2C = C = HN$, in Ar (Collins and Pimentel 1984). Xe-induced intersystem crossing from the initially generated singlet NH to the triplet might account for this difference.

Apart from NH, and despite an early report (Reiser et al. 1966) of the electronic spectra of several aryl nitrenes in organic glasses, nitrenes in general have until recently seemed difficult to stabilize in solidified gas matrices. Simple alkyl nitrenes, generated photolytically from the corresponding azides, rearrange very readily (possibly even in a concerted manner) to imine products (Stolkin et al. 1977, Dunkin and Thomson 1980b, Braathen et al. 1984), e.g.,

$$R_3C-N_3 \rightarrow [R_3C-N] \rightarrow R_2C = NR \quad (R = H, Me).$$

Further photolysis of the imine is sometimes observed (Braathen et al. 1984), but no acyclic alkyl nitrene has been detected in these matrix experiments. The driving force of the nitrene–imine rearrangement is such that it has been exploited to generate highly strained bridgehead imines (see section 4).

Since the first matrix-isolation study of the photolysis of phenyl azide (Chapman and LeRoux 1978, Chapman et al. 1979), it is now well-established (Donnelly et al. 1985) that the major matrix photoproducts of phenyl azides lacking *ortho* substituents are 1-aza-1,2,4,6-cycloheptatetraenes, e.g. (30), characterized by strong IR absorptions in the $1910-1880 \, cm^{-1}$ region. Although triplet phenylnitrene (28, X = H) and *p*-tolylnitrene (28, X = Me) have been detected in matrices by ESR (Chapman et al. 1978, Kuzaj et al. 1986), the stabilized concentrations were probably low, and IR studies have failed to identify either nitrenes or bicyclic azirines, e.g. (29), as intermediates in the formation of the ring-expansion products, e.g. (30). Ring expansion takes place

efficiently, even when the phenyl ring bears a powerful electron attracting group, such as CN (Donnelly et al. 1985) or NO_2 (Dunkin and Sweeney 1986). When the phenyl azide has two *ortho* substituents, however, as in pentafluorophenyl azide (Dunkin and Thomson 1982) or 2,6-dimethylphenyl azide (Dunkin et al. 1985), the ring expansion is blocked. In these cases, the matrix-isolated nitrenes have been observed directly by IR spectroscopy. The nitrenes do not have an easily recognizable group frequency, but their reactions with CO, yielding the corresponding isocyanates (RNCO), confirmed their identities. The nitrene–CO reaction does not take place so readily as that between carbenes and CO.

Annealing at temperatures up to about 40 K failed to induce thermal addition of CO to the triplet ground-state nitrenes; isocyanate formation occurred only on further UV-irradiation. This observation suggests that the isocyanates are formed by reaction of excited singlet state nitrenes with CO.

The matrix IR studies of aryl azide photolyses raised considerable doubts over the earlier report (Reiser et al. 1966) of the electronic spectra of aryl nitrenes in organic glasses. Once 2,6-disubstituted phenyl nitrenes had been authenticated by matrix IR studies, however, it became possible to compare their electronic spectra in both solidified-gas matrices at 10–20 K and organic glasses at 77 K. Figure 2 shows UV-visible absorption spectra of 1-aza-1,2,4,6-cycloheptatetraene (30, R = H) and 2,6-dimethylphenylnitrene in Ar matrices at 12 K, recorded in the author's own laboratory (Dunkin and Sweeney 1986). The two spectra are quite different, reflecting the different electronic and molecular structures of the two species. The nitrene spectrum has relatively weak structured absorptions near 300, 400 and 500 nm, as well as a strong absorption with $\lambda_{max} = 246$ nm. Pentafluorophenylnitrene exhibits a similar absorption pattern (Dunkin and Sweeney 1986). Leyva et al. (1986) have carried out photolyses of PhN_3 and a number of 2,6-disubstituted phenyl azides in glasses at 77 K, and have published electronic absorption, fluorescence and fluorescence excitation spectra of the products at wavelengths down to about 280 nm. The absorption spectra of 2,6-dimethylphenylnitrene in both 77 K glasses and 12 K

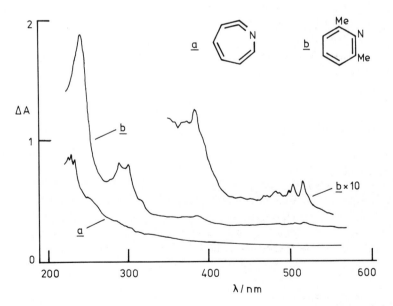

Fig. 2. UV–visible spectra of: (a) 1-aza-1,2,4,6-cycloheptatetraene, and (b) 2,6-dimethylphenyl-nitrene in Ar at 12 K.

Ar matrices are very similar, as are those of pentafluorophenylnitrene. A slightly enhanced resolution of vibrational structure in the Ar-matrix spectra, as expected, is the most significant difference. All the nitrene spectra show the characteristic visible absorptions near 400 and 500 nm. Remarkably, in 77 K glasses, even the phenyl azide photoproduct has these characteristic nitrene absorptions, and is thus quite distinct from the major photoproduct of PhN$_3$ in Ar or N$_2$ matrices. There is thus compelling evidence to suggest a strong solvent or temperature dependence in the photolysis of phenyl azide, the reasons for which have yet to be elucidated.

Although azirine intermediates such as (29) have not been detected in matrix photolyses of phenyl azides, IR studies of the photolyses of azidonaphthalenes, e.g. (31), in Ar and N$_2$ at 12 K (Dunkin and Thomson 1980a) revealed tricyclic azirines, e.g. (32) and (34), as intermediates in the overall formation of azacycloheptatetraene products, e.g. (33) and (35) (scheme 4). The azirines, identified by means of characteristic IR bands at 1736–1708 cm^{-1}, readily underwent photo-isomerization to azacycloheptatetraenes, in contrast to the observed stability of the analogous cyclopropenes (15) (West et al. 1986). As might by now be expected, nitrenes were not detected in these IR studies of azidonaphthalenes.

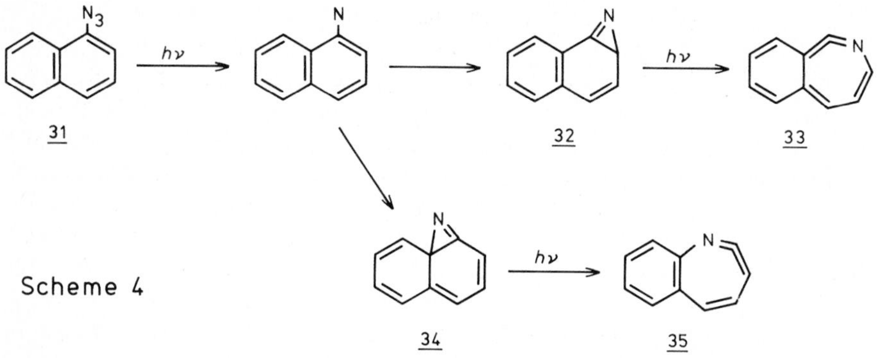

Scheme 4

Heteroaryl azides have also been subjected to combined matrix isolation and flow pyrolysis or photolysis experiments (Wentrup et al. 1980, Kuzaj et al. 1986). Typical of the results is the formation of the strained carbodiimide (36) [$v(N=C=N)_{as} = 2010$ cm^{-1}].

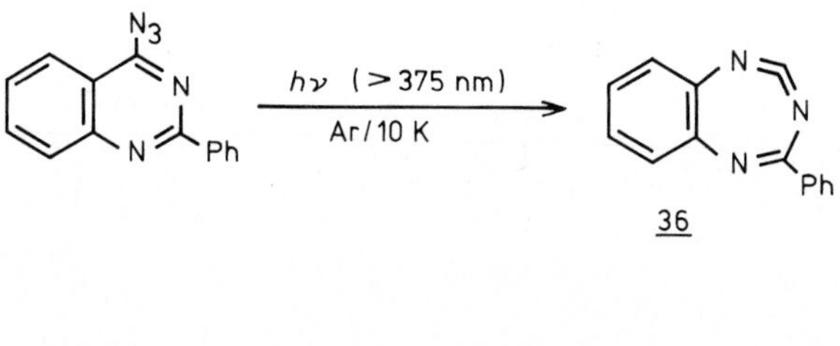

3.3. Radicals and biradicals

Organic chemists usually reserve the term *radical* for any doublet species with a single unpaired electron located on one or more carbon or hetero atoms, and this restricted usage is adopted here. Radicals are thus distinguished from biradicals – species with two more or less separate radical centres – and other hypovalent species such as carbenes and nitrenes. Radicals also differ from carbenes by being generally much less reactive and thus much more easily detected in fluid phases, especially by ESR. For radical studies, ESR is a powerful technique, because it is both sensitive and structurally informative. Moreover, ESR signals are not obscured by common solvents, either liquid or frozen. Consequently, there is a huge literature on ESR studies of organic radicals; and matrix isolation, by sheer dint of numbers of publications, has contributed less prominently to the advancement of radical chemistry than it has to other important areas. Nevertheless, real benefits derive from spectroscopic investigations of radicals in solidified-gas matrices. The three principal advantages are:

(1) IR spectra of radicals can be obtained;
(2) extremely reactive precursor species (e.g. F atoms) can be employed without reacting with the solvent, and
(3) singlet as well as paramagnetic products can be observed directly from both thermally and photochemically induced radical reactions.

A variety of methods for generating radicals in matrices have been developed. Detachment of H atoms by vacuum-UV photolysis of toluene (Miller and Andrews 1981) and phenol (Pullin and Andrews 1982), during condensation with Ar, gave benzyl and phenoxyl radicals, respectively, which were observed by absorption, emission, or excitation spectroscopy. More surprisingly, because cage recombination might be expected to reverse the reaction very efficiently, iodine atoms have been detached photochemically from matrix-isolated $PhC \equiv CI$ and $CF_2 = CFI$, and the resulting phenylethynyl (Kasai and McBay 1984) and perfluorovinyl (Kasai 1986) radicals examined by ESR. The most extensive series of publications concerned with photochemical generation of radicals in matrices, however, has come from Pacansky and co-workers (Pacansky et al. 1977, 1978, 1981, Pacansky and Coufal 1979, 1980a,b,c, Pacansky and Dupuis 1982, Pacansky and Brown 1983, Pacansky and Guitierrez 1983). These workers showed that, in the photolysis of matrix-isolated diacyl peroxides, a large proportion of the resulting pairs of radicals remains separated by the molecules of CO_2 that are also eliminated. Subsequent annealing often allows the CO_2 to diffuse away, permitting radical recombination or disproportionation. A recent example (Pacansky and Brown 1983) is provided by the photolysis of (37) in Ar matrices,

$$\underset{37}{Me\overset{\overset{\displaystyle O}{\|}}{C}-O-O-\overset{\overset{\displaystyle O}{\|}}{C}Ph} \xrightarrow{h\nu} \underset{38}{Me\overset{\overset{\displaystyle O}{\|}}{O}CPh} + Me\cdot + CO_2 + Ph\cdot.$$

The observed photoproducts were methyl benzoate ($\underline{38}$), methyl radical, phenyl radical, and CO_2. It is significant that although ($\underline{38}$), from recombination of $PhCO_2$ and CH_3 radicals, was amongst the products, no trace of phenyl acetate, from recombination of $MeCO_2$ and Ph radicals, was detected. This demonstrates dramatic lifetime differences between $PhCO_2$ and $MeCO_2$. Photolysis of diacyl peroxides has enabled matrix IR spectra to be obtained for a series of radicals, which, besides methyl and phenyl, include ethyl, *n*-propyl, *iso*-propyl, *n*-butyl, *iso*-butyl, *n*-pentyl, and *neo*-pentyl, as well as a number of deuteriated derivatives. In certain cases, interesting structural information for these radicals has been deduced. For example, analysis of the spectra suggests that, whereas rotation about the C–C bond in the ethyl radical is fairly free, the barrier for rotation of the methyl groups in the *iso*-propyl radical is sufficient to allow two conformers to co-exist in Ar matrices (Pacansky and Coufal 1979, 1980a,b,c, Pacansky and Dupuis 1982).

Another way of circumventing problems due to cage recombination is to deposit radicals in matrices after gas-phase generation by flow pyrolysis. The allyl radical, $CH_2 = CH-\dot{C}H_2$, e.g., has been isolated in Ar matrices following flow pyrolysis of allyl halides, hexa-1,5-diene, and several other precursors, and its IR, UV and ESR spectra were obtained (Mal'tsev et al. 1982a, 1984b, Maier et al. 1983). Similarly the *t*-butyl radical has been generated by pyrolysis of azoisobutane and 2-nitrosoisobutane (Pacansky and Chang 1981, Pacansky et al. 1982); its IR spectrum suggests a C_{3v} structure with longer than normal C–H bonds *trans* to the unpaired electron. On photolysis ($\lambda < 280$ nm) in Ar matrices, the *t*-butyl radical undergoes isomerization to the *iso*-butyl radical and H-atom elimination, yielding isobutene (Pacansky et al. 1982).

Two more methods for generating radicals in matrices deserve to be mentioned here, although they are discussed more fully in other chapters (see chs. 4 and 7). Both involve atom reactions. The first is the addition of atoms to unsaturated molecules, the second is the abstraction of H or halogen atoms. For example, H atoms generated in a gas-phase discharge, when co-deposited with $PhC \equiv CH$ in Ar at 20 K, gave the α-styryl radical, $Ph\dot{C} = CH_2$, and a second product derived from addition of an H atom to the aromatic ring (Andrews and Kelsall 1987). Similarly generated F atoms have been observed by matrix IR spectroscopy, to add to C_2H_2 (Jacox 1980), C_2H_4 (Jacox 1981a), C_2F_4 (Jacox 1984a), and benzene (Jacox 1982b) (scheme 5). As an alternative to addition, F atoms can also react with organic molecules by abstraction of H atoms; CH_3OH, CH_3CHO, and ethylene oxide, e.g., all gave radical matrix products in this way (scheme 5) (Jacox 1981b, 1982a). Co-condensation of Li atoms with halomethanes, $CHFX_2$ (X = Cl, Br, I), and excess Ar resulted in abstraction of halogen atoms, and the formation of fluorohalomethyl radicals, $\dot{C}HFX$ (X = Cl, Br, I), for which IR spectra were obtained (Prochaska et al. 1979).

Biradicals have so far been less extensively studied by matrix-isolation spectroscopy than radicals. There is considerable theoretical interest in these

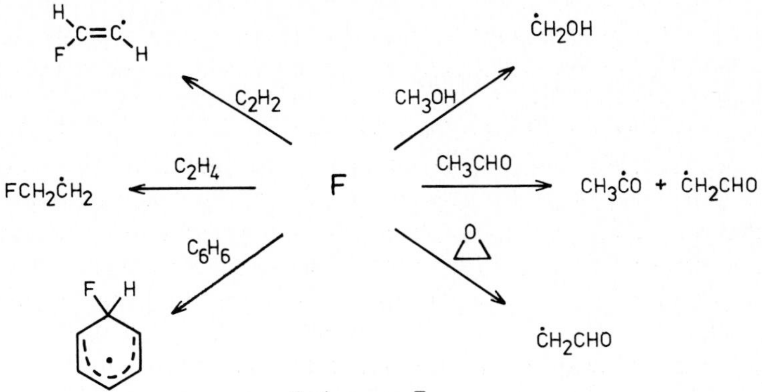

Scheme 5

species, however, especially with regard to whether the ground states are singlets or triplets. *m*-quinomethane (39), and a naphthalene analogue of (39) (Rule et al. 1979, 1982), the localized cyclobutadiyl (40) (Jain et al. 1984), and the delocalized cyclobutadiyl (41) (Snyder and Dougherty 1984) have all been

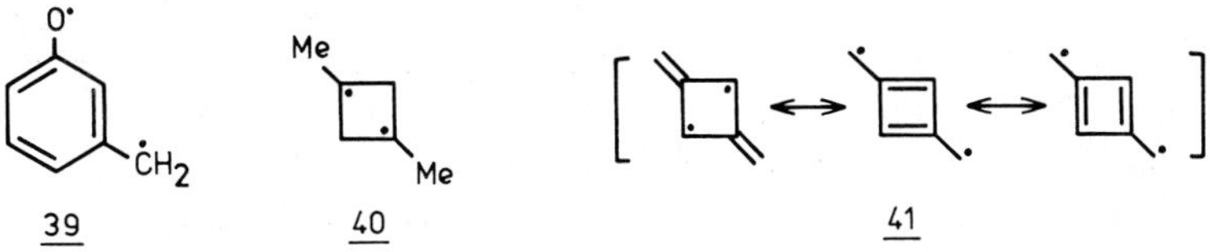

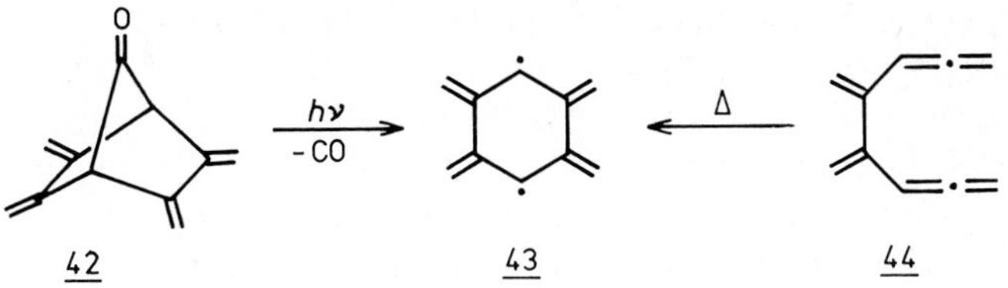

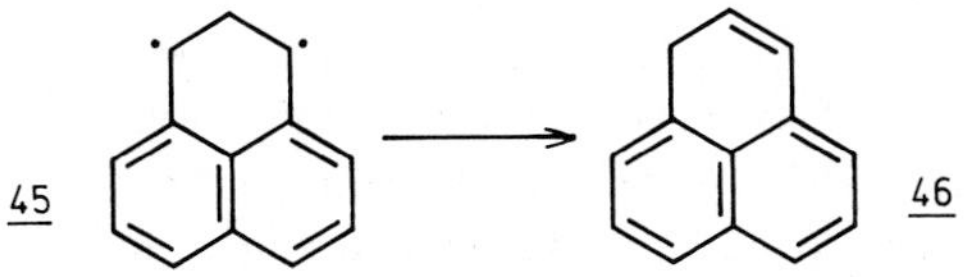

generated photolytically in 2-methyltetrahydrofuran and other glasses at temperatures of 10–77 K, and all seem to have triplet ground states. Tetramethylene-1,4-cyclohexanediyl (43) has been trapped in Ar and adamantane matrices following photolysis of the bicyclic precursor (42) or flow pyrolysis of polyene (44) (Roth et al. 1987). Contrary to a theoretical prediction of a singlet ground state for (43), it exhibits a triplet ESR spectrum in adamantane matrices, and its UV spectrum also suggests a triplet ground state. The matrix UV spectrum of 2,3-dimethylene-1,4-cyclohexanediyl (Roth et al. 1980) is likewise consistent with calculations for a triplet ground state. The thermal isomerization of triplet biradical (45) to (46) and the similar isomerization of a brominated derivative have been studied in polyethylene at 10–160 K by UV, IR and ESR spectroscopy (Fisher and Michl 1987). Below about 100 K, isomerization rates were independent of temperature. Both external and internal heavy atom effects were noted. The results indicate a spin-forbidden proton tunnelling mechanism for the isomerization.

4. Molecules with strained single and multiple bonds

During the past ten years or so, one of the most spectacular successes in organic chemistry has been the synthesis, by Maier and co-workers, of a derivative of tetrahedrane. Matrix experiments played a crucial role in the development of the synthesis (Maier et al. 1978, 1981d). Photolysis of tetra-t-butylcyclopentadienone (47) in Ar matrices led exclusively to the criss-cross adduct (48) (scheme 6). On prolonged photolysis (48) decomposed either to the ketene (49) and subsequently to alkyne (52), or by loss of CO, yielding tetra-t-butyltetrahedrane (51). As might be expected, matrix IR spectra were insufficient

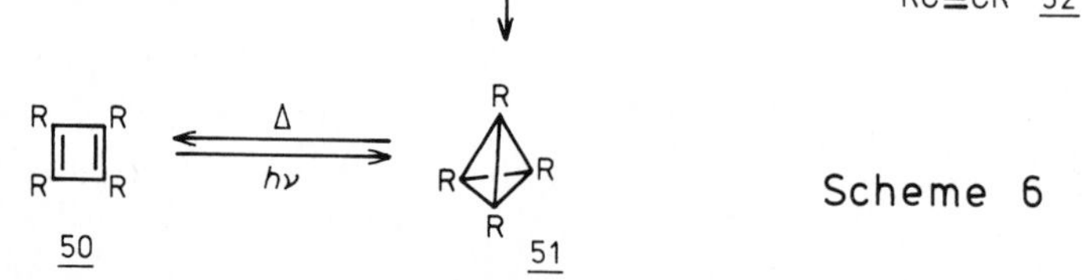

Scheme 6

on their own to identify the tetrahedrane product. Fortunately, however, similar photoreactions of (47) could be carried out in a frozen solvent at 77 K or in liquid solutions between –130°C and room temperature, thus permitting ^{1}H- and ^{13}C-NMR monitoring. The tetrahedrane skeleton is calculated to have about 550 kJ mol^{-1} of strain energy, and the generation of a derivative was achieved despite gloomy theoretical predictions of inevitable failure. Since the t-butyl group was expected to lower the thermostability of the skeletal C–C bonds, it was surprising that the tetra-*t*-butyl derivative (<u>51</u>) could be formed at all. Even more surprising was the subsequent discovery that (<u>51</u>) could be isolated from the solution photoproduct mixture by chromatography and was stable up to about $+130°C$, whereupon it rearranged to tetra-*t*-butylcyclobutadiene (<u>50</u>). UV-irradiation of (<u>50</u>), either in Ar at 10 K or in solution at room temperature, regenerated (<u>51</u>). Once the tetrahedrane was available as a stable solid, it was characterized very fully by photoelectron spectroscopy (Heilbronner et al. 1980), by ^{13}C- and ^{1}H-NMR, IR and Raman spectroscopy and by mass spectrometry (Maier et al. 1981e, Loerzer et al. 1983), and ultimately by an X-ray crystal structure analysis (Irngartinger et al. 1984). Its identity has thus been placed apparently beyond controversy. It is illustrative of the limitations of current matrix techniques, however, that this could not have been the case had (<u>51</u>) not been astonishingly stable.

So far as the author is aware, tetra-*t*-butyltetrahedrane (<u>51</u>) remains the only derivative of tetrahedrane which has been directly observed. Maier's group have made attempts to synthesize dimethyltetrahedrane (Maier and Reisenauer 1981a), tetramethyltetrahedrane (Maier and Reisenauer 1981b, Maier et al. 1981b,c), bis(trimethylsilyl)tetrahedrane (Maier et al. 1982a), and tetrahedrane itself (Lage et al. 1982), principally by matrix photolysis of likely precursors. Up until now, unfortunately, all these attempts have failed.

The introduction of strained bonds into molecules can be achieved in gas-phase dehalogenation reactions brought about by alkali metal atoms. Matrix-isolated [2.2.1]propellane (<u>53</u>), with a highly strained central single bond, has been generated in this manner (Walker et al. 1982); so too has the even more strained [2.1.1]propellane (Wiberg et al. 1983). When the two halogen atoms in the precursor are vicinally related, a strained double bond can be formed, as for example in bicyclo[2.2.0]hex-1(4)-ene (<u>54</u>) (Wiberg et al. 1986) and adamantene (<u>55</u>) (Conlin et al. 1979), both of which have been isolated in Ar matrices and characterized by IR spectroscopy.

Adamantene ($\underline{55}$) is a member of the class of highly reactive and theoretically interesting molecules with bridgehead double bonds which violate Bredt's rule. A more extreme example is the tricyclodecene ($\underline{56}$), which has been isolated in Ar at 10 K, after flow pyrolysis of a β-lactone precursor (Radziszewski et al. 1986). The double bond in ($\underline{56}$) is pyramidalized at both ends, and the resulting strain is apparent from the unusually low frequency $C = C$ stretch at 1557 cm^{-1} and from a broad UV absorption with λ_{max} near 245 nm, evidence of a low-lying $\pi\pi^*$ state. Anti-Bredt molecules containing bridgehead $C = N$ bonds are accessible through photolysis of bridgehead azides. Several research groups (Dunkin et al. 1983, Michl et al. 1983, Sheridan and Ganzer 1983, Radziszewski et al. 1984, 1985a,b) have reported matrix studies aimed at characterizing strained imines of this type. For instance, the azidobicyclo[2.2.2]octane ($\underline{57}$), when photolysed in Ar or N_2 at 12 K, yielded the anti-Bredt imine ($\underline{58}$) (Dunkin et al. 1983), with $\nu(C = N)$ some 60 cm^{-1} lower than normal at 1603–1601 cm^{-1}. Most of the strained bridgehead imines examined so far have $\nu(C = N)$ bands near 1600 cm^{-1}. The azaadamantene ($\underline{59}$) is therefore quite exceptionally strained,

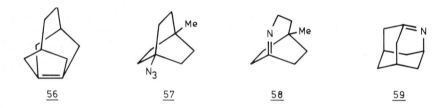

| $\underline{56}$ | $\underline{57}$ | $\underline{58}$ | $\underline{59}$ |

having $\nu(C = N)$ at 1451 cm^{-1} (Ar matrix) and an electronic absorption in the visible region (Radziszewski et al. 1984). This molecule was found to be sufficiently reactive to form a $2 + 2$ cycloadduct with CO_2 at 36 K. The most detailed studies of bridgehead imines have been carried out by Michl's group, using elegant combinations of IR, Raman, UV-visible, circular dichroism and ESR spectroscopy, and Ar, N_2, 3-methylpentane and polyethylene matrices. They have, for example, identified the E and Z isomers of 2-azabicyclo[3.2.1]oct-1-ene (Radziszewski et al. 1985b), and managed to inter-convert the enantiomeric forms of 4-azahomoadamant-3-ene by irradiation with circularly polarized light (Radziszewski et al. 1985a). In two cases they have also been able to detect bridgehead nitrene intermediates (Radziszewski et al. 1984, 1985a), which are thus the first non-aromatic organic nitrenes to be directly observed.

Among the early successes of the matrix spectroscopy of reactive organic species were the experiments conducted by Chapman and co-workers in which benzyne ($\underline{61}$, R = H) was generated from a variety of precursors (Dunkin 1980). The matrix IR spectrum of benzyne coupled with a set of force-field calculations (Laing and Berry 1976), suggested that the molecule has a cycloalkyne structure,

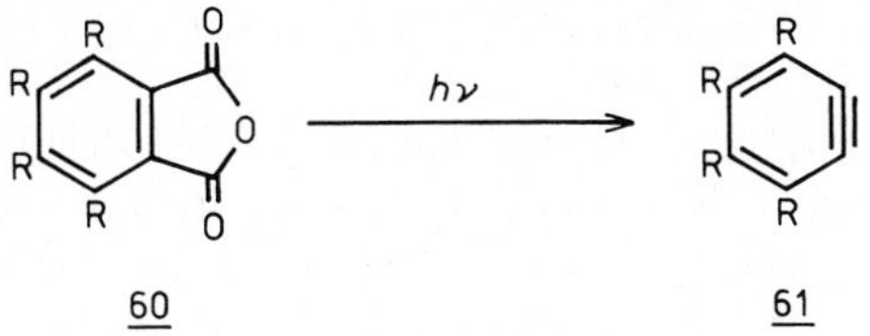

60   61

i.e. with a genuine triple bond. The subsequent discovery (Dunkin and MacDonald 1979) that phthalic anhydride (60, R = H) provided an extremely convenient precursor for benzyne facilitated the acquisition of the matrix IR spectrum of its deuteriated analogue (61, R = D). Agreement between the experimental matrix IR spectrum of the deuteriated benzyne and that predicted from the preferred force-field was reasonable for bands below 2000 cm^{-1}, but much less satisfactory for the critical bands above 2000 cm^{-1}. More recently, a corrected force field, however, has succeeded in reconciling all the data (Nam and Levoi 1985) and still suggest the presence of a triple bond. Attempts to utilize the aromatic anhydride route for the matrix preparation of the analogous pyridines unfortunately failed, leading instead to wholesale decomposition (Dunkin and MacDonald 1982). Matrix photolysis of the bisdiazoketone (62) led firstly to the cyclopropenone (63) and then, by loss of CO, to acenaphthyne (64) (Chapman et al. 1981). The $v(C \equiv C)$ IR band of (64) has not been definitely

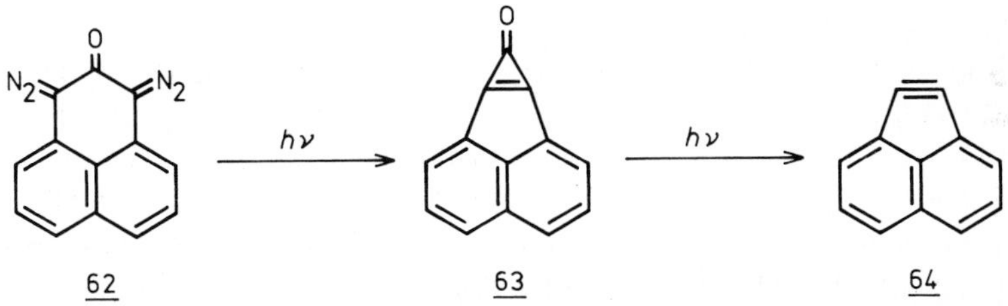

62   63   64

assigned, but an absorption at 1930 cm^{-1} (Ar at 15 K) could possibly correspond to this vibration. The attempted generation of cyclopentyne by analogous matrix photodecomposition of a bisdiazocyclohexanone proceeded via the expected cyclopropenone and loss of CO to a rearranged allenic product, but the hoped for cycloalkyne, which was a likely intermediate, was not detected (Chapman et al. 1981). Cyclic allenes and ketenimines, in which the cumulene systems are incorporated into small or medium sized rings, have carbon atoms of normally linear geometry forced into bent arrangements, and are thus related to cycloalkynes. The reader is referred to section 3 for a discussion of these species.

5. Organosilicon compounds

An area of distinct growth in recent years has been the matrix study of organisilicon molecules. It is of interest to note that a number of silicon analogues of simple oxygen-containing organic molecules have been generated in matrices by reaction of silane or silaalkanes with oxygen atoms derived from ozone photolysis. Strictly speaking not all of these are organic molecules, of course, but they include silanone ($H_2Si = O$), silanol (SiH_3OH), silanoic acid ($HSiO_2H$), silicic acid ($HO \cdot SiO \cdot OH$), methylsilanone ($MeSiHO$), and dimethyl-silanone (Me_2SiO) (Withnall and Andrews 1985a,b, 1986). The matrix IR spectra show that silanone $v(Si = O)$ bands occur in the region 1202 cm^{-1} (H_2SiO) to 1210 cm^{-1} ($Me_2Si = O$), while the $v(Si = O)$ frequency increases to 1249 cm^{-1} for silanoic acid and to 1270 cm^{-1} for silicic acid.

The largest body of work on organosilicon species in matrices so far has been devoted to the characterization, particularly by IR spectroscopy, of molecules containing $C = Si$ bonds (Raabe and Michl 1985, Maier 1986). The first approach to silenes (silaethylenes) was the photolysis of 1-diazo-2,2-dimethyl-2-silapropane ($\underline{65}$, $R^1 = R^2 = R^3 = Me$, $R^4 = H$) and its diazirine isomer in Ar matrices, which was reported simultaneously by Chapman et al. (1976) and Chedekel et al. (1976) (scheme 7). Elimination of N_2 gave first the carbene ($\underline{66}$, $R^1 = R^2 = R^3 = Me$, $R^4 = H$), which was identified by dimerization to E-1,2-bis(trimethylsilyl)ethylene on warming to room temperature, and secondly 1,1,2-trimethylsilene ($\underline{68}$, $R^1 = R^2 = R^3 = Me$, $R^4 = H$). Molecular vibrations with significant $v(Si = C)$ contributions were not assigned by these pioneering workers, but identification of the silene was confirmed by trapping it with methanol and by its thermal dimerization to stereoisomers of the 1,3-disilacy-clobutane ($\underline{69}$, $R^1 = R^2 = R^3 = Me$, $R^4 = H$). More recently, a similar matrix photolysis of ($\underline{65}$, $R^1 = SiMe_3$, $R^2 = R^3 = Me$, $R^4 = CO_2Et$) was shown (Sekiguchi et al. 1985) to yield a silaacrylate ($\underline{68}$, $R^1 = SiMe_3$, $R^2 = R^3 = Me$, $R^4 = CO_2Et$). Flow pyrolysis of monosilacyclobutanes ($\underline{67}$) has provided a second route to matrix-isolated silenes (scheme 7), which has been exploited by Mal'tsev, Nefedov and co-workers (Mal'tsev et al. 1979, 1982b, 1984a, Nefedov et al. 1980, Khabashesku et al. 1983), and by Gusel'nikov et al. (1980). The latter group have extended the reaction to silathietanes also (Gusel'nikov et al. 1983). For $Me_2Si = CH_2$, $v(Si = C)$ has been assigned (Gusel'nikov et al. 1980, Nefedov et al. 1980) to an IR band at 1004–1001 cm^{-1}. In the case of $Me_2Si = CD_2$ and $(CD_3)_2Si = CD_2$, on the other hand, $v(Si = C)$ corresponds to two bands in each IR spectrum: at 895 and 1118 cm^{-1} and at 866 and 1112 cm^{-1}, respectively (Mal'tsev et al. 1984a). The higher frequency band of each pair has the larger contribution from $v(Si = C)$. Maier et al. (1981a, 1984b) and Reisenauer et al. (1982) have developed a third, independent route to silenes, including the parent compound itself ($\underline{68}$, $R^1 = R^2 = R^3 = R^4 = H$), which involves flow pyrolysis of silabicyclo[2.2.2]octadienes ($\underline{70}$, $X = H, CF_3$) (scheme 7). Matrix IR spectra for a series of silenes ($CH_2 = SiR^2R^3$, with $R^2, R^3 = H, D, Me, Cl$) have thus been

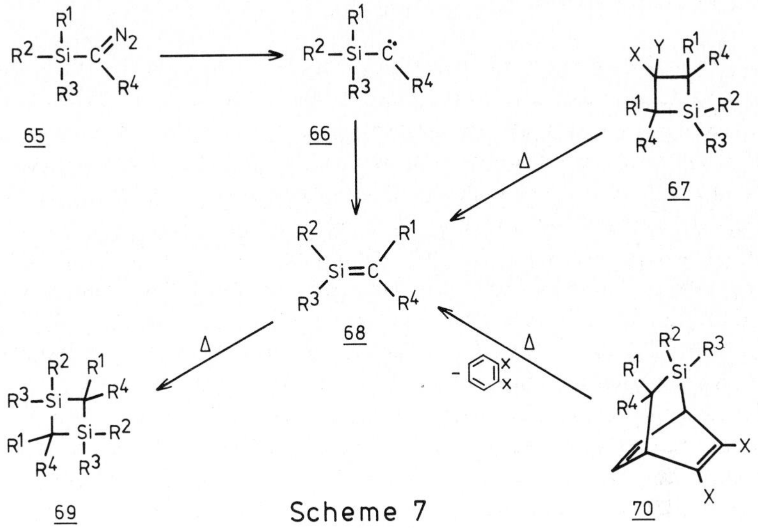

Scheme 7

obtained, and the general tendency for $v(Si = C)$ to mix with other vibrations has been confirmed. Recently, the techniques of photoselection and IR linear dichroism measurements have been employed (Arrington et al. 1984, Vančik et al. 1985, Raabe et al. 1986) to aid assignment of the IR absorptions of 1-methylsilene (73), particularly the $v(Si = C)$ band at 988 cm^{-1}.

A remarkable chemical property of some of the matrix-isolated silenes examined is a facile reversible isomerization to the corresponding silylenes, the silicon analogues of carbenes. 1-Methylsilene (73), e.g., disappeared after a short irradiation with 254 nm light, yielding dimethylsilylene (72), which has a broad UV absorption with λ_{max} near 450 nm (Reisenauer et al. 1982, Maier et al. 1984c). Subsequent irradiation at longer wavelengths reversed the reaction. Drahnak et al. (1979, 1981), Arrington et al. (1983, 1984), Vančik et al. (1985), and Raabe et al. (1986) have found several alternative precursors for dimethylsilylene, e.g. [Me$_2$Si]$_6$ and the diazide (71) (scheme 8). It was shown (Drahnak

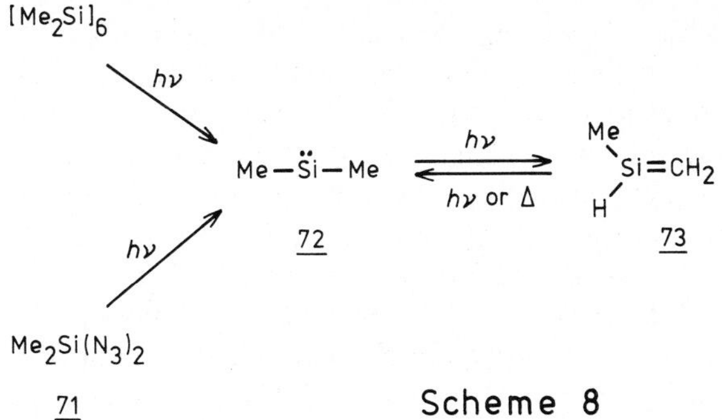

Scheme 8

et al. 1979, 1981) that in 3-methylpentane matrices 1-methylsilene (73), as well as its photochemical rearrangement, also underwent a thermal rearrangement to dimethylsilylene (72) at about 100 K.

Other important goals of matrix organosilicon chemistry have been syntheses of silabenzene and related silaarenes. Flow pyrolysis of any of the precursors (74), (75) or (76, X = H) (scheme 9) was found to yield a common product, which could be trapped in Ar matrices (Maier et al. 1980, 1982b, 1984a). Besides C–H stretches at 3060–3030 cm^{-1}, eighteen IR absorptions were observed for this product, including two weak bands at 2244 and 2219 cm^{-1}, assigned as Si–H stretches. UV absorptions at 212, 272 and 320 nm were also found to belong to the product, which was therefore identified as silabenzene (77). Flow pyrolysis of a methyl derivative of one of the silabenzene precursors (76, X = Me) similarly gave silatoluene (79), which has a well-resolved vibronic progression beginning at 322 nm (Kreil et al. 1980). The UV spectra of both (77) and (79) seem to confirm earlier predictions (Dewar et al. 1975, Schlegel et al. 1978) of aromatic character for silabenzenes. Irradiation of (77) with 320 nm light led to its disappearance. The change was accompanied by a significant shift in the ν(Si–H) band to 2142 cm^{-1}, indicating a change in geometry at the silicon atom from trigonal to tetrahedral, and thus suggesting that photo-isomerization to Dewar silabenzene (78) had occurred. Subsequently irradiation at 240 nm regenerated silabenzene. More recently, 1,4-disilabenzene has been isolated in Ar matrices from flow pyrolysis of 1,4-disilacyclohexa-2,5-diene (Maier et al. 1985b). Its UV spectrum has absorptions at 275, 340 and 408 nm, the last having

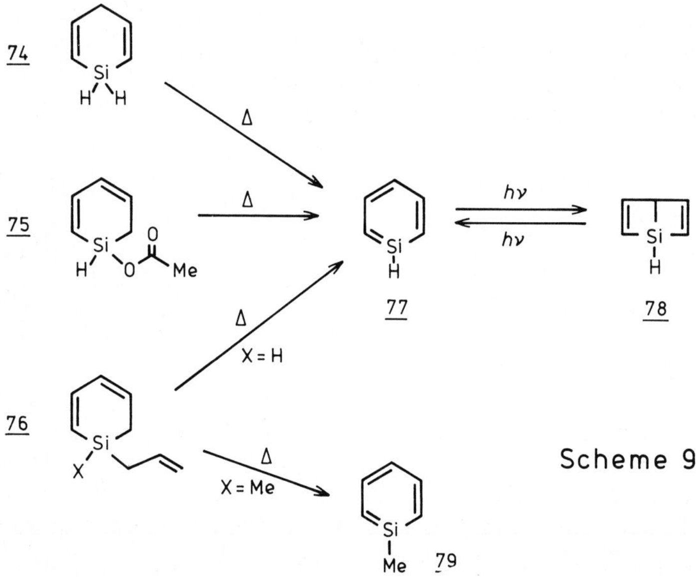

Scheme 9

clear vibrational structure. The UV spectra of benzene, silabenzene and disilabenzene form a series, all with similar sequences of absorptions, but shifted to longer wavelengths with increasing silicon substitution.

6. Carbonyl, thiocarbonyl and related compounds

6.1. Carbonyl compounds

Compared with the vast literature on the photochemistry of carbonyl compounds at more normal temperatures, there have been very few reports of matrix-isolation studies of these processes. Hitherto, the principal aims of matrix chemists working with organic carbonyl compounds have been to characterize spectroscopically a number of interesting and unstable ketones, lactones or ketenes. Despite an early paper (Chapman and McIntosh 1971) describing a partial IR spectrum of cyclopentadienone ($\underline{80}$, R = H) at 77 K and its reported occurrence as a product of certain matrix reactions (Bell and Dunkin 1983, Chapman and Hess 1984), a thorough spectroscopic study of this important molecule has only recently been made. Maier et al. (1985a) generated cyclopentadienone in Ar matrices from a variety of precursors, and obtained both IR $[\nu(C=O)$ 1727, 1724 cm$^{-1}]$ and UV spectra. The tetrafluoro derivative ($\underline{80}$, R = F) $[\nu(C=O)$ 1780, 1756, 1733 cm$^{-1}]$ has been trapped in Ar matrices after thermolysis of the anhydride ($\underline{81}$) (Grayston et al. 1980), and the tetrachloro derivative ($\underline{80}$, R = Cl) $[\nu(C=O)$ 1778, 1758 cm$^{-1}]$ was the major final product in the matrix photo-oxidation of tetrachlorodiazocyclopentadiene (Dunkin et al. 1986b). The highly strained 7-norbornadienone ($\underline{82}$) was formed in matrix photolysis of trienone ($\underline{83}$) (Birney and Berson 1985, 1986) and independently from two azo precursors (LeBlanc and Sheridan 1985). Strong carbonyl bands for ($\underline{82}$) were found by Birney and Berson at 1846 and

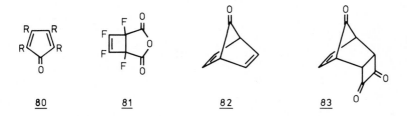

$\underline{80}$	$\underline{81}$	$\underline{82}$	$\underline{83}$

1795 cm^{-1} (shifted to 1861 and 1808 cm^{-1} in the d$_6$-analogue), and by LeBlanc and Sheridan at 1861 and 1790 cm^{-1}. The discrepancies in the $\nu(C=O)$ frequencies are probably attributable to site effects, but the high values of the frequencies, the existence of two carbonyl bands, and the observed shift on deuterisation all helped to confirm the identity of ($\underline{82}$). In polyethylene,

3-methylpentane or other organic solvents, (82) is stable up to about 200 K, thus allowing characterization by ^{1}H-NMR at 183–195 K. On being warmed above 200 K, or on photolysis, however, (82) decomposes to benzene and CO.

2-benzopyran-3-one (84), an unstable *o*-quinonoid isomer of coumarin, has been generated along with ketene (85) by photolysis of bis-lactone (86) in Ar at 10 K or *n*-propyl ether at 77 K (Bleasdale et al. 1983). It exhibited a series of electronic absorptions at 509, 497, 472, 463, 440, 417 and 395(sh) nm, and a $v(C = O)$ band at 1758 cm^{-1}. Flow pyrolysis of (86) probably also gave (84) initially, but the major product, isocoumarin, resulted from further rearrangement, presumably via (85). An *o*-quinonoid structure occurs again in the photoenol (87), which has been observed following photolysis (250–300 nm) of *o*-tolualdehyde in Ar, N$_2$, Xe and CO matrices at 12 K (Gebicki and Krantz 1984). *o*-tolualdehyde has *syn* and *anti* conformers of similar stability, but only the *syn* form underwent the isomerization. Photoenol (87) has a broad UV absorption with λ_{max} at 393 nm (Xe matrix).

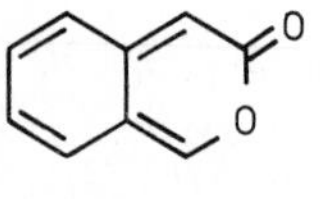

84 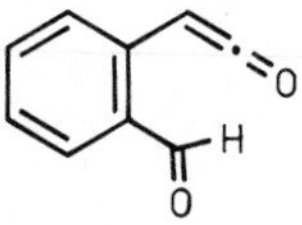   85 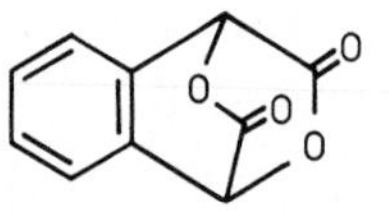   86 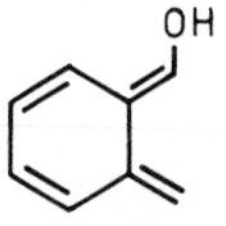   87

6.2. *Thiocarbonyl compounds*

The versatility of methylene and vinylene carbonates and their sulphur analogues as photoprecursors for unstable carbonyl and thiocarbonyl compounds has been demonstrated, particularly by Torres et al. (1981, 1982a,b, 1983c). For instance, methylene trithiocarbonate (88) eliminated CS$_2$ when photolysed (254 nm) in Ar and N$_2$ matrices, leaving thioformaldehyde, CH$_2$S, [$v(C = S)$ 1052 cm^{-1} in Ar] (Torres et al. 1982a). The vinylene thiocarbonate (89), on the other hand, under similar conditions lost CO, giving the previously unknown thioglyoxal (90) (Torres et al. 1983c). Thioglyoxal was found to be stable to further irradiation at 270 nm or longer wavelengths, but with 210 nm light slowly disappeared. One of the ultimate products was possibly the valence tautomer, 1,2-oxathiete (91), but a definite assignment was precluded owing to the weakness of its IR spectrum and the simultaneous occurrence of other decomposition pathways. Matrix photolysis (270 nm) of vinylene dithiocar-

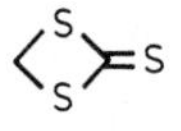

88 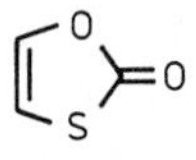   89 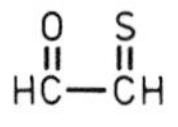   90    91

bonate (92) gave *s-cis* dithioglyoxal (93), which existed in photoequilibrium with its valence tautomer, dithiete (94), and its *s-trans* conformer (95) (scheme 10) (Torres et al. 1982b). Further UV irradiation resulted in a hydrogen shift, giving the thioketene (96) [$\nu(C=C=S)$ 1750 cm^{-1}), and finally elimination of a sulphur atom. Benzodithiete, the benzannelated analogue of (94), has been thermally generated from *o*-phenylene dithiocarbonate and two other precursors, and its Ar-matrix IR and gas-phase photoelectron spectra have been obtained (Breitenstein et al. 1982).

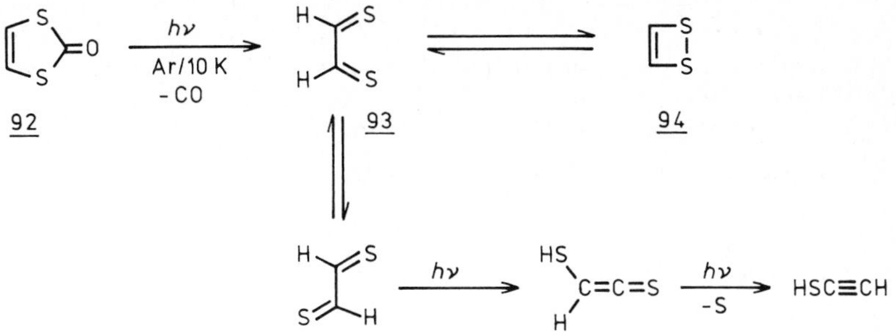

Scheme 10

Flow pyrolysis of the series of diallyl compounds (97, X, Y = O, S) resulted in elimination of 1,5-hexadiene and gave *p*-benzoquinone and its monothio and dithio analogues, which were detected by gas-phase photoelectron or Ar-matrix IR spectroscopy (Bock et al. 1983). An IR and UV study of the photo-enolization of monothioacetylacetone with 300 nm light (Gebicki and Krantz 1981a,b) showed that the predominant product was the H-bonded thioenol (98), which was formed together with a minor amount of the less stable conformer (99). The proportion of (99) could be increased by irradiation with 350 nm light, but reversion to (98) occurred on subsequent irradiation at 300 nm or on warming. These matrix results are claimed to settle a controversy over the enolization of monothioacetylacetone in solution, in which hitherto it has been believed that both an enol and a thioenol were formed.

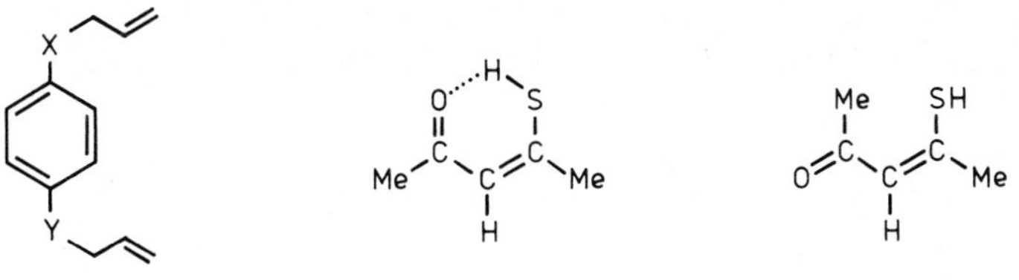

References

Adrian, F.J., J. Bohandy and B.F. Kim, 1984, J. Chem. Phys. **81**, 3805.

Allamandola, L.J., D. Lucas and G.C. Pimentel, 1978, Rev. Sci. Instrum. **49**, 913.

Andrews, L., and B.J. Kelsall, 1987, J. Phys. Chem. **91**, 1435.

Andrews, L., and C.K. Kohlmiller, 1982, J. Phys. Chem. **86**, 4548.

Arrington, C.A., R. West and J. Michl, 1983, J. Am. Chem. Soc. **105**, 6176.

Arrington, C.A., K.A. Klingensmith, R. West and J. Michl, 1984, J. Am. Chem. Soc. **106**, 525.

Baird, M.S., I.R. Dunkin, N. Hacker, M. Poliakoff and J.J. Turner, 1981, J. Am. Chem. Soc. **103**, 5190.

Bally, T., and S. Masamune, 1980, Tetrahedron **36**, 343.

Barnes, A.J., W.J. Orville-Thomas, A. Müller and R. Gaufrès, eds, 1981, Matrix Isolation Spectroscopy (Reidel, Dordrecht).

Bass, A.M., and H.P. Broida, eds, 1960, Formation and Trapping of Free Radicals (Academic Press, New York).

Beeler, A.J., A.M. Orendt, D.M. Grant, P.W. Cutts, J. Michl, K.W. Zilm, J.W. Downing, J.C. Facelli, M.S. Schindler and W. Kutzelnigg, 1984, J. Am. Chem. Soc. **106**, 7672.

Bell, G.A., and I.R. Dunkin, 1983, J. Chem. Soc., Chem. Commun., p. 1213.

Bell, G.A., and I.R. Dunkin, 1985, J. Chem. Soc. Faraday Trans II, **81**, 725.

Bell, G.A., I.R. Dunkin and C.J. Shields, 1985, Spectrochim. Acta A **41**, 1221.

Birney, D.M., and J.A. Berson, 1985, J. Am. Chem. Soc. **107**, 4553.

Birney, D.M., and J.A. Berson, 1986, Tetrahedron **42**, 1561.

Bleasdale, D.A., D.W. Jones, G. Maier and H.P. Reisenauer, 1983, J. Chem. Soc., Chem. Commun., p. 1095.

Bock, H., S. Mohmand, T. Hirabayashi, G. Maier and H.P. Reisenauer, 1983, Chem. Ber. **116**, 273.

Braathen, G.O., P. Klaeboe, C.J. Nielsen and H. Priebe, 1984, J. Mol. Struct. **115**, 197.

Breitenstein, M., R. Schulz and A. Schweig, 1982, J. Org. Chem. **47**, 1979.

Cartland, H.E., and G.C. Pimentel, 1986a, J. Phys. Chem. **90**, 1822.

Cartland, H.E., and G.C. Pimentel, 1986b, J. Phys. Chem. **90**, 5485.

Cesaro, S.N., H. Frei and G.C. Pimentel, 1983, J. Phys. Chem. **87**, 2142.

Chapman, O.L., 1979, Pure Appl. Chem. **51**, 331.

Chapman, O.L., and C.J. Abelt, 1987, J. Org. Chem. **52**, 1219.

Chapman, O.L., and T.C. Hess, 1979, J. Org. Chem. **44**, 962.

Chapman, O.L., and T.C. Hess, 1984, J. Am. Chem. Soc. **106**, 1842.

Chapman, O.L., and J.-P. LeRoux, 1978, J. Am. Chem. Soc. **100**, 282.

Chapman, O.L., and C.L. McIntosh, 1971, J. Chem. Soc., Chem. Commun., p. 770.

Chapman, O.L., and R.S. Sheridan, 1979, J. Am. Chem. Soc. **101**, 3690.

Chapman, O.L., C.-C. Chang, J. Kolc, M.E. Jung, J.A. Lowe, T.J. Barton and M.L. Tumey, 1976, J. Am. Chem. Soc. **98**, 7844.

Chapman, O.L., R.S. Sheridan and J.-P. LeRoux, 1978, J. Am. Chem. Soc. **100**, 6245.

Chapman, O.L., R.S. Sheridan and J.-P. LeRoux, 1979, Recl. Trav. Chim. Pays-Bas **98**, 334.

Chapman, O.L., J. Gano, P.R. West, M. Regitz and G. Maas, 1981, J. Am. Chem. Soc. **103**, 7033.

Chapman, O.L., R.J. McMahon and P.R. West, 1984, J. Am. Chem. Soc. **106**, 7973.

Chedekel, M.R., M. Skoglund, R.L. Kreeger and H. Schechter, 1976, J. Am. Chem. Soc. **98**, 7846.

Cirelli, G., A. Russu, R. Wolf, M. Rudin, A. Schweiger and H.H. Günthard, 1982, Chem. Phys. Lett. **92**, 223.

Collins, S.T., and G.C. Pimentel, 1984, J. Phys. Chem. **88**, 4258.

Conlin, R.T., R.D. Miller and J. Michl, 1979, J. Am. Chem. Soc. **101**, 7637.

Cradock, S., and A.J. Hinchcliffe, 1975, Matrix Isolation (Cambridge University Press, Cambridge, UK).

Dewar, M.J.S., D.H. Lo and C.A. Ramsden, 1975, J. Am. Chem. Soc. **97**, 1311.

Donnelly, T., I.R. Dunkin, D.S.D. Norwood, A. Prentice, C.J. Shields and P.C.P. Thomson, 1985, J. Chem. Soc., Perkin Trans. **2**, 307.

Dows, D.A., E. Whittle and G.C. Pimentel, 1955, J. Chem. Phys. **23**, 1475.

Drahnak, T.J., J. Michl and R. West, 1979, J. Am. Chem. Soc. **101**, 5427.

Drahnak, T.J., J. Michl and R. West, 1981, J. Am. Chem. Soc. **103**, 1845.

Dunkin, I.R., 1980, Chem. Soc. Rev. **9**, 1.

Dunkin, I.R., 1986, Spectrochim. Acta A **42**, 649.

Dunkin, I.R., and G.A. Bell, 1985, Tetrahedron **41**, 339.

Dunkin, I.R., and J.G. MacDonald, 1979, J. Chem. Soc., Chem. Commun., p. 772.

Dunkin, I.R., and J.G. MacDonald, 1982, Tetrahedron Lett. **23**, 4839.

Dunkin, I.R., and C.J. Shields, 1986, J. Chem. Soc., Chem. Commun., p. 154.

Dunkin, I.R., and D. Sweeney, 1986, unpublished results.

Dunkin, I.R., and P.C.P. Thomson, 1980a, J. Chem. Soc., Chem. Commun., p. 499.

Dunkin, I.R., and P.C.P. Thomson, 1980b, Tetrahedron Lett. **21**, 3813.

Dunkin, I.R., and P.C.P. Thomson, 1982, J. Chem. Soc., Chem. Commun., p. 1192.

Dunkin, I.R., C.J. Shields, H. Quast and B. Seiferling, 1983, Tetrahedron Lett. **24**, 3887.

Dunkin, I.R., T. Donnelly and T.S. Lockhart, 1985, Tetrahedron Lett. **26**, 359.

Dunkin, I.R., D. Griller, A.S. Nazran, D.J. Northcott, J.M. Park and A.H. Reddoch, 1986a, J. Chem. Soc., Chem. Commun., p. 435.

Dunkin, I.R., G.A. Bell, F.G. McLeod and A. McCluskey, 1986b, Spectrochim. Acta A **42**, 567.

Facelli, J.C., A.M. Orendt, A.J. Beeler, M.S. Solum, G. Depke, K.D. Malsch, J.W. Downing, P.S. Murthy, D.M. Grant and J. Michl, 1985, J. Am. Chem. Soc. **107**, 6749.

Facelli, J.C., A.M. Orendt, M.S. Solum, G. Depke, D.M. Grant and J. Michl, 1986, J. Am. Chem. Soc. **108**, 4268.

Facelli, J.C., D.M. Grant and J. Michl, 1987, Acc. Chem. Res. **20**, 152.

Fisher, J.J., and J. Michl, 1987, J. Am. Chem. Soc. **109**, 583.

Frei, H., 1983, J. Chem. Phys. **79**, 748.

Frei, H., and G.C. Pimentel, 1983a, J. Chem. Phys. **78**, 3698.

Frei, H., and G.C. Pimentel, 1983b, J. Chem. Phys. **79**, 3307.

Frei, H., and G.C. Pimentel, 1985, Ann. Rev. Phys. Chem. **36**, 491.

Frei, H., L. Fredin and G.C. Pimentel, 1981, J. Chem. Phys. **74**, 397.

Ganzer, G.A., R.S. Sheridan and M.T.H. Liu, 1986, J. Am. Chem. Soc. **108**, 1517.

Gebicki, J., and A. Krantz, 1981a, J. Am. Chem. Soc. **103**, 4521.

Gebicki, J., and A. Krantz, 1981b, J. Chem. Soc., Chem. Commun., p. 486.

Gebicki, J., and A. Krantz, 1984, J. Chem. Soc., Perkin Trans. **2**, 1623.

Grayston, M.W., W.D. Saunders and D.M. Lemal, 1980, J. Am. Chem. Soc. **102**, 413.

Gusel'nikov, L.E., V.V. Volkova, V.G. Avakyan and N.S. Nametkin, 1980, J. Organomet. Chem. **201**, 137.

Gusel'nikov, L.E., V.V. Volkova, V.G. Avakyan, N.S. Nametkin, M.G. Voronkov, S.V. Kirpichenko and E.N. Suslova, 1983, J. Organomet. Chem. **254**, 173.

Hallam, H.E., ed., 1973, Vibrational Spectroscopy of Trapped Species (Wiley, London).

Hayes, R.A., T.C. Hess, R.J. McMahon and O.L. Chapman, 1983, J. Am. Chem. Soc. **105**, 7786.

Heilbronner, E., T.B. Jones, A. Krebs, G. Maier, K.-D. Malsch, J. Pocklington and A. Schmelzer, 1980, J. Am. Chem. Soc. **102**, 564.

Horspool, W.M., 1983, Ann. Rep. Prog. Chem. Sect. C, **80**, 87.

Irngartinger, H., A. Goldmann, R. Jahn, M. Nixdorf, H. Rodewald, G. Maier, K.-D. Malsch and R. Emrich, 1984, Angew. Chem. Int. Ed. Engl. **23**, 993.

Jacox, M.E., 1980, Chem. Phys. **53**, 307.

Jacox, M.E., 1981a, Chem. Phys. **58**, 289.

Jacox, M.E., 1981b, Chem. Phys. **59**, 213.

Jacox, M.E., 1982a, Chem. Phys. **69**, 407.

Jacox, M.E., 1982b, J. Phys. Chem. **86**, 670.

Jacox, M.E., 1984a, J. Phys. Chem. **88**, 445.

Jacox, M.E., 1984b, J. Phys. Chem. Ref. Data **13**, 945.

Jain, R., G.J. Snyder and D.A. Dougherty, 1984, J. Am. Chem. Soc. **106**, 7294.

Jonkman, H.T., and J. Michl, 1978, J. Chem. Soc., Chem. Commun., p. 751.

Jonkman, H.T., J. Michl, R.N. King and J.D. Andrade, 1978, Anal. Chem. **50**, 2078.

Kasai, P.H., 1986, J. Phys. Chem. **90**, 5034.

Kasai, P.H., and H.C. McBay, 1984, J. Phys. Chem. **88**, 5932.

Kesselmayer, M.A., and R.S. Sheridan, 1986a, J. Am. Chem. Soc. **108**, 99.

Kesselmayer, M.A., and R.S. Sheridan, 1986b, J. Am. Chem. Soc. **108**, 844.

Khabashesku, V.N., E.G. Baskir, A.K. Mal'tsev and O.M. Nefedov, 1983, Izv. Akad. Nauk SSSR, Ser. Khim., p. 238.

Knudsen, A.K., and G.C. Pimentel, 1983, J. Chem. Phys. **78**, 6780.

Kreil, C.L., O.L. Chapman, G.T. Burns and T.J. Barton, 1980, J. Am. Chem. Soc. **102**, 841.

Kühne, H., 1978, Ber. Bunsenges. Phys. Chem. **82**, 15.

Kuzaj, M., H. Lüerssen and C. Wentrup, 1986, Angew. Chem. Int. Ed. Engl. **25**, 480.

Laganis, E.D., D.S. Janik, T.J. Curphey and D.M. Lemal, 1983, J. Am. Chem. Soc. **105**, 7457.

Lage, H.W., H.P. Reisenauer and G. Maier, 1982, Tetrahedron Lett. **23**, 3893.

Laing, J.W., and R.S. Berry, 1976, J. Am. Chem. Soc. **98**, 660.

LeBlanc, B.F., and R.S. Sheridan, 1985, J. Am. Chem. Soc. **107**, 4554.

Lee, Y.-P., and G.C. Pimentel, 1981a, J. Chem. Phys. **74**, 4851.

Lee, Y.-P., and G.C. Pimentel, 1981b, J. Chem. Phys. **75**, 4241.

Leyva, E., M.S. Platz, G. Persy and J. Wirz, 1986, J. Am. Chem. Soc. **108**, 3783.

Loerzer, T., R. Machinek, W. Lüttke, L.H. Franz, K.-D. Malsch and G. Maier, 1983, Angew. Chem. Int. Ed. Engl. **22**, 878.

Lund, P.A., Z. Hasan, P.N. Schatz, J.H. Miller and L. Andrews, 1982, Chem. Phys. Lett. **91**, 437.

Maier, G., 1986, Pure Appl. Chem. **58**, 95.

Maier, G., and H.P. Reisenauer, 1981a, Chem. Ber. **114**, 3916.

Maier, G., and H.P. Reisenauer, 1981b, Chem. Ber. **114**, 3959.

Maier, G., and H.P. Reisenauer, 1986, Angew. Chem. Int. Ed. Engl. **25**, 819.

Maier, G., S. Pfriem, U. Schäfer and R. Matusch, 1978, Angew. Chem. Int. Ed. Engl. **17**, 520.

Maier, G., G. Mihm and H.P. Reisenauer, 1980, Angew. Chem. Int. Ed. Engl. **19**, 52.

Maier, G., G. Mihm and H.P. Reisenauer, 1981a, Angew. Chem. Int. Ed. Engl. **20**, 597.

Maier, G., M. Schneider, G. Kreiling and W. Mayer, 1981b, Chem. Ber. **114**, 3922.

Maier, G., W. Mayer, H.-A. Freitag, H.P. Reisenauer and R. Askani, 1981c, Chem. Ber. **114**, 3935.

Maier, G., S. Pfriem, U. Schäfer, K.-D. Malsch and R. Matusch, 1981d, Chem. Ber. **114**, 3965.

Maier, G., S. Pfriem, K.-D. Malsch, H.-O. Kalinowski and K. Dehnicke, 1981e, Chem. Ber. **114**, 3988.

Maier, G., M. Hoppe, H.P. Reisenauer and C. Krüger, 1982a, Angew. Chem. Int. Ed. Engl. **21**, 437.

Maier, G., G. Mihm and H.P. Reisenauer, 1982b, Chem. Ber. **115**, 801.

Maier, G., H.P. Reisenauer and T. Sayraç, 1982c, Chem. Ber. **115**, 2192.

Maier, G., T. Sayraç and H.P. Reisenauer, 1982d, Chem. Ber. **115**, 2202.

Maier, G., H.P. Reisenauer, B. Rohde and K. Dehnicke, 1983, Chem. Ber. **116**, 732.

Maier, G., G. Mihm, R.O.W. Baumgärtner and H.P. Reisenauer, 1984a, Chem. Ber. **117**, 2337.

Maier, G., G. Mihm and H.P. Reisenauer, 1984b, Chem. Ber. **117**, 2351.

Maier, G., G. Mihm, H.P. Reisenauer and D. Littmann, 1984c, Chem. Ber. **117**, 2369.

Maier, G., L.H. Franz, H.-G. Hartan, K. Lanz and H.P. Reisenauer, 1985a, Chem. Ber. **118**, 3196.

Maier, G., K. Schöttler and H.P. Reisenauer, 1985b, Tetrahedron Lett. **26**, 4079.

Mal'tsev, A.K., V.N. Kobashesku and O.M. Nefedov, 1979, Izv. Akad. Nauk SSSR, Ser. Khim., p. 2152.

Mal'tsev, A.K., V.A. Korolov and O.M. Nefedov, 1982a, Izv. Akad. Nauk SSSR, Ser. Khim., p. 2415.

Mal'tsev, A.K., V.N. Khabashesku and O.M. Nefedov, 1982b, J. Organomet. Chem. **226**, 11.

Mal'tsev, A.K., V.N. Khabashesku and O.M. Nefedov, 1984a, J. Organomet. Chem. **271**, 55.

Mal'tsev, A.K., V.A. Korolov and O.M. Nefedov, 1984b, Izv. Akad. Nauk SSSR, Ser. Khim., p. 555.

Mal'tsev, A.K., P.S. Zuev and O.M. Nefedov, 1985a, Izv. Akad. Nauk SSSR, Ser. Khim., p. 957.

Mal'tsev, A.K., P.S. Zuev and O.M. Nefedov, 1985b, Izv. Akad. Nauk SSSR, Ser. Khim., p. 2159.

McCarty Jr, M., and G.W. Robinson, 1959, J. Am. Chem. Soc. **81**, 4472.

McMahon, R.J., and O.L. Chapman, 1986, J. Am. Chem. Soc. **108**, 1713.

McMahon, R.J., and O.L. Chapman, 1987, J. Am. Chem. Soc. **109**, 683.

McMahon, R.J., O.L. Chapman, R.A. Hayes, T.C. Hess and H.-P. Krimmer, 1985, J. Am. Chem. Soc. **107**, 7597.

Meyer, B., 1971, Low Temperature Spectroscopy (Elsevier, New York).

Michl, J., 1983, Int. J. Mass. Spectrom. Ion Phys. **53**, 255.

Michl, J., and E.W. Thulstrup, 1986, Spectroscopy with Polarized Light (VCH, Weinheim, FRG).

Michl, J., G.J. Radziszewski, J.W. Downing, K.B. Wiberg, F.H. Walker, R.D. Miller, P. Kovacic, M. Jawdosiuk and V. Bonačić-Koutecký, 1983, Pure Appl. Chem. **55**, 315.

Miller, J.H., and L. Andrews, 1981, J. Mol. Spectrosc. **90**, 20.

Milligan, D.E., and M.E. Jacox, 1964, J. Chem. Phys. **41**, 2838.

Moore, C.B., and G.C. Pimentel, 1964, J. Chem. Phys. **41**, 3504.

Moskovits, M., and G.A. Ozin, eds, 1976, Cryochemistry (Wiley, New York).

Murai, H., J. Ribo, M. Torres and O.P. Strausz, 1981, J. Am. Chem. Soc. **103**, 6422.

Nam, H.-H., and G.E. Levoi, 1985, Spectrochim. Acta A **41**, 67.

Nefedov, O.M., A.K. Mal'tsev, V.N. Khabashesku and V.A. Korolev, 1980, J. Organomet. Chem. **201**, 123.

Orendt, A.M., J.C. Facelli, D.M. Grant, J. Michl, F.H. Walker, W.P. Dailey, S.T. Waddell, K.B. Wiberg, M. Schindler and W. Kutzelnigg, 1985, Theor. Chim. Acta **68**, 421.

Otteson, D., and J. Michl, 1984, J. Org. Chem. **49**, 866.

Pacansky, J., and D.W. Brown, 1983, J. Phys. Chem. **87**, 1553.

Pacansky, J., and J.S. Chang, 1981, J. Chem. Phys. **74**, 5539.

Pacansky, J., and H. Coufal, 1979, J. Chem. Phys. **71**, 2811.

Pacansky, J., and H. Coufal, 1980a, J. Chem. Phys. **72**, 3298.

Pacansky, J., and H. Coufal, 1980b, J. Chem. Phys. **72**, 5285.

Pacansky, J., and H. Coufal, 1980c, J. Mol. Struct. **60**, 255.

Pacansky, J., and M. Dupuis, 1982, J. Am. Chem. Soc. **104**, 415.

Pacansky, J., and A. Gutierrez, 1983, J. Phys. Chem. **87**, 3074.

Pacansky, J., D.E. Horne, G.P. Gardini and J. Bargon, 1977, J. Phys. Chem. **81**, 2149.

Pacansky, J., G.P. Gardini and J. Bargon, 1978, Ber. Bunsenges. Phys. Chem. **82**, 19.

Pacansky, J., D.W. Brown and J.S. Chang, 1981, J. Phys. Chem. **85**, 2562.

Pacansky, J., J.S. Chang and D.W. Brown, 1982, Tetrahedron **38**, 257.

Perutz, R.N., 1985a, Ann. Rep. Progr. Chem. C **82**, 157.

Perutz, R.N., 1985b, Chem. Rev. **85**, 97.

Platz, M.S., 1984, in: Azides and Nitrenes, ed. E.F.V. Scriven (Academic Press, Orlando) ch. 7.

Prochaska, F.T., B.W. Keelan and L. Andrews, 1979, J. Mol. Spectrosc. **76**, 142.

Pullin, D., and L. Andrews, 1982, J. Mol. Struct. **95**, 181.

Raabe, G., and J. Michl, 1985, Chem. Rev. **85**, 419.

Raabe, G., H. Vančik, R. West and J. Michl, 1986, J. Am. Chem. Soc. **108**, 671.

Radziszewski, J.G., J.W. Downing, C. Wentrup, P. Kaszynski, M. Jawdosiuk, P. Kovacic and J. Michl, 1984, J. Am. Chem. Soc. **106**, 7996.

Radziszewski, J.G., J.W. Downing, M. Jawdosiuk, P. Kovacic and J. Michl, 1985a, J. Am. Chem. Soc. **107**, 594.

Radziszewski, J.G., J.W. Downing, C. Wentrup, P. Kaszynski, M. Jawdosiuk, P. Kovacic and J. Michl, 1985b, J. Am. Chem. Soc. **107**, 2799.

Radziszewski, J.G., T.-K. Yin, F. Miyake, G.E. Renzoni, W.T. Borden and J. Michl, 1986, J. Am. Chem. Soc. **108**, 3544.

Reisenauer, H.P., G. Mihm and G. Maier, 1982, Angew. Chem. Int. Ed. Engl. **21**, 854.

Reisenauer, H.P., G. Maier, A. Riemann and R.W. Hoffmann, 1984, Angew. Chem. Int. Ed. Engl. **23**, 641.

Reiser, A., H. Wagner and G. Bowes, 1966, Tetrahedron Lett., p. 2635.

Rosengren, K., and G.C. Pimentel, 1965, J. Chem. Phys. **43**, 507.

Roth, W.R., M. Biermann, G. Erker, K. Jelich, W. Gerhartz and H. Görner, 1980, Chem. Ber. **113**, 586.

Roth, W.R., R. Langer, M. Bartmann, B. Stevermann, G. Maier, H.P. Reisenauer, R. Sustmann and W. Müller, 1987, Angew. Chem. Int. Ed. Engl. **26**, 256.

Rule, M., A.R. Matlin, E.F. Hilinski, D.A. Dougherty and J.A. Berson, 1979, J. Am. Chem. Soc. **101**, 5098.

Rule, M., A.R. Matlin, D.E. Seeger, E.F. Hilinski, D.A. Dougherty and J.A. Berson, 1982, Tetrahedron **38**, 787.

Sander, W., 1986, Angew. Chem. Int. Ed. Engl. **25**, 255.

Schlegel, H.B., B. Coleman and J. Jones Jr, 1978, J. Am. Chem. Soc. **100**, 6499.

Sekiguchi, A., W. Ando and K. Honda, 1985, Tetrahedron Lett. **26**, 2337.

Sheridan, R.S., and G.A. Ganzer, 1983, J. Am. Chem. Soc. **105**, 6158.

Sheridan, R.S., and M.A. Kesselmayer, 1984, J. Am. Chem. Soc. **106**, 436.

Snyder, G.J., and D.A. Dougherty, 1985, J. Am. Chem. Soc. **107**, 1774.

Sodeau, J.R., and E.K.C. Lee, 1978, Chem. Phys. Lett. **57**, 71.

Solum, M.S., J.C. Facelli, J. Michl and D.M. Grant, 1986, J. Am. Chem. Soc. **108**, 6464.

Stolkin, I., T.-K. Ha and H.H. Günthard, 1977, Chem. Phys. **21**, 327.

Strub, H., A.J. Beeler, D.M. Grant, J. Michl, P.W. Cutts and K.W. Zilm, 1983, J. Am. Chem. Soc. **105**, 3333.

Toriyama, K., M. Iwasaki, K. Nunome and H. Muto, 1981, J. Chem. Phys. **75**, 1633.

Torres, M., A. Clement and O.P. Strausz, 1980a, J. Org. Chem. **45**, 2271.

Torres, M., E.M. Lown, H.E. Gunning and O.P. Strausz, 1980b, Pure Appl. Chem. **52**, 1623.

Torres, M., J. Ribo, A. Clement and O.P. Strausz, 1981, Nouv. J. Chim. **5**, 351.

Torres, M., I. Safarik, A. Clement and O.P. Strausz, 1982a, Can. J. Chem. **60**, 1187.

Torres, M., A. Clement, O.P. Strausz, A.C. Weedon and P. de Mayo, 1982b, Nouv. J. Chim. **6**, 401.

Torres, M., J. Ribo, A. Clement and O.P. Strausz, 1983a, Can. J. Chem. **61**, 996.

Torres, M., J.L. Bourdelande, A. Clement and O.P. Strausz, 1983b, J. Am. Chem. Soc. **105**, 1698.

Torres, M., A. Clement and O.P. Strausz, 1983c, Nouv. J. Chim. **7**, 269.

Tseng, K.L., and J. Michl, 1977, J. Am. Chem. Soc. **99**, 4840.

Vaccani, S., H. Kühne, A. Bauder and H.H. Günthard, 1977, Chem. Phys. Lett. **50**, 187.

Vančik, H., G. Raabe, M.J. Michalczyk, R. West and J. Michl, 1985, J. Am. Chem. Soc. **107**, 4097.

Walker, F.H., K.B. Wiberg and J. Michl, 1982, J. Am. Chem. Soc. **104**, 2056.

Wentrup, C., 1984, in: Azides and Nitrenes, ed. E.F.V. Scriven (Academic Press, Orlando) ch. 8.

Wentrup, C., C. Thétaz, E. Tagliaferri, H.J. Lindner, B. Kitschke, H.-W. Winter and H.P. Reisenauer, 1980, Angew. Chem. Int. Ed. Engl. **19**, 566.

Wentrup, C., G. Gross, A. Maquestiau and R. Flammang, 1983, Angew. Chem. Int. Ed. Engl. **22**, 542.

West, P.R., O.L. Chapman and J.-P. LeRoux, 1982, J. Am. Chem. Soc. **104**, 1779.

West, P.R., A.M. Mooring, R.J. McMahon and O.L. Chapman, 1986, J. Org. Chem. **51**, 1316.

Whittle, E., D.A. Dows and G.C. Pimentel, 1954, J. Chem. Phys. **22**, 1943.

Wiberg, K.B., F.H. Walker, W.E. Pratt and J. Michl, 1983, J. Am. Chem. Soc. **105**, 3638.
Wiberg, K.B., M.G. Matturro, P.J. Okarma, M.E. Jason, W.P. Dailey, G.J. Burgmaier, W.F. Bailey and P. Warner, 1986, Tetrahedron **42**, 1895.
Withnall, R., and L. Andrews, 1985a, J. Am. Chem. Soc. **107**, 2567.
Withnall, R., and L. Andrews, 1985b, J. Phys. Chem. **89**, 3261.
Withnall, R., and L. Andrews, 1986, J. Am. Chem. Soc. **108**, 8118.
Zilm, K.W., and D.M. Grant, 1981, J. Am. Chem. Soc. **103**, 2913.
Zilm, K.W., R.T. Conlin, D.M. Grant and J. Michl, 1978, J. Am. Chem. Soc. **100**, 8038.
Zilm, K.W., R.T. Conlin, D.M. Grant and J. Michl, 1980, J. Am. Chem. Soc. **102**, 6672.

CHAPTER 9

Inorganic and Organometallic Photochemistry

Robin N. PERUTZ

Department of Chemistry
University of York
York YO1 5DD
United Kingdom

Chemistry and Physics of
Matrix-Isolated Species
L. Andrews and M. Moskovits, eds

Contents

1. Introduction and general principles

Photolysis of matrix-isolated molecules is a particularly effective route to unstable molecules and ions. Matrix isolation offers the opportunity of characterising the molecular and electronic structure of both primary and secondary photoproducts. The twin aims of defining the photochemical pathway and discovering the structure of the photoproducts have motivated much of the work reviewed here. While some of the molecules first observed by matrix isolation have subsequently proved stable enough to be observed by other static methods [e.g., HOF (Appelman 1973), $Cr(CO)_5N_2$ (Burdett et al. 1978b)] many more remain unique to the matrix environment. Perhaps the most striking advances of recent years have been made in techniques complementary to matrix isolation, which allow unstable molecules to be characterised in solution or in the gas phase, often by time-resolved methods. These experiments have vindicated most of the conclusions of the matrix experiments. Meanwhile, matrix isolation remains by far the most powerful general method of obtaining *structural* information on highly unstable polyatomics in their ground electronic states.

Control of photolytic reactions may be achieved through wavelength selectivity. The short-wavelength limit for exciting the guest molecules in a matrix is set by the onset of exciton absorption by the host material. For noble-gas matrices, this limit lies well into the vacuum ultraviolet (Sonntag 1977). Photochemical reactions have been induced by radiation from the vacuum UV to the vibrational infrared, almost always via single-photon processes. Successful experiments have also been reported using X- and γ-irradiation, but the matrix material must then be intimately involved in the reaction.

The success of matrix isolation depends on the use of a number of different spectroscopies to characterise the photoproducts. While infrared spectroscopy remains the mainstay for characterisation, UV/visible absorption and emission are essential in order to have rational direction of the photolysis wavelength. Laser-induced fluorescence offers the opportunity of obtaining far more highly resolved emission and excitation spectra than are obtainable with broad-band sources [see chapter by Bondybey (ch. 5)]. The energies of vibrational transitions which are infrared forbidden may be obtained by laser-induced fluorescence or by Raman spectroscopy (Downs and Hawkins 1983).

Paramagnetic molecules may be detected by electron-spin resonance, if they have odd numbers of unpaired electrons and low orbital angular momentum. With even numbers of unpaired electrons or high orbital momentum, they may become ESR silent. Magnetic circular dichroism is an alternative technique for spotting paramagnetic molecules, which is appropriate to those species which are ESR silent. It has the particular advantage of direct correlation with the UV/visible spectrum. Variation of temperature and magnetic field allows the optical determination of magnetisation curves, and, in favourable cases, of g-values. Most of these methods will be illustrated later in this chapter. Further methods such as photoelectron spectroscopy or inelastic neutron scattering are discussed in chapter 1 of the present volume or by Perutz (1986).

In section 2 the photoprocesses are considered that may be observed in matrices. In the succeeding section a series of case histories will be used to illustrate the principles of such matrix experiments and the conclusions which may be derived from them. Throughout, connections are stressed between matrix-isolation results and experiments on unstable molecules conducted in other phases. Photochemical reactions of organometallics will feature prominently, reflecting both my own interest and the current advances in studying unstable organometallics by matrix isolation and time-resolved methods. Matrix isolation provides the structural basis for much of the kinetic work.

The use of photochemistry in studying metal atom reactions is covered in the chapter by Hauge et al. (ch. 10); infrared photochemistry is discussed in the chapter by Frei and Pimentel (ch. 6) and elsewhere (Lotta et al. 1984, Frei and Pimentel 1985). For more comprehensive reviews of photochemistry involving matrix-isolated atoms and small molecules, the reader is referred to Perutz (1985a,b). The review by Hitam et al. (1984b) lists photoprocesses of organometallics in full. Case histories of $Fe(CO)_4$ (Poliakoff 1978, Poliakoff and Weitz 1987) and $Cr(CO)_5$ (Turner et al. 1977, rather dated) are also available.

2. Photoprocesses in matrices

This section is divided into two major parts, according to whether we are examining:
   (a) photolysis of isolated molecules in inert matrices (i.e., normally unimolecular reactions), or
   (b) photolysis in reactive or doped matrices, or of aggregates.
In the second group, the reactions are typically bimolecular. Further subdivisions are made according to the photoprocesses.

2.1. Isolated molecules in inert matrices

2.1.1. Simple fragmentation
Noble gases are unreactive towards the majority of molecules and their photoproducts. Often, nitrogen or methane may also be used as inert hosts.

Such isolated molecules may fragment on photolysis yielding small atoms, especially H and F, or molecules such as N_2, CO, CO_2, C_2H_4. This method forms the key to synthesis of many important unstable molecules [see also chapters by Andrews (ch. 2) and Jacox (ch. 4)],

$$UF_6 \xrightarrow{h\nu/Ar} UF_5 + F \quad \text{(Paine et al. 1976)} \tag{1}$$

$$FC(O)N_3 \xrightarrow{h\nu/Ar} FNCO + N_2 \xrightarrow{h\nu/Ar} NF + CO \tag{2}$$

$$\text{(Gholivand and Willner 1984)}$$

$$Fe(CO)_5 \underset{h\nu_2/Ar}{\overset{h\nu_1/Ar}{\rightleftharpoons}} Fe(CO)_4 + CO \quad \text{(Poliakoff 1978)} \tag{3}$$

$$(\eta^5\text{-}C_5H_5)Rh(C_2H_4)_2 \underset{h\nu_2/Ar}{\overset{h\nu_1/Ar}{\rightleftharpoons}} (\eta^5\text{-}C_5H_5)Rh(C_2H_4) + C_2H_4 \tag{4}$$

$$\text{(Haddleton and Perutz 1985, Haddleton et al. 1988).}$$

After photolysis smaller atoms or molecules may diffuse a short distance through the matrix so preventing a back reaction. Others remain in the same cage and a complex is formed between the two photofragments, as is the case for $Fe(CO)_4 \cdots CO$ [eq. (3)]. Such complexes are detected by the perturbations induced in the IR spectra. However, if there is no activation energy for the back reaction, in-cage recombination may ensue thermally even at 10 K and no permanent photoproduct is formed. This cage effect prevents the use of photofragmentation for the release of heavier atoms or molecules, unless there is a mechanism by which it may be circumvented (see below). Thus photolysis of $Mn_2(CO)_{10}$ fails to generate the expected $Mn(CO)_5$ radicals and ICl is not a source of iodine atoms in matrices. When the photofragments remain close together, the back reaction may often be effected photochemically by exciting a product molecule selectively. Such photochemical reversal has become a good test for the production of coordinatively unsaturated organometallics [examples are given in eqs. (3) and (4)].

Sometimes, surprisingly large fragments may be eliminated, showing that in-cage recombination is not just the effect of slower diffusion. In the cases below, even PCl_3 or 3-bromopyridine can be eliminated.

$$LW(CO)_5 \xrightarrow[\text{Ar or } CH_4]{h\nu} W(CO)_5 + L \quad (L = 3\text{-bromopyridine, } PCl_3, \text{ etc.})$$

$$\text{(Boxhoorn et al. 1980, McHugh et al. 1979)} \tag{5}$$

Experiments with polarised photoselection have shown that there is a pseudo-rotation mechanism available to metal pentacarbonyls, which allows the ejected ligand to escape from the vacant site without diffusion, thus obviating the cage effect [fig. 1 (Turner et al. 1977, Burdett et al. 1978a); see also section 2.2.2]. A more detailed discussion of the cage effect is given by Perutz (1985a).

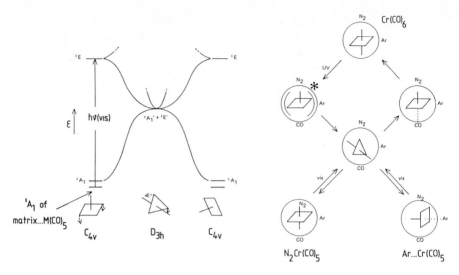

Fig. 1. The pseudo-rotation mechanism for avoiding the cage effect in $LM(CO)_5$ ($M = Cr, Mo, W$) molecules. The diagram at the left shows how excitation of square-pyramidal (spy) $M(CO)_5$ causes crossing to a trigonal bipyramidal state which decays back to spy $M(CO)_5$. The spy $M(CO)_5$ is regenerated either in the original orientation or rotated through 120°. The diagram at the right shows how this mechanism can effect ligand exchange. In this example, interconversion is shown between $N_2Cr(CO)_5$, $Ar \cdots Cr(CO)_5$ and $Cr(CO)_6$. After each reorientation, $Cr(CO)_5$ reacts with the atom or molecule closest to its vacant site. Here the three possibilities are N_2, Ar and CO. Another example is shown in scheme 2 without showing the orientation. [Adapted with permission from Turner et al. (1977; Pure and Appl. Chem. **49**, 271).]

2.1.2. Bond making during fragmentation

When bonds are made during photofragmentation, there is usually a substantial barrier to any recombination reaction, so that in-cage combination is prevented. Examples include H–H, C–H, C–F, halogen–halogen and H–halogen bond formation,

$$(\eta^5\text{-}C_5H_5)_2TaH_3 \xrightarrow{\ h\nu,\, Ar\ } (\eta^5\text{-}C_5H_5)_2TaH + H_2 \quad \text{(Baynham et al. 1985)}$$

$$\text{(6)}$$

$$(\eta^5\text{-}C_5H_5)_2W(CH_3)H \xrightarrow[\ Ar\ ]{\ h\nu\ } (\eta^5\text{-}C_5H_5)_2W + CH_4$$

$$\text{(Chetwynd-Talbot et al. 1982),} \quad \text{(7)}$$

$$X\text{-}CO\text{-}NF_2 \xrightarrow[\ Ar\ ]{\ h\nu\ } FNCO + XF \quad (X = H,\ NF_2,\ CF_3)$$

$$\text{(Gholivand and Willner 1984).} \quad \text{(8)}$$

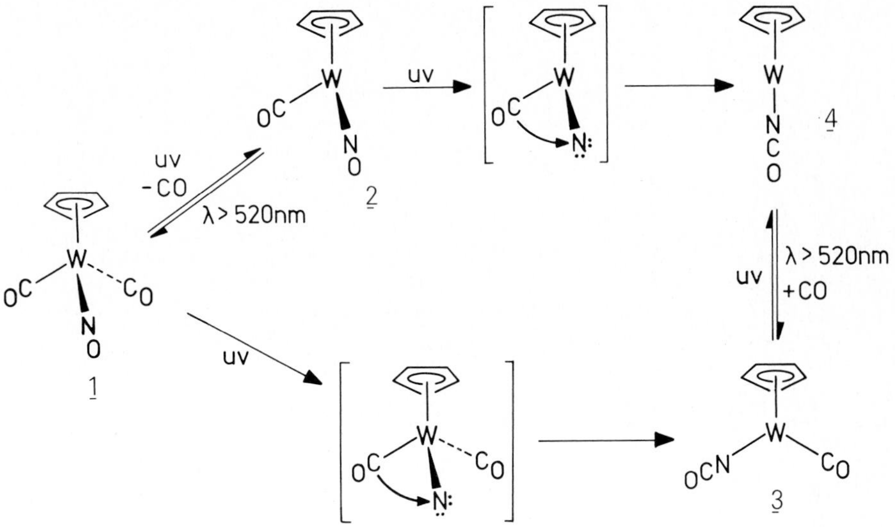

Scheme 1. Matrix photochemistry of $(\eta^5\text{-}C_5H_5)W(CO)_2NO$. Rearrangement to isocyanato complexes. Three photoproducts are observed, 2, 3 and 4. The upper pathway is preferred by the authors, but the lower rearrangements cannot be excluded. [Adapted with permission from Hitam et al. (1984a; J. Chem. Soc. Chem. Commun., p. 471; Copyright 1984, Royal Society of Chemistry).]

The examples of eqs. (6) and (7) show bond formation in the smaller molecule (e.g., H_2), which was eliminated from a transition-metal complex. It is also possible to induce bond-formation or rearrangement in both fragments as is clear from eq. (8). This principle applies equally to organometallics. Thus scheme 1 shows formation of an isocyanato complex from a metal(carbonyl)nitrosyl, presumably via a nitrene (Hitam et al. 1984a).

In another example, we see the stepwise change from singly bonded arsenide to doubly bonded arsenide [eq. (9)] involving a change in geometry at arsenic as the π-bond is formed (Mahmoud et al. 1984; $Cp = \eta^5\text{-}C_5H_5$)

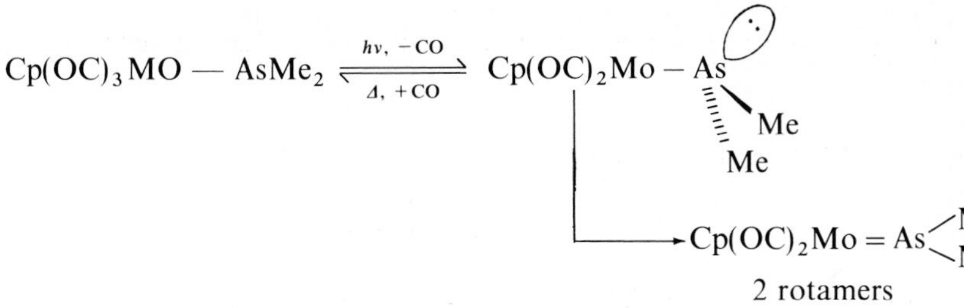

$$Cp(OC)_3MO - AsMe_2 \underset{\Delta, +CO}{\overset{h\nu, -CO}{\rightleftharpoons}} Cp(OC)_2Mo - As \begin{matrix} \\ Me \\ Me \end{matrix} \qquad (9)$$

$$Cp(OC)_2Mo = As \begin{matrix} Me \\ Me \end{matrix}$$

2 rotamers

Stepwise coordination of an organic ligand can also be observed as is illustrated for the η^1-allyl → η^3-allyl conversion in eq. (10) (Hitam et al. 1985),

$$\text{Cp(OC)}_3\text{W} \underset{\Delta, +CO}{\overset{\lambda < 370 \text{ nm}, -CO}{\rightleftharpoons}} \text{Cp(OC)}_2\text{W} \rightleftharpoons \text{Cp(OC)}_2\text{W} \tag{10}$$

2.1.3. Isomerisation

Photolysis provides a most effective method of generating an unstable isomer of a matrix-isolated molecule. In some examples, it is possible to revert to the original isomer by changing the photolysis wavelength to coincide with the absorption of the new isomer. Recent examples include:

$$\text{ONONO}_2 \xrightarrow[\lambda = 436 \text{ nm}]{\text{Ar}} \text{O}_2\text{N–NO}_2 \quad \text{(Bandow et al. 1984)}, \tag{11}$$

$$\text{H}_2\text{C} = \text{SiH}_2 \underset{254 \text{ nm}}{\overset{\lambda < 320 \text{ nm}}{\rightleftharpoons}} \text{H}_3\text{CSiH} \quad \text{(Maier et al. 1984a,b)}, \tag{12}$$

$$\xrightarrow[\text{Ar}]{\lambda < 290 \text{ nm}} \quad \text{(Haddleton and Perutz 1986)}. \tag{13}$$

Such reactions may proceed by an intramolecular route or involve photofragmentations followed by recombination. Equation (12) shows a key difference between carbenes and silylenes. This ability of silenes to isomerise to silylenes sets them up for quite different reactivity and has proved very controversial. The iridium complex [eq. (13)] unexpectedly isomerises to a metal vinyl hydride, a reaction which was unprecedented among stable ethylene complexes. However, a very similar reaction had been observed in matrices for Fe(C$_2$H$_4$) [Kafafi et al. 1985, see also chapter by Hauge et al. (ch. 10)]. Indeed, it soon proved possible to observe the unstable isomer in solution, and further examples of such reactions have been discovered since.

A group of intramolecular isomerisations which can be singled out are the rotamerisations, in which one conformer is converted to another. Many examples of this type of reaction can be induced by irradiation in the vibrational infrared, especially among C$_2$ molecules such as 2-fluoroethanol [see chapter by Frei and Pimentel (ch. 6)]. Another example which harks back to the original IR-induced isomerisation of HONO is that of HSNO (Müller et al. 1984; see section 3.1),

$$\text{trans-HSNO} \underset{\lambda > 2000 \text{ nm}}{\overset{\lambda = 585 \text{ nm}}{\rightleftharpoons}} \text{cis-HSNO}. \tag{14}$$

2.1.4. Photo-ionization

Neutral molecules usually have ionization energies above 6 eV in the gas phase, with the consequence that photo-ionization requires vacuum ultraviolet (VUV) radiation. Stabilization of ions in an Ar matrix by electrostatic attraction and by solvation may be as high as 1.4 eV for $(C_2H_4)^+$ and even 3.4 eV for H^+, but high energy sources remain necessary for ionization (Gedanken et al. 1973). Nevertheless, it is possible to form ions from, e.g., HCl with radiation of 10.2 eV, although its ionization energy is 12.7 eV. Examples of ionization with VUV and γ-radiation are given here, and examples of co-deposition with alkali metals in section 2.2.5. Open discharge methods and electron bombardment are reviewed in the chapters by Knight (ch. 7) and Andrews (ch. 2). In contrast to neutral molecules, any anions are likely to have very low ionization energies, with the result that matrix-isolated ions are usually neutralised by visible or near-UV irradiation.

It is not possible to measure charge directly in a matrix, so identification of ions depends on distinctive spectroscopic characteristics. Unfortunately, it is very rare to be able to identify both anion and cation conclusively. For instance, VUV irradiation of ethyne in argon gives C_2^-, an ion with conspicuous absorption and emission spectra. No cation is detected, probably because C_2^- is such a strong absorber that it is detected at exceptionally low concentration. As indicated in a general sense above, C_2^- is removed by visible irradiation.

Other non-metal ions which have been studied intensively in matrices include HX_2^-, X_3^- (X = Cl, Br), HAr_n^+ and halomethyl ions (see, e.g., Jacox 1978, Andrews 1979, Bondybey and Pimentel 1972, Van Ijzendoorn 1983, Bondybey et al. 1971, Andrews and Prochaska 1979, Perutz 1985b for a summary). Metal carbonyl ions also exhibit their charge conspicuously, since their $\nu(CO)$ bands shift ca. 100 cm^{-1} per unit of charge. VUV photolysis of matrix-isolated metal carbonyls causes ion production, probably via photo-ionization and dissociative electron capture, e.g.,

$$2Cr(CO)_6 \xrightarrow{\text{VUV, Ar}} [Cr(CO)_6]^+ + [Cr(CO)_5^-] + CO. \qquad (15)$$

$$(1990\,cm^{-1}) \qquad\qquad (2095\,cm^{-1}) \quad (1839, 1858\,cm^{-1})$$

It should be remembered that excess electrons are very mobile in solid argon, so there is no need for the two precursor molecules to be close together. Although the stoichiometry of $[Cr(CO)_5]^-$ has been established by isotopic substitution, the cation is too weak an absorber to verify the assignment in the same way (Breeze et al. 1981).

The most reliable method of detecting radical-ions, namely ESR, has been applied both to small molecules [see chapter by Knight (ch. 7)] and to metal carbonyls. In the latter case, research has been concentrated on the use of γ-radiation, e.g., to form the remarkable 19-electron $[HCo(CO)_4]^-$ from

HCo(CO)$_4$ (Fairhurst et al. 1983). This ion is unusual among electron-addition products to metal carbonyls, in retaining its structure without distortion or ligand loss.

2.2. Bimolecular photochemical processes: reactive, doped and concentrated matrices

In section 2.1, the photoprocesses available to an isolated molecule in an inert matrix were considered. If the matrix itself is reactive, or if an inert matrix is doped with a suitable reagent, bimolecular processes become possible. Reactive matrices include N$_2$, CO, O$_2$ and CH$_4$, but noble gases, especially krypton and xenon may also be reactive towards some substrates. The same molecules together with many others such as H$_2$, O$_3$, N$_2$O, HI and F$_2$, may be used as dopant reagents together with an inert host such as argon. Bimolecular processes also occur if the concentration of guest in the matrix is allowed to rise so that pairs of molecules are "isolated" as nearest neighbours in the matrix.

In such reactions, the choice of matrix ratios becomes particularly important. A pure matrix (e.g., N$_2$) has the attraction of giving much sharper IR spectra than a heavily doped matrix (e.g., Ar + 10% N$_2$). On the other hand, reaction of one guest molecule with several molecules of N$_2$ may occur in the pure matrix, whereas the reaction can be controlled in the doped matrix. The choice of matrix ratio is discussed in more detail elsewhere (Perutz 1985a). The question of the stoichiometry of the reaction can be put on a more quantitative basis either by isotopic labelling or by investigating how the absorbance, A, of the photoproduct varies with the matrix ratio, R. If log A is plotted against R, the stoichiometry of reaction should be given by the slope. This method has been applied with moderate success by Almond et al. (1985) to the reaction of W(CO)$_6$ with N$_2$O in methane matrices.

It is also important to consider the nature of the absorber in such reactions. Is it the guest which absorbs the light as in eq. (16) (Church et al. 1981), or is it the dopant as in

$$\text{HMn(CO)}_5 \xrightarrow[h\nu]{\text{CO}} \text{Mn(CO)}_5 + \text{HCO} \qquad (16)$$

the photolysis of HI/ethane/Xe matrices? In the latter case photolysis of HI generates hydrogen atoms which react with ethane to form C$_2$H$_5$ on annealing (Toriyama et al. 1980, Wakahara et al. 1982). Alternatively, a complex may be formed between guest and dopant in the ground state as in Me$_2$S·Cl$_2$ (Machara and Ault 1987), or in the excited state as for (Fe·CH$_4$)* (Ozin et al. 1986). The last situation is similar to the description "harpooning" given to the Xe/Cl$_2$ reaction (Fajardo and Apkarian 1986).

In sections 2.2.1–2.2.6, examples are given of the photoprocesses which occur in doped, reactive and concentrated matrices. This is followed, in section 2.3, by a discussion of the role of molecular complexes, prior to, during and succeeding photolysis.

2.2.1. Photo-induced addition

Simple addition reactions may be promoted photochemically in matrices, possibly via excitation of molecular complexes. Examples employing UV radiation include the addition of HCl to AlCl to form $HAlCl_2$ and the formation of $OGeF_2$ from GeO and F_2 (Schnöckel 1978, 1981). However, some reactions of this type, such as the addition of F_2 to C_2H_4, can even be stimulated by IR excitation [see chapter by Frei and Pimentel (ch. 6)].

Addition has also been observed among metallocenes. For instance, the 16-electron complex, chromocene, may be converted to Cp_2CrCO by irradiation in a CO matrix. However, there is no need to look only to reactive matrices for addition reactions. Many of the fragmentation reactions of transition-metal complexes are reversed by irradiation at longer wavelength, as is shown in the examples of eqs. (3) and (4) for addition of CO or C_2H_4.

2.2.2. Fragmentation followed by reaction

The restrictions on photofragmentation caused by in-cage recombination were highlighted in section 2.1.1. One of the simpler applications of a reactive matrix is directed at avoidance of the cage effect. Diazomethane, for example, is not photolysed in an inert matrix to give CH_2 because of the cage effect, but it is photolysed in a CO-doped N_2 matrix to give ketene, CH_2CO (DeMore et al. 1959). $HMn(CO)_5$ loses CO on mercury arc photolysis in Ar matrices, but apparently not H atoms, probably because of in-cage recombination. However, if a carbon monoxide matrix is used instead, $HMn(CO)_4$ is no longer generated on photolysis, since it reacts instantly with CO. Hydrogen atom loss is now observed [eq. (16)] because the expelled atoms are trapped by CO forming the formyl radical (Church et al. 1981, 1983; see section 3.2).

Reaction with CO can also be used to demonstrate decoordination of organic ligands, which is inhibited by the cage effect in inert matrices. For instance, stepwise decoordination of cycloheptatriene from $(\eta^6\text{-}C_7H_8)Mo(CO)_3$ is observed in a CO matrix (Hooker and Rest 1982),

$$(\eta^6\text{-}C_7H_8)Mo(CO)_3 \xrightarrow{h\nu} (\eta^4\text{-}C_7H_8)Mo(CO)_4 \xrightarrow{h\nu} Mo(CO)_6 + C_7H_8.$$

$$(17)$$

One of the commonest uses of a reactive matrix in organometallic chemistry is to effect substitution. We have seen above that a polyene can be substituted by CO. Probably the most extensive series of substitution reactions have been observed with chromium hexacarbonyl (Perutz and Turner 1975a,b, Almond et al. 1985, Burdett et al. 1978a,b, Sweany 1985),

$$Cr(CO)_6 \xrightarrow{L, h\nu_1} LCr(CO)_5 + CO, \tag{18}$$

with L = Ar, Kr, Xe, CH_4, CO_2, N_2O, N_2, H_2, $Cr(CO)_6 \cdots OC$. When mixed

R.N. Perutz

matrices are employed, e.g., Ne + Xe or Ar + N_2, a series of reversible substitutions may be observed by selective photolysis (scheme 2). This is a group of highly unconventional ligands, but the distinct characteristics of each complex can be established through a combination of IR and UV spectroscopy.

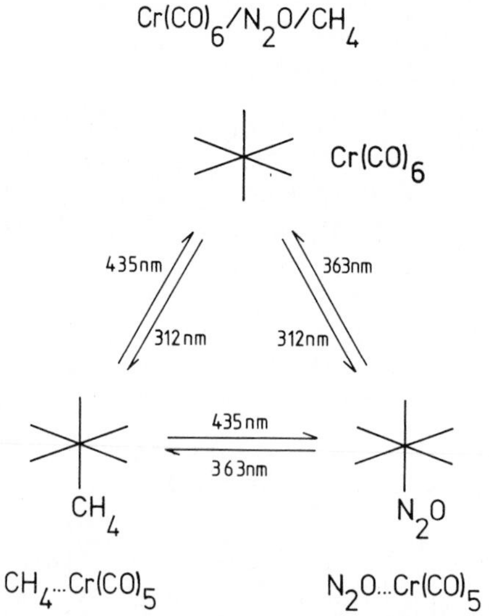

$$Cr(CO)_6/N_2O/CH_4$$

Scheme 2. Interconversion of $Cr(CO)_6$, $CH_4\cdots Cr(CO)_5$ and $N_2O\cdots Cr(CO)_5$ in an N_2O-doped CH_4 matrix. The reactions proceed by the photochemical mechanism of fig. 1.

At the time of the original experiments, it was not possible to establish whether unperturbed $Cr(CO)_5$ could be matrix-isolated. $Cr(CO)_5$ adopts a square pyramidal structure in all matrices. In the least perturbing matrix, Ne, its absorption maximum was at 624 nm, and its principal $\nu(CO)$ bands at 1977 and 1944 cm^{-1}. It has since become possible to measure spectra of $Cr(CO)_5$ in the gas phase; the resulting values of 620 nm, for the absorption maximum, and 1980 and 1948 cm^{-1} for the $\nu(CO)$ bands, are remarkably close to those for a Ne matrix (Breckenridge and Stewart 1986, Seder et al. 1986). Thus "$NeCr(CO)_5$" really is identical to gas-phase $Cr(CO)_5$. Many of the other $LCr(CO)_5$ molecules with exotic ligands have also been detected by solution flash photolysis (L = N_2, H_2, alkane) or by photolysis in liquid xenon (L = N_2, H_2, Xe) (Kelly et al. 1983, Simpson et al. 1983, Church et al. 1984a, Upmacis et al. 1985, Turner et al. 1983).

Oxidative addition of reagents such as H_2 or, more remarkably, methane can also take place on matrix photolysis,

$$Fe(CO)_5 \xrightarrow{h\nu, \, H_2, \, Ar} Fe(CO)_4(H)_2 + CO \quad \text{(Sweany 1981),} \qquad (19)$$

$$\text{CpRh(L)(CO)} \xrightarrow{h\nu,\ CH_4} \text{CpRh(CO)(H)(CH}_3) \quad L = C_2H_4, \text{CO}$$

$$\text{(Haddleton 1986, Rest et al. 1987)} \quad (20)$$

Here H_2 binds to $Fe(CO)_4$ as a dihydride, in contrast to the dihydrogen complex formed with $Cr(CO)_5$. Notice that these reactions relate closely to the simple addition processes of section 2.2.1.

2.2.3. Atom transfer reactions

One of the most successful approaches to matrix photochemistry has been the method of incorporating a reagent, RX, acting as photochemical source of an atom X (e.g., H, C, O, F, S) which can be transferred to a suitable substrate A. By controlling the source reagent and the photolysis wavelength, it is even possible to select the electronic state of the atom,

$$\text{RX} + \text{A} \xrightarrow{h\nu,\ Ar} \text{R} + \text{XA}, \quad (I)$$

or

$$\text{RX} \cdots \text{A} \xrightarrow{h\nu,\ Ar} \text{R} + \text{XA}. \quad (II)$$

A simple demonstration reaction of type I, where there is no evidence for complex formation is the reaction of PF_3 with OCS. Here OCS acts as a photochemical source of S atoms (Hawkins et al. 1985),

$$\text{OCS} + \text{PF}_3 \xrightarrow{Ar,\ h\nu} \text{OC} + \text{SPF}_3. \quad (21)$$

Another particularly successful atom transfer reagent is ozone. Recent research has exposed the important role of complex formation (reaction type II) in these reactions as in the example of ICl (Hawkins et al. 1984, fig. 2),

$$\text{O}_3 + \text{ICl} \xrightarrow{Ar} \text{O}_3 \cdot \text{ICl} \xrightarrow{450-750\ nm} \text{OICl} + \text{O}_2. \quad (22)$$

Both OCS and O_3 can act as sources of excited-state (^1D) atoms as well as ground-state (^3P) atoms. When OCS is photolysed in pure methane, the products include CH_3SH and $CH_2 = S$, formed by reaction of $S(^1D)$ with methane. On the other hand, replacement of solid methane by methane/argon mixtures eliminates these excited-state products, leaving CS_2 as the only sulphur-containing product. It appears that the lifetime of excited S atoms in CH_4/Ar matrices is too short to effect insertion (Hawkins et al. 1985). Some control of the production of excited oxygen atoms from ozone may be achieved by wavelength selection. At shorter wavelengths both $O(^1D)$ and $O(^3P)$ are formed, whereas visible radiation gives exclusively $O(^3P)$. Reactions of ozone are discussed in more detail in the chapter by Andrews (ch. 2).

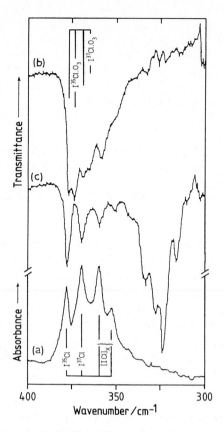

Fig. 2. The infrared spectrum in the I–Cl stretching region of argon matrices containing: (a) ICl alone; (b) $ICl + {}^{16}O_{3-x}{}^{18}O_x$ (50% ^{18}O enriched) showing bands of complexes and isolated ICl; (c) same as (b) but after photolysis ($340 < \lambda < 600$ nm). The effect of photolysis has been to convert the $ICl \cdot O_3$ complex to OICl and O_2ICl, leaving only uncomplexed ICl behind. Both products contribute to the new bands in the 310–340 cm^{-1} region. [Reproduced with permission from Hawkins et al. (1984, J. Am. Chem. Soc., **106**, 3076; Copyright 1984, American Chemical Society).]

These methods have not been used extensively in transition-metal chemistry, but their potential is indicated by the hydrogen-atom transfer reaction of eq. (23) (Morton and Preston 1984a,b),

$$Ni(CO)_4 + HI \xrightarrow{\ Kr,\ h\nu\ } Ni(CO)_2CHO + HNi(CO)_3 + CO + I. \qquad (23)$$

A comprehensive account of atom-transfer reactions and their mechanisms is given by Perutz (1985a). Further examples will also be found in the succeeding sections. The effect of complex formation is discussed in section 2.3.

2.2.4. Reactions of dimers

As a matrix becomes more concentrated, the proportion of nearest-neighbour pairs among the guest molecules increases. Often, such dimers have a significantly perturbed IR spectrum and may exhibit distinct photochemistry. Indeed, they represent a special case of complex formation prior to photolysis. Matrix isolation complements gas-phase beam methods as an approach to studying spectra of dimers, but is particularly suitable for studying their photochemistry. Hawkins and Downs (1984a) have examined the photochemistry of cis-$(NO)_2$, which is shown to yield N_2O and ground-state oxygen atoms. The oxygen atoms go on to react with $(NO)_2$ to form both isomers of N_2O_3,

$$cis\text{-}(NO)_2 \xrightarrow{\lambda < 320 \text{ nm}} N_2O + O(X^3P) \xrightarrow{(NO)_2} ONONO + O_2NNO.$$

(24)

In contrast, monomeric NO is photostable. The photogenerated oxygen atoms can also be transferred to other substrates incorporated into the matrix, such as CO, and S atoms formed simultaneously from OCS. The ground-state interaction between NO molecules is weak, yet the photochemistry is quite distinct. Unfortunately, little is known about the nature of the excited states involved in the reaction. Moving to a slightly stronger interaction, we note that N_2O_4 is also photosensitive, not only undergoing isomerisation (section 2.1.3) but forming a variety of other nitrogen oxides including cis-$(NO)_2$ (Bandow et al. 1984). Further reactions of cis-$(NO)_2$ will feature in section 3.1.

2.2.5. Photo-electron transfer reactions and heterolytic reaction

Of the methods of generating ions in matrices mentioned in section 2.1.4, one of the earliest to be developed was the method of co-condensation with an alkali-metal atom. Electron transfer may often be achieved with low-energy photolysis in the visible region, since the resulting ion-pair is stabilised by electrostatic attraction (see Kasai 1971, Perutz 1985a). The use of low-energy radiation avoids the concomitant production of radicals which is inevitable in the use of high-energy radiation. Co-condensation of alkali atoms with metal hexacarbonyls, followed by photolysis with visible light, provides an alternative route to $Cr(CO)_5^-$ [Breeze et al. 1981, see eq. (15)]. Whereas many metal carbonyl anions have been characterised by ESR following γ-radiolysis in matrices or crystal lattices, the IR experiments remain the only description of $Cr(CO)_5^-$. Like isoelectronic $Mn(CO)_5$ (section 3), it proves to be square pyramidal.

Heterolytic reaction on photolysis usually requires a polar solvent, and has rarely been observed in matrices. Reactions to metal carbonylnitrosyls were first thought to be heterolytic, but are now described as isomerisations with partial intramolecular electron transfer (Hitam et al. 1984b). Recently, evidence has been published of another heterolytic reaction in the presence of CO as a trapping reagent. Like many other metal carbonyls $Cp(OC)Fe(\mu\text{-}CO)_2Co(CO)_3$

loses CO on photolysis in argon matrices. However, on photolysis in CO matrices, heterolytic cleavage occurs with concomitant uptake of CO, forming an ion-pair (Fletcher et al. 1984),

$$Cp(OC)Fe(\mu\text{-}CO)_2Co(CO)_3 \xrightarrow{\text{367 nm, CO}} [CpFe(CO)_3]^+ [Co(CO)_4]^-.$$

(25)

2.2.6. Photochemical reaction with dioxygen

As techniques have improved, it has become possible to tackle reactions of increasing complexity, such as those of small molecules with O_2. These photo-oxidations are often complicated by the formation of a series of photoproducts, which are themselves photosensitive. Solid oxygen itself absorbs appreciably at wavelengths as long as 300 nm, and guests in solid O_2 are liable to form contact pairs with distinct charge-transfer (CT) bands. For instance, cis-2-butene shows a maximum at 234 nm and undergoes photolysis at wavelengths as long as 322 nm, giving trans-2-butene, CH_3CHO, CO_2, CO and O_3 (Hashimoto and Akimoto 1987). The reaction mechanism involves excitation to a charge-transfer state which may react to form photoproducts, or decay either to $^3[\text{alkene}] + {}^3[O_2]$ or $^3[\text{alkene}] + {}^1[O_2]$. The major pathway to oxygenated products proceeds directly from the CT state. Additional complications may arise from reactions involving more than one oxygen molecule.

Lee and co-workers have studied the photo-oxidation of sulphur-containing molecules in solid oxygen by IR spectroscopy. While SO_2 monomer proves to be photostable, SO_2 dimers are oxidised to SO_3; H_2S gives a variety of products including H_2O, SO_2, SO_3, H_2SO_4, $H_2O\cdot SO_2$ and, probably $H_2O\cdot SO_3$ (Sodeau and Lee 1980, Tso and Lee 1984). Unfortunately, no UV spectra have been recorded of these matrices, so the role of charge-transfer complexes remains unexplored.

Transition-metal systems form strong complexes with dioxygen, which may bond either in the end-on, bent ($\eta^1\text{-}O_2$), or in the side-ways ($\eta^2\text{-}O_2$) mode. This provides the opportunity of studying the photochemistry of very specific complexes. Thus co-condensation of Fe(tpp) (tpp = tetraphenylporphyrin) with O_2 yields Fe(tpp)($\eta^1\text{-}O_2$), which reacts on laser photolysis at 406.7 nm to form the oxo complex, OFe(tpp) (Bajdor and Nakamoto 1984).

Considerable research work has been devoted to photo-oxidation of metal hexacarbonyls, particularly in dilute O_2/Ar or O_2/CH_4 matrices. Under these circumstances it is possible to selectively convert $M(CO)_6$ to $Ar\cdots M(CO)_5$ or $CH_4\cdots M(CO)_5$ (M = Cr, Mo, W) in the first step. Dioxygen only becomes involved during the subsequent step in which the metal pentacarbonyl is excited. The reaction sequence (scheme 3) involves first an $\eta^2\text{-}O_2$ complex, then cis and trans-dioxo complexes, and finally binary metal oxides. The distinction between dioxygen and dioxo complexes is complicated because both absorb in the

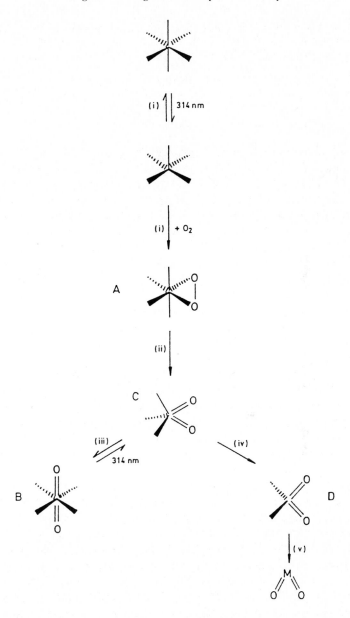

Scheme 3. Photo-oxidation at the wavelengths indicated of $M(CO)_6$ (M = Cr, Mo, W) in O_2-doped matrices. An unmarked line represents a bond to a CO ligand. (i) Cr, 544 nm; Mo, 403 nm; W, 435 nm; (ii) Cr, W, 367 nm; Mo, 314 nm; (iii) Mo and W only, Mo 403 nm or anneal, W anneal; (iv) Cr, Mo, W 314 nm, Cr anneal as alternative; (v) Cr, Mo, W(?) 314 nm. [Reproduced with permission from Almond et al. (1986, Inorg. Chem **25**, 19; Copyright 1986, American Chemical Society).]

$750-1000 \text{ cm}^{-1}$ region, but they can be distinguished by the shifts of $[M]^{16}O^{18}O$ relative to $[M]^{16}O^{16}O$ and $[M]^{18}O^{18}O$ (Almond et al. 1986, Crayston et al. 1984). Intriguingly, intermediate B, $trans$-$(O)_2M(CO)_4$, is also formed by photolysis of $M(CO)_6$ in CO_2 or N_2O-doped matrices (Almond et al. 1985), before reacting further to form oxides such as CrO_2. Thus, it is now possible to trace the pathway of photo-oxidation of a metal carbonyl in a series of steps through to the final metal oxide.

2.3. Intermolecular complex formation in matrix photochemistry

Complex formation plays a crucial role throughout matrix photochemistry. Mutual perturbation of the molecular vibrations of nearest neighbours often forms the basis of studies of stable molecules in matrices. When the interaction between the participants is stronger, charge-transfer (CT) bands may appear in the electronic spectrum. Photochemistry may ensue from the CT excited state. Even if the ground-state interaction is weak (e.g., in XeCl) excited-state interactions may be much stronger, so formation of excimers and exciplexes is significant.

If we return to the example of the $ICl \cdot O_3$ complex, we recall that OICl is formed only from the precursor complex and not from isolated molecules [eq. (22)]. Recombination reactions, as discussed in section 2.1.1, are photochemical reactions specific to the complex. For instance, UV photolysis of $Fe(CO)_5$ in Ar generates the $Fe(CO)_4 \cdots CO$ complex, and near IR or visible photolysis of $Fe(CO)_4 \cdots CO$ regenerates $Fe(CO)_5$. Similarly $Xe \cdots Cr(CO)_5$ may be converted to $Ar \cdots Cr(CO)_5$ in a mixed Ar/Xe matrix (scheme 2). Thus, we may argue that intermolecular complexes form the link between the unimolecular reactions of section 2.1 and the bimolecular reactions of section 2.2. The same relationship between complexes in precursors and photoproducts applies to dimers. The reaction to form $(N_2O + O)$ is specific to cis-$(NO)_2$ [eq. (24)]. Conversely, the linear dimer $(HCN)_2$ is formed specifically by photolysis of sym-tetrazine in a matrix (Pacansky 1977, Knoezinger and Wittenbeck 1982), and $(HO_2)_2$ can be synthesised by photo-oxidation of glyoxal (Diem et al. 1982, Tso and Lee 1985).

Exciplexes derived from rare-gas halides now have great importance in laser systems – they may also be observed in matrices. Early studies showed that XeX (X = F, Cl) emission in matrices was strongly red-shifted with respect to the gas phase. The vibrational parameters were little different in the ground state, but strongly perturbed in the excited state (Ault and Andrews 1976, Goodman and Brus 1976). Large matrix shifts are typical of the most ionic molecules, so this is an indication of charge separation in the excited state. Recent experiments on the XeCl and Xe_2Cl system have again shown very large red shifts of emission compared to the gas phase. $Xe_2^+Cl^-$ emits at 573 nm with a lifetime of ca. 225 ns whether in Ar, Kr or Xe host. These exciplexes were made by two-photon absorption of matrix-isolated HCl or Cl_2 at 308 nm (Fajardo and Apkarian 1986).

3. Case histories

In section 2, various types of photochemical process available to matrix-isolated molecules were outlined. In this section I will use case histories to bring out more about the detailed structural and electronic information which can be deduced from these experiments, and to indicate how the matrix results link with those from other methods.

3.1. Sulphur–nitrogen–oxygen compounds

The extensive possibilities for studying isomerism are exhibited particularly spectacularly in the chemistry of S–N–O compounds. Only one acid derivative is stable in the gas phase and this compound, HNSO, exists mainly in the *cis* form. It is synthesised by reaction of NH_3 and $SOCl_2$, but detailed gas-phase studies are hampered by its instability and tendency to polymerise on condensation. Early matrix-isolation work by Tchir and Spratley (1975a,b,c) using ^2H and ^{15}N substitution established the photochemical conversion of *cis*-HNSO into *trans*-HNSO with $\lambda > 300$ nm.* When shorter wavelength radiation was used, other products were formed which were identified as *cis*-HOSN, *trans*-HSNO, *cis*-HSNO and SNO. Growth curves showed that *cis*-HOSN was the primary product of this reaction and HSNO was formed in a secondary reaction. Later work by Müller et al. (1984) demonstrated that wavelength-selective photolysis could be used to interconvert *cis* and *trans* isomers of HSNO. Mid-infrared radiation is also effective in *cis* → *trans* interconversion [eq. (14)], as for HONO. (Detailed work with IR lasers has yet to be reported.)

Recently, the Zürich group (Nonella et al. 1987) has shown that yet another isomer may be generated by initial conversion of *cis*-HNSO to *cis*-HOSN and HSNO, followed by irradiation at 365 nm. This wavelength regenerates *cis*-HNSO and its *trans* isomer, but also causes production of a new molecule assigned as *trans*-HONS. Subsequent long-wavelength irradiation removes *trans*-HONS and reforms *cis*-HNSO directly (scheme 4). The assignment to *trans*-HONS rather than *trans*-HOSN depends on the shifts observed on isotopic substitution and on detailed force-field calculations. In keeping with the assignment, elaborate SCF–CI calculations find a potential minimum at the *trans* geometry for HONS, but not for HOSN. In those compounds for which both *cis* and *trans* isomers are identified, the stretching frequency to hydrogen is slightly lower in

---

* There remains some doubt as to the validity of the original assignment of the long-wavelength photoproduct to *trans*-HNSO (Huber 1987).

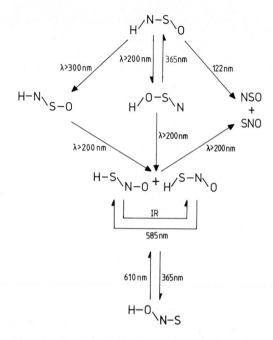

Scheme 4. Photochemistry of HNSO and its photoproducts in argon matrices.

the *cis*-isomer, suggesting internal hydrogen bonding. However, the role of H-bonding does not appear to have been investigated in the calculations. Little is known of the mechanism of the isomerizations, but the specificity of the long-wavelength irradiation argues for intramolecular processes.

Vacuum-ultraviolet photolysis of HNSO can be used to form the triatomics NSO and SNO (Tchir and Spratley 1975c). SNO is also formed together with SN_2O_2 by sulphur atom transfer to *cis*-$(NO)_2$ induced by photolysis of OCS,

$$NO \xrightarrow{\text{UV, Ar, OCS}} NO + SNO, \tag{26}$$

$$\underset{O\quad O}{\overset{N\cdots N}{\underset{\|}{\overset{\|}{}}}} \xrightarrow{\text{UV, Ar, OCS}} \underset{S}{\overset{S}{}}{>}N{-}N{\overset{}{\underset{O}{\diagdown}}} \underset{\lambda > 375\,nm}{\overset{\text{Nernst glower}}{\rightleftharpoons}} \underset{O}{\overset{N}{\diagup}}\underset{S}{}\underset{O}{\overset{N}{\diagdown}}. \tag{27}$$

The sulphur-atom method has been used to examine the vibrational characteristics of SNO on isotopic substitution. The molecule exhibits Fermi resonance between $v(NO)$ and $2v(NS)$, so there are always two bands in the NO-stretching region confusing the interpretation of the spectrum. The other produyct, SN_2O_2, like its oxygen analogue N_2O_3, undergoes isomerisation, probably as indicated in eq. (27) (Hawkins and Downs 1984b).

The methyl-substituted derivative, CH_3SNO, also exhibits *cis–trans* isomerism. A stationary state is set up on photolysis, and the corresponding position of equilibrium varies with photolysis wavelength. By studying $v(NO)$ for the two isomers, their relative extinction coefficients can be determined, and hence the proportions of each isomer may be discovered. By assuming that the proportions on deposition reflect the room-temperature distribution in the gas phase, it is also possible to determine the difference in free energy between the isomers (Müller and Huber 1984).

3.2. Photochemistry of $HMn(CO)_5$, $CH_3Mn(CO)_5$ and $Mn_2(CO)_{10}$

The $Mn(CO)_5$ radical is isolobal with the methyl radical and occupies as crucial a niche in organometallic chemistry as does the methyl radical in organic chemistry. For many years evidence for its existence was deduced from kinetic studies, but the radical itself eluded detection. Success was claimed on several occasions, only to be disproved later. As it turned out, it had been made in a CO matrix by photolysis of $Mn(CO)_4NO$, but the spectra were not good enough to prove its identity or its structure. The commonest source of $Mn(CO)_5$ in solution is $Mn_2(CO)_{10}$, yet it was only realised recently that another photo-product of $Mn_2(CO)_{10}$ exists and plays a vital role in determining reaction products, namely $Mn_2(CO)_9$. The photochemical reactivities of the closely related carbonyls $HMn(CO)_5$ and $CH_3Mn(CO)_5$ were studied in the early stages of matrix isolation of metal carbonyls. However, only in the last few years has it become possible to determine the structure of their photoproducts.

3.2.1. Manganese pentacarbonyl

Photolysis of $HMn(CO)_5$ in a CO matrix investigated with IR detection causes production of HCO and a new carbonyl, identical to that formed by similar photolysis of $Mn(CO)_4NO$. By prolonged photolysis of $HMn(CO)_5$ in a mixed $^{13}CO/^{12}CO$ matrix, an equilibrium is set up and ^{13}CO is incorporated into $HMn(CO)_5$ and its photoproduct. The resulting spectrum may be fitted in frequency and intensity to a set of force constants using a CO-factored force field. Notice that this spectrum is the superposition of the spectra of each of the possible isotopomers according to the probabilities of their occurrence. Unlike most force-constant problems where there are usually more force constants than observed bands, this type of problem is overdetermined. By these means Church et al. (1981) established that the product was square-pyramidal $Mn(CO)_5$ (fig. 3). Moreover from the intensity pattern of the $v(CO)$ modes of $Mn(^{12}CO)_5$ an axial-equatorial bond angle of 96° may be estimated. By use of plane-polarised light, the authors generated preferentially-oriented $Mn(CO)_5$, which exhibits dichroic IR spectra. These experiments confirmed the identity of the $v(CO)$ modes and established the direction of the photo-active transition moment of $HMn(CO)_5$ (fig. 4). They showed evidence for reorientation of $Mn(CO)_5$ by polarised photolysis, probably via a Berry pseudo-rotation (see fig. 1). The

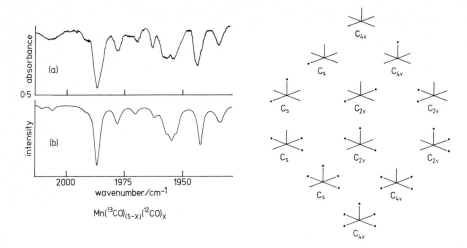

Fig. 3. (a) IR spectrum in the $v(CO)$ region obtained by prolonged photolysis of $HMn(CO)_5$ in a CO matrix containing ^{12}CO (55%) and ^{13}CO (45%). All the bands shown are due to $Mn(^{13}CO)_{5-x}(^{12}CO)_x$ ($x = 0-5$). (b) Spectrum predicted for C_{4v} $Mn(CO)_5$ with an axial-equatorial bond angle of 96°. The spectrum is fitted using a CO-factored force field. The final simulation is formed from the sum of contributions in appropriate proportions from each of the isotopomers shown symbolically at the right. The ^{13}CO labels are indicated by a dot. [Adapted with permission from Church et al. (1981; J. Am. Chem. Soc. **103**, 7515; Copyright 1981, American Chemical Society).]

results of the matrix experiments also established that $Mn(CO)_5$ has a near-IR absorption at 798 nm.

Initially, observation of $Mn(CO)_5$ was confirmed to CO matrices, in which any CO-loss products immediately recombined to reform $HMn(CO)_5$ [see eq. (16)]. Subsequently, use of a more sensitive detection method, ESR, showed that $Mn(CO)_5$ could also be formed by UV photolysis in argon matrices (see below). Alternatively, it can be made in sufficient yield for IR detection in argon matrices by using ArF laser irradiation at 193 nm. Presumably the higher energy photons supply enough translational energy for the hydrogen atoms to diffuse out of the cage (Church et al. 1983).

The ESR measurements (Symons and Sweany 1982) not only allowed an independent verification of the C_{4v} structure of $Mn(CO)_5$, but provided an estimate of 60% d_z^2 and 8% 4s character of the SOMO. They demonstrated that the ESR spectrum in solution would show negligible ^{55}Mn hyperfine coupling rendering detection very difficult. Unfortunately, the matrix spectra were rather broad and exhibited an unusual intensity pattern. By using a Kr matrix and γ-irradiation of $HMn(CO)_5$ in a sealed tube at 77 K, Fairhurst et al. (1984) were able to record an exceedingly sharp spectrum with a normal intensity pattern (fig. 5). It now became possible to observe ^{13}C hyperfine coupling on

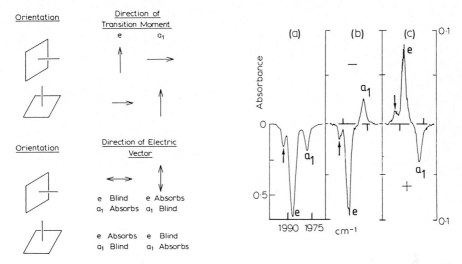

Fig. 4. The schematic diagram at the left shows the relationship: (top) between the transition moment direction and the orientation of square-pyramidal $Mn(CO)_5$; (bottom) between the direction of the electric vector of linearly polarised light and the orientation. At the right are shown: (a) normal spectrum of $Mn(CO)_5$ in $\nu(CO)$ region obtained by unpolarised UV photolysis of $HMn(CO)_5$; (b) Polarised spectrum obtained after prolonged irradiation at $\lambda = 330$ nm with photolysis polariser horizontal. The spectrum is presented as the difference (spectrum with detection polariser vertical) − (spectrum with polariser horizontal). (c) Similar difference spectrum after prolonged photolysis with polariser horizontal. Notice that the e mode has opposite polarisation to the a_1 mode. The weak band marked with an arrow has been shown to be a matrix splitting of the e mode. The effect of prolonged irradiation at 830 nm is to reorient $Mn(CO)_5$, probably via a pseudo-rotation mechanism (see fig. 1), and to reverse the polarisation. The symmetry of the photo-active transition moment is shown to be $e(x,y)$. [Reproduced with permission from Church et al. (1981, J. Am. Chem. Soc. **103**, 7515; Copyright 1981, American Chemical Society).]

enriched samples and distinguish coupling to the unique carbon from coupling to the equatorial carbons. Very remarkably each peak of $Mn(^{12}CO)_5$ was surrounded by a weak decet of satellites, which were shown to arise from coupling to ^{83}Kr ($I = \frac{9}{2}$ and 11% natural abundance). The isotopic data gave an estimate of the spin population of each C(2p) orbital as ca. 5%, with a similar spin density delocalised onto the unique krypton. Thus matrix-isolation work has provided a very detailed structural picture of $Mn(CO)_5$.

3.2.2. Hydridomanganese tetracarbonyl and methylmanganese tetracarbonyl

The major product of UV photolysis of $HMn(CO)_5$ in an argon matrix is not $Mn(CO)_5$. It has at least three $\nu(CO)$ bands and a weak $\nu(MnH)$ band; like other coordinatively unsaturated metal carbonyls, it reacts with CO to reform the precursor on long-wavelength photolysis. This much evidence is suggestive of $HMn(CO)_4$, but does not indicate whether its structure is based on a substituted

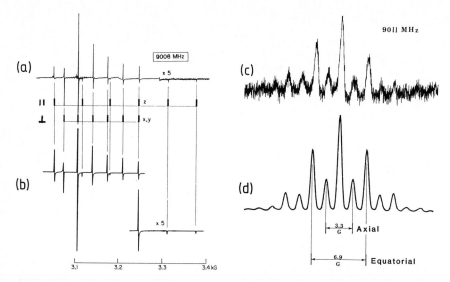

Fig. 5. ESR spectra of $Kr \cdots Mn(CO)_5$ obtained by γ-radiolysis of $HMn(CO)_5$ in Kr at 77 K. (a) Spectrum of $Kr \cdots Mn(^{12}CO)_5$ showing six parallel and six perpendicular components, arising from hyperfine coupling to ^{55}Mn ($I = \frac{5}{2}$). (b) Computer simulation of the spectrum. Note that the satellites due to coupling to ^{83}Kr are observed only at very much higher expansion. (c) The $m_I = \frac{5}{2}$ parallel transition of $Kr \cdots Mn(CO)_5$ obtained using a sample of $HMn(CO)_5$ enriched to 30% with ^{13}C ($I = \frac{1}{2}$). (d) Simulation of enriched spectrum (c) showing coupling to axial ^{13}CO and equatorial ^{13}CO. The marked bands arise from C_{4v} and C_s isomers of $Mn(^{12}CO)_4(^{13}CO)$, respectively. The remaining bands arise principally from $Mn(^{12}CO)_5$ and $Mn(^{12}CO)_3(^{13}CO)_2$ – see also fig. 3. [Reproduced with permission from Fairhurst et al. (1984, Organometallics **3**, 1389; Copyright 1984, American Chemical Society).]

square pyramid (C_s symmetry) or a trigonal bipyramid (C_{2v} symmetry). Its visible absorption is more revealing: like that of $Cr(CO)_5$ (section 2.2.2) it is shifted dramatically (2500 cm^{-1}) in a methane matrix, strong evidence for the square-pyramidal structure with Ar or CH_4 interacting with the LUMO via the sixth coordination site. Confirmation of the square-pyramidal C_s structure is also obtained by very detailed isotopic studies [cf. $Mn(CO)_5$]. However, the authors show that this distinction is not always possible and the converse, namely proof of a C_{2v} structure by ^{13}CO labelling, will not be feasible (Church et al. 1983).

Square pyramidal $HMn(CO)_4$ should have two isomers, with the hydride axial or equatorial. Selective photolysis of C_s $HMn(CO)_4$ does indeed generate the other isomer, C_{4v} $HMn(CO)_4$. By choosing the appropriate photolysis wavelength, the direction of isomerisation can be reversed. The entire matrix photochemistry of $HMn(CO)_5$ is summarised in scheme 5.

Similar studies of $CH_3Mn(CO)_5$ photolysis in Ar and CH_4 matrices showed that the major photolysis product, $CH_3Mn(CO)_4$, again exists in two isomers

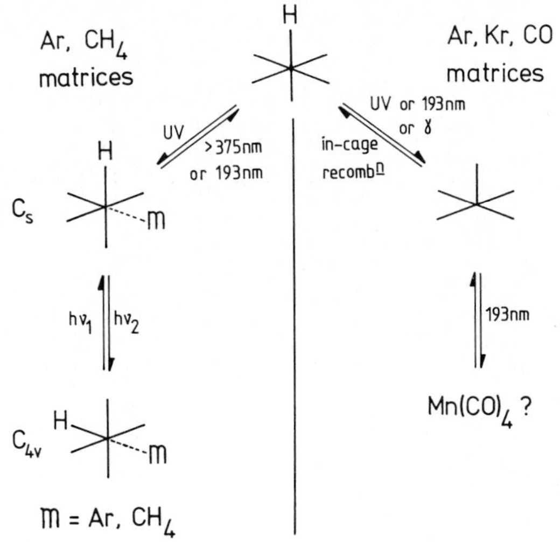

H

Ar, CH$_4$
matrices

Ar, Kr, CO
matrices

UV or 193nm
or γ

UV
>375nm
or 193nm

in-cage
recomb$^{\square}$

H

C_s ----m

hν_1 hν_2

193nm

C_{4v} H ----m

Mn(CO)$_4$?

m = Ar, CH$_4$

Scheme 5. Photochemistry of HMn(CO)$_5$ in matrices. [Adapted with permission from Church et al. (1983, Inorg. Chem. **22**, 3259; Copyright 1983, American Chemical Society).]

based on substitution of a square pyramid. Once more the sixth coordination position is occupied by the matrix atom or molecule acting as a "token" ligand and causing substantial perturbation of the optical spectrum (Horton-Martin et al. 1986). Use of a CO matrix does not result in formation of Mn(CO)$_5$ or CH$_3$C(O)Mn(CO)$_5$, but in formation of CH$_3$Mn(CO)$_4$ again (McHugh and Rest 1980). This time we may surmise that it undergoes an "isocarbonyl" interaction with the oxygen of a CO in the sixth site, but confirmation via visible absorption spectroscopy is lacking. A better defined pair of complexes, *cis-* and *trans*-CH$_3$Mn(CO)$_4$N$_2$ may be generated using a nitrogen matrix. Recently, Turner et al. (1987) employed partial isotopic substitution of the methyl group (CHD$_2$ groups) to obtain "isolated" CH-stretching frequencies free of perturbation by Fermi resonance or interaction between CH stretching vibrations. By this method, they show a downward shift of v(CH) on reaction with nitrogen, corresponding to a lengthening of the C–H bond of 0.0011–0.0015 Å.

3.2.3. Dimanganese–enneacarbonyl

Matrix photolysis of Mn$_2$(CO)$_{10}$ (Hepp and Wrighton 1983) first upset the conventional wisdom that the only significant photoprocess in this molecule was Mn–Mn bond-fission, by showing that photolysis in a 3-methylpentane glass generated Mn$_2$(CO)$_9$ with an extraordinarily low-frequency IR band due to a bridging carbonyl (1760 cm^{-1}). These authors argued that loss of CO accounted for as many as 30% of the reactive excited states in solution. Dunkin et al. (1984) followed these experiments up with an investigation of the photolysis of Mn$_2$(CO)$_{10}$ in argon matrices. Again, no Mn(CO)$_5$ was observed, presumably because of prompt recombination, but the spectrum of $Mn_2(CO)_9$ was very similar to that observed in the glasses. When photolysed with polarised light all the bands of the terminal carbonyls of $Mn_2(CO)_9$ proved to be dichroic, but the bridging mode was not. Both its low frequency and the lack of dichroism can be explained by an off-axis $\eta^1-\eta^2$ bridging carbonyl,

$$-\!\!\stackrel{|\diagup}{\underset{\diagup\,|}{Mn}}\!\!-\!\!\stackrel{\diagdown\diagup}{\underset{\diagup\diagdown}{Mn}}\!\!-\quad\xrightarrow{h\nu}\quad \stackrel{}{\underset{|}{\diagup\!\!\!>\!Mn}}\!\!\stackrel{\displaystyle C\!\!\stackrel{\textstyle O}{\diagup\!\!\!\diagup}}{-\!\!\!-\!\!\!-}\!\!\stackrel{|\diagup}{Mn}\!\!\diagdown \tag{28}$$

This structure serves to relieve the electronic unsaturation of both manganese centres.

Matrix studies have also been carried out on the heterobimetallic complex $MnRe(CO)_{10}$. The photoproduct $MnRe(CO)_9$ proves to be bridged, presumably like $Mn_2(CO)_9$. By using $MnRe(CO)_{10}$ labelled with ^{13}CO at the Re end of the molecule, it proved possible to show that CO was lost only from the Mn end (Firth et al. 1986, 1987).

3.2.4. Detection of manganese carbonyl intermediates in solution

The first convincing direct observation of $Mn(CO)_5$ in solution came in 1978 shortly before the matrix work. Waltz et al. (1978) generated $Mn(CO)_5$ by pulse radiolysis and showed that it dimerised extremely rapidly (ca. $10^9\ dm^3\ mol^{-1}\ s^{-1}$) and had an absorption maximum of 830 nm in EtOH (cf. 798 nm in CO matrix). Flash photolysis experiments (Wegman et al. 1981) showed that a further intermediate was generated, but it was not appreciated at that stage that this was a primary photoproduct. The combination of picosecond work (Rothberg et al. 1982) and millisecond measurements (Yesaka et al. 1983) proved that this second intermediate was a primary photoproduct, but with a quite different reactivity from $Mn(CO)_5$. Whereas $Mn(CO)_5$ decayed by second-order kinetics in microseconds, the second transient reacted with CO and phosphines in milliseconds (rate constant $\sim 10^7\ dm^3\ mol^{-1}\ s^{-1}$): it was clearly $Mn_2(CO)_9$. The final link-up with matrix isolation came with the advent of flash photolysis with IR detection (or time-resolved IR spectroscopy). Church et al. (1984b) were able to detect both $Mn(CO)_5$ and $Mn_2(CO)_9$ by IR spectroscopy in hydrocarbon solution. The IR bands of $Mn(CO)_5$ and $Mn_2(CO)_9$, including the key $1760\ cm^{-1}$ bridging mode, were in excellent agreement with the matrix values; their decay kinetics were consistent with the previous measurements by UV detection. The photolysis of $MnRe(CO)_{10}$ has also been studied by time-resolved IR spectroscopy. The photoproducts are $Mn(CO)_5$, $Re(CO)_5$ and $MnRe(CO)_9$. As in the matrix, $MnRe(CO)_9$ has a bridged structure; moreover, it is again formed specifically by loss of CO at the manganese end of the molecule (Firth et al. 1986, 1987).

In conclusion, matrix isolation has provided detailed structural characterisation for $Mn(CO)_5$, $Mn_2(CO)_9$ and the two isomers of $XMn(CO)_4$ ($X = CH_3$, H). Much has been deduced about their electronic structure and photochemical reactivity. The thermal reactivity is derived from time-resolved methods in solution. The coarse spectral characteristics agree excellently with the matrix data but cannot be used to deduce the detailed structure of such unstable molecules.

3.3. Photochemistry of $(\eta^5\text{-}C_5H_5)_2MH_n$: the unstable metallocenes of molybdenum, tungsten and rhenium

Metallocenes, parallel ring *bis*(η^5-cyclopentadienyl)metal complexes are stable for all the first-row transition metals from vanadium to nickel and have valence electron counts from 15 to 20. Yet second- and third-row transition metals only form isolable metallocenes with the kinetically stable eighteen-electron configuration, the only exception being the marginally stable rhodocene. Instead, we find a variety of bent-ring derivatives and dinuclear complexes mostly with eighteen-electron configurations. The simplest of these are the hydrides

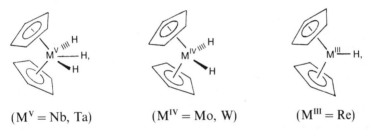

$(M^V = Nb, Ta)$ $\qquad$ $(M^{IV} = Mo, W)$ $\qquad$ $(M^{III} = Re)$

In 1972, Green and Giannotti (1972) first reported that Cp_2WH_2 ($Cp = \eta^5\text{-}C_5H_5$) behaved as a source of Cp_2W on photolysis, reacting like a carbene to insert into C–H bonds of arenes, and even of Me_4Si, forming, e.g., $Cp_2W(C_6H_5)H$. Very similar reactions could be induced thermally with $Cp_2W(CH_3)H$ (Berry et al. 1979, 1980, Cooper et al. 1979). These C–H activation reactions proved to be unique to the tungsten systems,

$$Cp_2WH_2 \xrightarrow[298\,K]{h\nu,\, C_6H_6} Cp_2W(Ph)H + H_2. \tag{29}$$

C–H bond activation by molybdenum occurs only in the absence of an added ligand, when a dinuclear product is formed (Berry et al. 1980). Could matrix isolation be used to characterise the reaction intermediates, even though there would be no intense $\nu(CO)$ modes?

3.3.1. Generation of unstable metallocenes
All of the hydrides are sufficiently volatile for matrix isolation and systematic studies showed that they are all photosensitive. A common product was

generated on photolysis of a variety of Cp_2WL and Cp_2WX_2 derivatives; this product could only be Cp_2W (scheme 6). The bands of the expelled ligand, either CO or C_2H_4, showed perturbations typical of isolation as a cage-pair or of weak complex. The IR spectrum of Cp_2W had the simplicity of a high-symmetry molecule, with bands close to those of ferrocene, and shifts similar to those of ferrocene on ring deuteration (fig. 6). The two exceptions were:

   (i) the low-frequency regions where the heavier tungsten atom reduces the frequency of the skeletal modes;
   (ii) the band at 3240 cm^{-1} which was the most intense in the spectrum.

The very small shift on total deuteration showed that this strange band was an electronic transition, not a vibrational transition at all (see below). More evidence for a parallel ring structure came from the UV/vis spectrum. Instead of the broad bands so commonly found for organometallics, an intense highly structured band in the blue gave a strong indication of high molecular symmetry and hence the parallel-ring structure. Experiments on Cp_2MoH_2 proceeded similarly, except that Cp_2Mo exhibited no electronic transition in the IR and

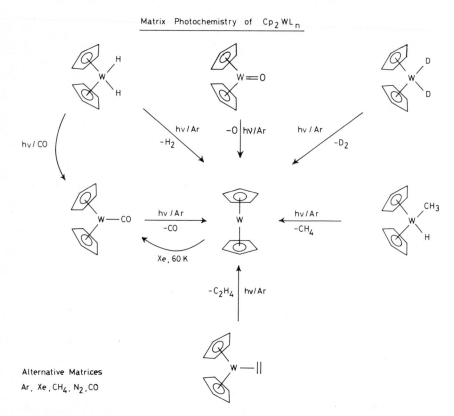

Scheme 6. Matrix photochemistry of Cp_2WL_n complexes showing the variety of routes to Cp_2W and its reaction with CO.

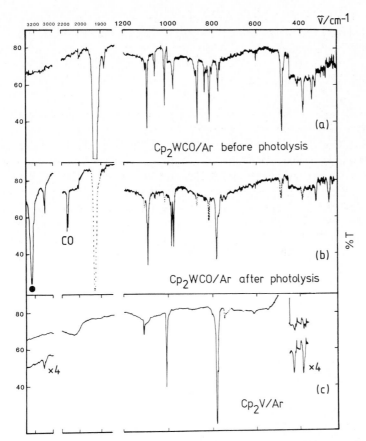

Fig. 6. IR spectra showing the production of tungstenocene in an Ar matrix: (a) Spectrum of Cp_2WCO in Ar before photolysis; (b) Spectrum after photolysis. The bands marked with a full line are due to Cp_2W and CO, those marked dotted are due to remaining Cp_2WCO. Notice the electronic transition at 3240 cm^{-1} (marked ●) and the splitting of the band of photo-ejected CO, caused by formation of $Cp_2W\cdots CO$ in addition to free CO. The spectrum of Cp_2W from other sources is essentially identical (see scheme 5). (c) Spectrum of Cp_2V in Ar, shown for comparison, the band marked with a dotted line is due to an impurity. The frequency pattern is very similar to Cp_2W but the intensity pattern is significantly different.

had broader vibrational bands resembling chromocene, indicative of a Jahn–Teller distorted ground state (Chetwynd-Talbot et al. 1982).

Unlike many reactions with metal carbonyls, these reductive eliminations are not readily reversed. However, Cp_2MoH_2 could be converted to Cp_2MoCO *stepwise* in a CO matrix,

$$Cp_2MoH_2 \xrightarrow{UV, -H_2} Cp_2Mo \xrightarrow{312\ nm,\ +CO} Cp_2MoCO. \qquad (30)$$

$$\xrightarrow{UV, -H_2 + CO}$$

The reaction with CO could also be observed thermally, so it was evident that bending of the rings presented only a small activation barrier (Chetwynd-Talbot et al. 1982),

$$Cp_2M \xrightarrow[60\,K]{Xe/CO} Cp_2MCO. \tag{31}$$

It was far from clear at the outset that Cp_2ReH would be photosensitive, and experiments in Ar matrices gave no photolysis products. However, two pathways were observed in a CO matrix, either H-atom loss with trapping by CO, or attack by CO with consequent ring slip. It later transpired that in-cage recombination, which must have inhibited production of Cp_2Re in argon, could also be avoided in nitrogen matrices. Both pathways were again in evidence [see eq. (32)], but emphasis is placed here on the formation of the metallocene,

$$\tag{32}$$

Evidence for rhenocene came from its IR spectrum and that of its d_{10}-analogue, and, once again, from its intense visible absorption with the vibrational fine structure (fig. 7a). The progression frequencies of all these optical bands are close to $300\,cm^{-1}$, in keeping with the symmetric (ring–metal–ring) stretching mode (Chetwynd-Talbot et al. 1983).

The trihydrides, Cp_2MH_3 (M = Nb, Ta), may be expected to exhibit either the photolysis pathways of Cp_2MH_2 (i.e., H_2 loss) or of Cp_2ReH (H-atom loss). Comparison of the spectra obtained on photolysis of Cp_2MH_3 and $Cp_2M(H)CO$ showed a common product with a clear $\nu(MH)$ band. The dominant route was loss of H_2. The behaviour of the carbonyls $Cp_2M(H)CO$ in argon matrices provided important information about the cage effect. Usually, metal carbonyls are much more photosensitive than metal hydrides, but $Cp_2Ta(H)CO$ proved a much poorer source of Cp_2TaH than did Cp_2TaH_3. The reason became clear from the lack of structure of the band of expelled CO. There was no sign of a weak $Cp_2TaH \cdots CO$ complex implying that product was formed only if CO succeeded in diffusing out of the cage. Unlike Cp_2W, there must be no barrier to recombination since Cp_2TaH retains the bent structure of its precursor. Photolysis of $Cp_2M(H)CO$ could also be expected to cause H-atom loss giving Cp_2MCO radicals (Baynham et al. 1985). This process was observed only in CO matrices [cf. $HMn(CO_5)$].

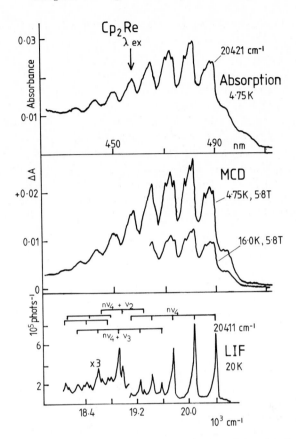

Fig. 7. Electronic spectra of Cp_2Re in a nitrogen matrix formed by photolysis of Cp_2ReH, illustrating the effect of using different methods to examine the lowest ligand-to-metal charge-transfer band. (a) Conventional absorption spectrum at 4.75 K showing the major progression of frequency $337\,cm^{-1}$ due to v'_{sym}(ring–Re–ring) and the fine structure due to coupling to matrix phonons in the excited state. The arrow points to the wavelength of the laser used for LIF. (b) MCD spectra at 4.75 K and 16.0 K both at 5.8 T. Notice that the form of the spectra is almost identical to (a) but the spectra are temperature dependent in intensity proving the paramagnetism of the ground state (see fig. 9). (c) Laser-induced fluorescence (LIF) spectrum. Notice the change of direction of the abscissa scale which is now in $10^3\,cm^{-1}$. The major vibrational progressions give v''_{sym}(ring–Re–ring) $(326\,cm^{-1})$ but other IR-forbidden modes may also be determined. The first band pin-points the (0,0) transition. [Reproduced with permission from Cox et al. (1984; Chem. Phys. Lett. **108**, 415).]

3.3.2. Electronic structure of unstable metallocenes

The matrix-isolation methods led, therefore, to the identification of three open-shell metallocenes, Cp_2Mo and Cp_2W with d^4 sixteen-electron configurations and Cp_2Re with a d^5 seventeen-electron configuration. In addition, the d^2

sixteen-electron complexes Cp_2NbH and Cp_2TaH were studied. These results showed conclusively that matrix isolation could be used successfully for the study of unstable organometallics without the advantage of $v(CO)$ modes. The clue to the electronic structure of the metallocenes lay in the unusual IR electronic transition, found only for Cp_2W, and more importantly, in the visible absorption bands (ligand-to-metal charge-transfer transitions). These absorption bands offered the opportunity of employing two further techniques: laser-induced fluorescence (LIF) and magnetic circular dichroism (MCD).

Laser-induced fluorescence has been observed successfully for both Cp_2Re and Cp_2Mo, yielding some of the most highly resolved electronic spectra recorded for an organometallic (fig. 7c, Cox et al. 1984). These spectra pinpoint the (0,0) transitions very precisely and provide values of $v(CpMCp)$ in the ground state. Considering the length of the progression, it is a surprise that the ground- and excited-state values of $v(CpReCp)$ differ by only 11 cm^{-1}.

Magnetic circular dichroism has been observed for all three new metallocenes mentioned above. The signals are very similar in pattern to the absorption spectra, but their intensity increases dramatically as the temperature is lowered (fig. 7b). The temperature dependence is proof of a paramagnetic ground state, as can be understood from the following argument. The lowest electronic state is split into two levels by the magnetic field, one of which absorbs left (LCP), the other right circularly polarised (RCP) light. As the temperature is lowered, the population of the lower state increases, so enhancing the differential absorbance of RCP versus LCP light (fig. 8).

Further information about the g-values of the guest in its ground state can be obtained by more detailed analysis of the MCD spectrum by either of two

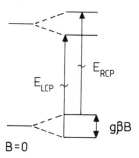

Fig. 8. Schematic diagram showing the principles of magnetic circular dichroism (MCD) for a molecule with a paramagnetic ground state. At the left the diagram shows ground and excited states without applied magnetic field, at the right in a magnetic field. The magnetic field splits ground and excited state according to the Zeeman effect. Left-circularly polarised (LCP) light is absorbed by one Zeeman state; right-circularly polarised (RCP) light by other one. The intensity of the MCD signal depends on the difference in absorption of LCP and RCP light, and hence on the Boltzmann population of the ground Zeeman state. The dependence of MCD intensity on B/T can be used to extract the g value (see figs. 7 and 9).

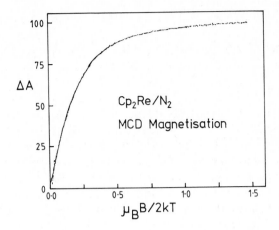

Fig. 9. MCD magnetisation curve obtained by measuring the differential absorbance at 478 nm as a function of B/T. The curve shows points from two sets of measurements, one at 1.8 K, the other at 6.5 K. The value of $g_\parallel$ in the ground electronic state may be obtained by fitting the data to a suitable function for a randomly oriented set of molecules (see table 1 and fig. 7).

methods. The first employs the ratio of integrated differential absorbance (from the MCD spectrum) to the integrated absorbance (from the conventional spectrum). This method gave reasonable estimates of $g_\parallel$, but suffered from the usual inaccuracy of band-area measurements. A better approach uses a detailed analysis of the magnetisation curves obtained by varying the magnetic flux density at a series of temperatures. These optically determined magnetisation curves are wavelength specific and, hence, better targeted than conventional magnetisation experiments (fig. 9).

We have employed these methods to establish that the new metallocenes have orbitally degenerate ground states: $^3E_2$ for the d^4 systems and $^2E_2$ for the d^5 complexes. Taking into account the high spin–orbit coupling, the $g_\parallel$ values may be predicted for the ionic limit. They are given together with the experimental values in table 1. The high angular momentum is only slightly reduced by

Table 1
Values of $g_\parallel$ for open-shell metallocenes.

Electron configuration	$g_\parallel$ (ionic limit)	$g_\parallel$ (experimental)	
d^4 $(e_{2g}^3 a_{1g}^1)$	4.0	Cp$_2$Mo	2.75
		Cp$_2$Re	5.34
d^5 $(e_{2g}^3 a_{1g}^2)$	6.0	Cp$_2^*$Re	5.07

covalency. For the third-row metallocenes, all Jahn–Teller coupling is quenched by spin–orbit coupling. The unusual IR electronic transition of Cp_2W arises from a vibronic transition between spin–orbit sub-states of the ground term (Cox et al. 1983, 1984). Such transitions have been observed for other complexes with similar ground states, such as ReF_6.

3.3.3. Comparisons with solution studies

Strong circumstantial evidence for Cp_2Mo and Cp_2W as reaction intermediates had been obtained well in advance of the matrix work (see above), but further connections with solution photochemistry have come more recently. It has proved possible to detect Cp_2Mo in solution by laser-flash photolysis of Cp_2MoH_2 (Perutz and Scaiano 1984). Its UV/visible spectrum proves remarkably close to that observed in the matrix and it turns out to be highly reactive. Its lifetime in the presence of CO is as short as 7 µs, and it reacts almost as rapidly with the Cp_2MoH_2 precursor when CO is absent.

The photochemistry of Cp_2ReH in solution has been, as yet, little explored. However, it certainly undergoes very similar CO addition reactions to those observed in matrices (Chetwynd-Talbot et al. 1983). Experiments designed to generate the Cp_2Re fragment result in dimers instead of the monomer (Pasman and Snel 1984). The most interesting link to the matrix work has come from a discovery prompted by the matrix work: photolysis of Cp^*ReH ($Cp^* = \eta^5$-C_5Me_5) in pentane yields the stable molecule Cp^*_2Re (Cloke and Day 1985). Extensive investigations of Cp^*_2Re in matrices, in solution and in frozen solution have shown that Cp^*_2Re has the same electronic ground state as Cp_2Re in these media. Its MCD magnetisation curve has been recorded as well as its ESR spectrum in frozen solution. The two methods give identical $g_{\parallel}$ values (table 1) within experimental error, confirming the power of the MCD method. The absorption and LIF spectra of Cp^*_2Re are even better resolved than those of Cp_2Re, probably because its conformation is locked into the eclipsed position (Bandy et al. 1988).

References

Almond, M.J., A.J. Downs and R.N. Perutz, 1985, Inorg. Chem. **24**, 275.
Almond, M.J., J.A. Crayston, A.J. Downs, M. Poliakoff and J.J. Turner, 1986, Inorg. Chem. **25**, 19.
Andrews, L., 1979, Ann. Rev. Phys. Chem. **30**, 79.
Andrews, L., and F.J. Prochaska, 1979, J. Chem. Phys. **70**, 4717.
Appelman, E.H., 1973, Acc. Chem. Res. **6**, 113.
Ault, B.S., and L. Andrews, 1976, J. Chem. Phys. **65**, 4192.
Bajdor, K., and K. Nakamoto, 1984, J. Am. Chem. Soc. **106**, 3045.
Bandow, H., H. Akimoto, S. Akiyama and T. Tezuka, 1984, Chem. Phys. Lett. **111**, 496.
Bandy, J.A., F.G.N. Cloke, G. Cooper, J.P. Day, R.B. Girling, R.G. Graham, J.C. Green, R. Grinter and R.N. Perutz, 1988, J. Am. Chem. Soc. **110**, 5039.
Baynham, R.F.G., J. Chetwynd-Talbot, P. Grebenik, R.N. Perutz and M.H.A. Powell, 1985, J. Organomet. Chem. **284**, 229.

Berry, M., K. Elmitt and M.L.H. Green, 1979, J. Chem. Soc. Dalton Trans., p. 1950.

Berry, M., N.J. Cooper, M.L.H. Green and S.J. Simpson, 1980, J. Chem. Soc. Dalton Trans., 29.

Bondybey, V.E., and G.C. Pimentel, 1972, J. Chem. Phys. **56**, 3832.

Bondybey, V.E., G.C. Pimentel and P.N. Noble, 1971, J. Chem. Phys. **55**, 540.

Boxhoorn, G., D.J. Stufkens and A. Oskam, 1978, Inorg. Chem. Acta **29**, 243.

Boxhoorn, G., A. Oskam, T. McHugh and A.J. Rest, 1980, Inorg. Chim. Acta **44**, L1.

Breckenridge, W.H., and G.M. Stewart, 1986, J. Am. Chem. Soc. **108**, 364.

Breeze, P.A., J.K. Burdett and J.J. Turner, 1981, Inorg. Chem. **20**, 3369.

Burdett, J.K., J.M. Grzybowski, R.N. Perutz, M. Poliakoff, J.J. Turner and R.F. Turner, 1978a, Inorg. Chem. **17**, 147.

Burdett, J.K., A.J. Downs, G.P. Gaskill, M.A. Graham, J.J. Turner and R.F. Turner, 1978b, Inorg. Chem. **17**, 523.

Chetwynd-Talbot, J., P. Grebenik and R.N. Perutz, 1982, Inorg. Chem. **21**, 3647.

Chetwynd-Talbot, J., P. Grebenik, R.N. Perutz and M.H.A. Powell, 1983, Inorg. Chem. **22**, 1675.

Church, S.P., M. Poliakoff, J.A. Timney and J.J. Turner, 1981, J. Am. Chem. Soc. **103**, 7515.

Church, S.P., M. Poliakoff, J.A. Timney and J.J. Turner, 1983, Inorg. Chem. **22**, 3259.

Church, S.P., F.W. Grevels, H. Hermann and K. Schaffner, 1984a, Inorg. Chem. **23**, 3830.

Church, S.P., H. Hermann, F.W. Grevels and K. Schaffner, 1984b, J. Chem. Soc., Chem. Commun., p. 785.

Cloke, F.G.N., and J.P. Day, 1985, J. Chem. Soc., Chem. Commun., p. 967.

Cooper, N.J., M.L.H. Green and R. Mahtab, 1979, J. Chem. Soc. Dalton Trans., p. 1557.

Cox, P.A., P. Grebenik, R.N. Perutz, M.D. Robinson, R. Grinter and D.R. Stern, 1983, Inorg. Chem. **22**, 3614.

Cox, P.A., P. Grebenik, R.N. Perutz, R.G. Graham and R. Grinter, 1984, Chem. Phys. Lett. **108**, 415.

Crayston, J.A., M.J. Almond, A.J. Downs, M. Poliakoff and J.J. Turner, 1984, Inorg. Chem. **23**, 3051.

DeMore, W.B., H.O. Pritchard, N. Davidson, 1959, J. Am. Chem. Soc. **81**, 5874.

Diem, M., T.-L. Tso and E.K.C. Lee, 1982, J. Chem. Phys. **76**, 6452.

Downs, A.J., and M. Hawkins, 1983, Adv. IR Ramans Spectrosc. **10**, 1.

Dunkin, I.R., P. Härter and C.J. Shields, 1984, J. Am. Chem. Soc. **106**, 7248.

Fairhurst, S.A., J.R. Morton and K.F. Preston, 1983, J. Magn. Resonance **55**, 453.

Fairhurst, S.A., J.R. Morton, R.N. Perutz and K.F. Preston, 1984, Organometallics **3**, 1389.

Fajardo, M.E., and V.A. Apkarian, 1986, J. Chem. Phys. **85**, 5660.

Firth, S., P.M. Hodges, M. Poliakoff and J.J. Turner, 1986, Inorg. Chem. **25**, 4608.

Firth, S., P.M. Hodges, M. Poliakoff, J.J. Turner and M.J. Therien, 1987, J. Organomet. Chem. **331**, 347.

Fletcher, S.C., M. Poliakoff and J.J. Turner, 1984, J. Organomet. Chem. **268**, 259.

Frei, H., and G.C. Pimentel, 1985, Ann. Rev. Phys. Chem. **36**, 491.

Gedanken, A., B. Raz and J. Jortner, 1973, J. Chem. Phys. **58**, 1178.

Gholivand, K., and H. Willner, 1984, Z. Naturforsch. b **39**, 1211.

Goodman, J., and L.E. Brus, 1976, J. Chem. Phys. **65**, 3808.

Green, M.L.H., and G. Giannotti, 1972, J. Chem. Soc., Chem. Commun., p. 1114.

Haddleton, D.M., 1986, J. Organomet. Chem. **308**, C15.

Haddleton, D.M., and R.N. Perutz, 1985, J. Chem. Soc., Chem. Commun., p. 1372.

Haddleton, D.M., and R.N. Perutz, 1986, J. Chem. Soc., Chem. Commun., p. 1734.

Haddleton, D.M., A. McCamley and R.N. Perutz, 1988, J. Am. Chem. Soc. **110**, 1810.

Hashimoto, S., and H. Akimoto, 1987, J. Phys. Chem. **91**, 1347.

Hawkins, M., and A.J. Downs, 1984a, J. Phys. Chem. **88**, 1527.

Hawkins, M., and A.J. Downs, 1984b, J. Phys. Chem. **88**, 3042.

Hawkins, M., L. Andrews, A.J. Downs and D.J. Drury, 1984, J. Am. Chem. Soc. **106**, 3076.

Hawkins, M., M.J. Almond and A.J. Downs, 1985, J. Phys. Chem. **89**, 3326.

Hepp, A.F., and M.S. Wrighton, 1983, J. Am. Chem. Soc. **105**, 5934.
Hitam, R.B., A.J. Rest, M. Herberhold and W. Kremnitz, 1984a, J. Chem. Soc., Chem. Commun., p. 471.
Hitam, R.B., A. Mahmoud and A.J. Rest, 1984b, Coord. Chem. Rev. **55**, 1.
Hitam, R.B., K.A. Mahmoud and A.J. Rest, 1985, J. Organomet. Chem. **291**, 321.
Hooker, R.H., and A.J. Rest, 1982, J. Chem. Soc. Dalton Trans., p. 2029.
Horton-Martin, A., M. Poliakoff and J.J. Turner, 1986, Organomet. **5**, 405.
Huber, J.R., 1987, personal communication.
Jacox, M.E., 1978, Rev. Chem. Intermed. **2**, 1.
Kafafi, Z.H., R.H. Hauge and J.L. Margrave, 1985, J. Am. Chem. Soc. **107**, 7550.
Kasai, P.H., 1971, Acc. Chem. Res. **4**, 329.
Kelly, J.M., C. Long and R. Bonneau, 1983, J. Phys. Chem. **87**, 3344.
Knoezinger, E., and R. Wittenbeck, 1982, Ber. Bunsenges. Phys. Chem. **86**, 742.
Lotta, T., J. Murto, M. Rasanen and A. Aspiala, 1984, J. Mol. Struct. **114**, 333.
Machara, N.P., and B.S. Ault, 1987, J. Phys. Chem. **91**, 2046.
Mahmoud, K.A., A.J. Rest, M. Luksza, K. Jörg and W. Malisch, 1984, Organomet. **3**, 501.
Maier, G., G. Mihm and H.P. Reisenauer, 1984a, Chem. Ber. **117**, 2351.
Maier, G., G. Mihm, H.P. Reisenauer and D. Littmann, 1984b, Chem. Ber. **117**, 2369.
McHugh, T.M., and A.J. Rest, 1980, J. Chem. Soc. Dalton Trans., p. 2323.
McHugh, T.M., A.J. Rest and J.R. Sodeau, 1979, J. Chem. Soc. Dalton Trans., p. 184.
Morton, J.R., and K.F. Preston, 1984a, Chem. Phys. Lett. **111**, 611.
Morton, J.R., and K.F. Preston, 1984b, J. Chem. Phys. **81**, 5775.
Müller, R.P., and J.R. Huber, 1984, J. Phys. Chem. **88**, 1605.
Müller, R.P., M. Nonella, P. Russegger and J.R. Huber, 1984, Chem. Phys. **87**, 351.
Nonella, M., J.R. Huber and T.-K. Ha, 1987, J. Phys. Chem. **91**, 5203.
Ozin, G.A., J.G. McCaffrey and J.M. Parnis, 1986, Angew. Chem. Int. Ed. Engl. **25**, 1072.
Pacansky, J., 1977, J. Phys. Chem. **81**, 2241.
Paine, R.T., R.S. McDowell, L.B. Asprey and L.H. Jones, 1976, J. Chem. Phys. **64**, 3081.
Pasman, P., and J.J.M. Snel, 1984, J. Organomet. Chem. **276**, 387.
Perutz, R.N., 1985a, Chem. Rev. **85**, 77.
Perutz, R.N., 1985b, Chem. Rev. **85**, 97.
Perutz, R.N., 1986, Ann. Rep. Prog. Chem., Sect. C, **82**, 157.
Perutz, R.N., and J.C. Scaiano, 1984, J. Chem. Soc., Chem. Commun., p. 457.
Perutz, R.N., and J.J. Turner, 1975a, Inorg. Chem. **14**, 262.
Perutz, R.N., and J.J. Turner, 1975b, J. Am. Chem. Soc. **97**, 4791.
Poliakoff, M., 1978, Chem. Soc. Rev. **7**, 527.
Poliakoff, M., and E. Weitz, 1987, Acc. Chem. Res. **20**, 408.
Rest, A.J., I. Whitwell, W.A.G. Graham, J.K. Hoyano and A.D. McMaster, 1987, J. Chem. Soc. Dalton Trans., p. 1181.
Rothberg, L.J., N.J. Cooper, K.S. Peters and V. Vaida, 1982, J. Am. Chem. Soc. **104**, 3536.
Schnöckel, H., 1978, J. Mol. Struct. **50**, 275.
Schnöckel, H., 1981, J. Mol. Struct. **70**, 183.
Seder, T.A., S.P. Church and E. Weitz, 1986, J. Am. Chem. Soc. **108**, 4721.
Simpson, M.B., M. Poliakoff, J.J. Turner, W.B. Maier and J.G. McLaughlin, 1983, J. Chem. Soc., Chem. Commun., p. 1355.
Sodeau, J.R., and E.K.C. Lee, 1980, J. Phys. Chem. **84**, 3358.
Sonntag, B., 1977, in: Rare Gas Solids, Vol. 2, eds M.L. Klein and J.A. Venables (Academic Press, New York) ch. 17.
Sweany, R.L., 1981, J. Am. Chem. Soc. **103**, 2410.
Sweany, R.L., 1985, J. Am. Chem. Soc. **107**, 2374.
Symons, M.C.R., and R.L. Sweany, 1982, Organomet. **1**, 834.

Tchir, P.O., and R.D. Spratley, 1975a, Can. J. Chem. **53**, 2311.

Tchir, P.O., and R.D. Spratley, 1975b, Can. J. Chem. **53**, 2318.

Tchir, P.O., and R.D. Spratley, 1975c, Can. J. Chem. **53**, 2331.

Toriyama, K., K. Nunome and M. Wasaki, 1980, J. Phys. Chem. **84**, 2374.

Tso, T.-L., and E.K.C. Lee, 1984, J. Phys. Chem. **88**, 2776.

Tso, T.-L., and E.K.C. Lee, 1985, J. Phys. Chem. **89**, 1618.

Turner, J.J., J.K. Burdett, R.N. Perutz and M. Poliakoff, 1977, Pure Appl. Chem. **49**, 271.

Turner, J.J., M.B. Simpson, M. Poliakoff, W.B. Maier and M.A. Graham, 1983, Inorg. Chem. **22**, 911.

Turner, J.J., M.A. Healy and M. Poliakoff, 1987, High-energy processes in Organometallic Chemistry, ACS Symposium, Vol. 333, p. 110.

Upmacis, R.K., G.E. Gadd, M. Poliakoff, M.B. Simpson, J.J. Turner, R. Whyman and A.F. Simpson, 1985, J. Chem. Soc., Chem. Commun., p. 27.

Van Ijzendoorn, L.J., L.J. Allamandola, F. Bass and J.M. Greenberg, 1983, J. Chem. Phys. **78** 7019.

Wakahara, A., T. Miyuzaki and K. Fueki, 1982, J. Phys. Chem. **86**, 1781.

Waltz, W.L., O. Hackelberg, L.M. Dorfman and A. Wojcicki, 1978, J. Am. Chem. Soc. **100**, 7259.

Wegman, R.W., R.J. Olsen, D.R. Gard, L.R. Faulkner and T.L. Brown, 1981, J. Am. Chem. Soc. **103**, 6089.

Yesaka, H., T. Kobayashi, K. Yasufuku and S. Nagakura, 1983, J. Am. Chem. Soc. **105**, 6249.

CHAPTER 10

Reactions of First-Row Transition Metal Atoms and Small Clusters in Matrices

Robert H. HAUGE and John L. MARGRAVE

Rice University
Department of Chemistry
P.O. Box 1892
Houston, TX 77251
USA

and

Zakya H. KAFAFI

Naval Research Laboratory
Optical Sciences Division
Code 6551
Washington, DC 20375
USA

Chemistry and Physics of
Matrix-Isolated Species
L. Andrews and M. Moskovits, eds

Contents

1. Introduction

Matrix-isolation studies of reactions of transition metal atoms and small clusters may be divided into three categories:

(1) Reactions between metal atoms which lead to metal clustering. The study of metal clusters in matrices is a very active area, which has been reviewed by Olsen and Klabunde (1987), Howard and Mile (1987), Moskovits (1986), Ozin and Andrews (1986), Weltner and Van Zee (1984) and Klabunde (1980), along with discussion presented within the present text.

(2) Interactions between metal atoms/clusters and ligands which result in the formation of metal–ligand complexes or adducts. The formation of metal–ligand species in matrices has been extensively reviewed in previous works (Barnes et al. 1981, Moskovits and Ozin 1976, Cradock and Hinchcliffe 1975, Hallam 1973), as well as in a number of reviews on matrix isolation (Perutz 1984a,b, Barnes 1980, Poliakoff et al. 1979, Moskovits and Ozin 1975).

(3) Reactions between metal atoms/clusters and ligand(s) which give rise to some form of bond activation of the reactant ligand. This chapter discusses the reactions between transition metal atoms and small ligands which ultimately result in bond insertion and molecular rearrangement of associated ligands. Interest has generally focused on the reactivity of simple but important molecules such as H_2O, CH_4, NH_3, and H_2. Other reactants have also been used such as diazomethane (Chang 1987, Chang et al. 1985, 1987a,b, 1988a,b), cyclopropane (Kline et al. 1988), ethylene, acetylene, ethylene oxide (Kafafi et al. 1987, Kline 1987), cyclopentadiene (Ball et al. 1985, Ball, 1987), and cyclo-hexadiene (Ball et al. 1986).

For metal-induced bond activation to occur spontaneously at 10 K on an inert surface or in an inert matrix, reaction barriers must be less than a few kJ/mol. It is of general interest, in fact, that matrix-isolation reaction studies, in the absence of other forms of excitation, provide an excellent test for the activated character of a specific reaction. On matrix surfaces or in matrices it may be assumed that the reactants interact through their respective ground states due to rapid energy exchange with the matrix surface boundary layer, with some exceptions for particularly metastable electronic and vibrational states.

Very little is known about the nature of the matrix boundary layer where reaction is occurring. It seems clear, however, that metal atoms diffuse significant distances at or near the surface during which an encounter with a second reactant is possible. This second reactant may also be mobile or immobilized but still available for reaction due to its proximity to the surface. The extent of reaction appears to be quite sensitive to the mobility of the reactants. For example, hydrogen reactions have been observed with only trace amounts of hydrogen present, apparently due to its high mobility (Hauge et al. 1987). One also observes the formation of metal dimers in the low parts per thousand molar range in argon matrices at ~ 10 K while very little dimer forms in a krypton matrix at the same temperature and concentration. It is also known that variation of matrix deposition temperature by only a few degrees strongly affects the extent of reaction. Warming to similar temperatures after deposition of matrices generally affects transition metal reactions to a much less extent.

The identification of the number of metal atoms or reactant molecules involved in the observed matrix reaction can be quite difficult. In fact, the matrix literature provides a number of examples of species identified as MA products which were later identified as MA_2 products or metal dimer products (Kafafi et al. 1985b). Others have documented the importance of careful concentration studies in determining the order of the observed matrix reactions. Moskovits (1986), in particular, has outlined methods for analyzing the product growth curves with variation of the molar ratios of reactants with respect to the host matrix. All of the respective methods rely on identification of the lowest order, one to one, MA product which serves as a reference for ascertaining the nuclearity of higher order reactions. In certain cases, one to one product formation does not occur or the product is not easily distinguished from the reactants and in these cases it can be very risky to attempt to define the nuclearity of the reaction simply from the slope of observed product versus metal concentration. For example, diatomic iron reactions (Kafafi et al. 1985b), with ethylene in an argon matrix show a one to one growth pattern in the atomic concentration range from three to ten parts per thousand simply because the concentration of the atomic iron–ethylene adduct saturates, thus causing the diatomic iron product to be first order in atomic iron. This was only discovered because a very slightly perturbed atomic iron monoethylene adduct was ultimately observed at even lower concentrations where its growth had not saturated.

Generally we have found that for transition metals in argon matrices at ~ 10 K that metal–matrix ratios less than two parts per thousand are necessary for atomic reactions to dominate the chemistry. However, in krypton matrices at 10 K atomic reactions appear to dominate up to ten parts per thousand. Once a one to one product is identified, dimetal and trimetal species can usually be identified by the order of their appearance as the metal concentration is increased. Higher nuclearity metal-cluster products are difficult to identify definitively from concentration studies only, because of the increasing similarity of their metal concentration dependence. Comparative reaction studies of the atom, diatom and triatom are of great interest, however, in that they provide an opportunity to study one-, two- and three-fold reaction sites.

It will be interesting in the future to contrast the reaction studies of metal species in matrices with their reactivity in cooled molecular beams. Reaction studies of metal clusters in molecular beams have the great advantage of being able to define with certainty the reacting metal species since they are directly monitored via their mass. Studies by Morse et al. (1985), Richtsmeier et al. (1985), and Whetten et al. (1985), have shown large variations of reactivity with cluster size. The most marked variations were observed for molecular hydrogen reactions. As yet, structural probes which indicate the nature of the molecular rearrangements that are occurring on these large cluster surfaces are not available. Collision-induced fragmentation does provide some insight into the chemistry, although indirectly. Thus, reaction studies of small metal-cluster species in matrices, which use the highly versatile FTIR instrumentation currently available, remain the best means of investigating molecular rearrangements. Tunneling reactions to form new products are also more likely for reactions in a matrix than for molecular-beam chemical studies since weakly bound complexes exist indefinitely in a matrix. For example, the experimental studies of Ismail et al. (1982), and subsequent theoretical studies by Tachibana et al. (1987) have indicated that the atomic silicon–water adduct converts to HSiOH over a period of hours via a tunneling mechanism.

Since this chapter will be limited to transition metal chemistry, studies of main group metals will not be discussed, although the same interesting questions regarding reaction mechanisms, activation energies, product geometries, product force constants and photochemical behavior are applicable. Emphasis will be on the first-row transition metal atoms reaction since the vast majority of reaction studies resulting in bond insertion chemistry have been carried out with these metals. This is undoubtedly due to their relative ease of vaporization. With the availability of pulsed laser vaporization the second- and third-row transition metals are likely to be increasingly studied.

2. Reactions of first-row transition metal atoms and small clusters in matrices

2.1. Reactions with water

Coordination complexes of water are of interest since water can not act as a π-electron acceptor but rather only as a σ- or π-electron donor. The bending frequency of water has been found to be very sensitive to the extent of electron donation. This is not surprising since removal of an electron from the $3a_1$ (σ lone pair) orbital of water causes water to become linear and results in a decrease in the bending frequency of ~ 700 cm^{-1}. This very large frequency change suggests that shifts of the bending mode of water can provide a sensitive measure of the extent of σ-electron donation.

Curtiss et al. (1987) and Curtiss and Pople (1985) have calculated the binding energies of water as well as the extent of electron transfer to some of the Group Ia and IIa metal atoms and find that they do correlate with the measured shifts of the bending mode of water within the respective groups. The shifts did not, however, correlate with the calculated binding energies between groups Ia and IIa, but did appear to reflect the extent of electron transfer. Thus, the measured shifts are perhaps best thought of as a relative measure of the σ-electron accepting capability, the so-called Lewis acidity, of the metal. It is considered a measure of σ-interaction since π-electron donation by water is not expected to affect the bending mode of water.

Measured shifts in the bending mode of water when it is adducted to first-row transition metal atoms are given in table 1. Sc, Ti, V (Kauffman et al. 1985a,b), and Ni (Park et al. 1988), were found to react spontaneously with water on the matrix surface and formed the respective hydroxy–metal hydrides. Also included are the values for the diatomic species. Note that the differences between the atomic and diatomic species for the same metal are reversed for chromium relative to the late transition metals. It is tempting to suggest that the electron-accepting d-orbitals of late transition metals have been destabilized relative to the atom and are thus poorer electron acceptors since only the partially filled anti-bonding d-orbital combinations are available for interaction with water.

Blomberg et al. (1986) have suggested a simplified dipole–induced-dipole interaction as being primarily responsible for the bonding between water and the polarizable metal species. Manceron et al. (1985) used a similar model to explain the marked enhancement of the intensity of the symmetric stretching mode of water when bound to lithium. It is interesting to speculate that careful measurements of the relative band intensities for the bound and free water molecule will provide a direct measure of the induced dipole and thus a measure of the dipole–induced-dipole interaction energy. Calculations of Blomberg

Table 1

Frequency shifts (in cm^{-1}) of the bending mode of water ($1593\ cm^{-1}$) when adducted to the respective metal atoms and diatoms.

Metal atom	$-\Delta v$	Metal diatom	$-\Delta v$
Ca	30.8	Ca_2	–
Cr	13.1	Cr_2	25.5
Mn	16.5	Mn_2	19.2
Fe	30.5	Fe_2	16.9
Co	29.0	Co_2	–
Cu	20.5	Cu_2	10.9
Zn	5.6	Zn_2	8.3

et al. (1986) suggest that the induced dipole is roughly equal to the water dipole for atomic nickel interacting with water. This suggests that the symmetric modes of water would be enhanced by a factor of four over free water for a C_{2v} geometry. Studies of the sensitivity of ligand vibrational transition moments to adduct formation is expected to provide new insights into ligand–metal bonding.

As mentioned above, nonactivated bond-insertion chemistry was seen for a number of the transition metals. It is perhaps not surprising that Sc, Ti, and V undergo spontaneous insertion since they are very electropositive elements with high reaction exothermicities of ~ 200 kJ/mol. However, the spontaneous reaction of atomic nickel was not expected and it was the only metal of the latter first-row transition metals which spontaneously reacted. The exothermic reaction energies are all of the order of ~ 100 kJ/mol for Cr, Mn, Fe, Co, and Ni. The reactivity of nickel is thus not due to the exothermicity of the reaction, but is most likely due to its low lying d^9s^1 electronic configuration. One might also expect Cu to react spontaneously since it exists in a $d^{10}s^1$ state; however, its reaction energy has decreased to ~ 15 kJ/mol. The bond insertion product, HMOH, for all of the above metal–water adduct species can be obtained by electronically exciting the metal–water adduct. Vibrational frequency values for the M–H and M–OH bonds are given in table 2 along with calculated force constants. Note that for HNiOH two sets of values were observed, which are thought to be due to different geometrical isomers. These force constants are plotted in fig. 1 as a function of atomic number.

It is interesting to note that both the M–H and M–OH bond force constants increase across the series. If a line is drawn between Ca and Mn, one sees that the

Table 2

Hydrido metal hydroxides stretching frequencies (in cm^{-1}) and force constants (in mdyn/Å) calculated from a diatomic model (argon matrix).

	H–M	D–M	M–OH	M–OD	k(H–M)	k(M–OH)	Δk^*
HCaOH	1232.8	887.3	574.8	576.5	0.92	2.47	1.55
HScOH	1485.1	1070.0	715.8	698.2	1.29	3.73	2.44
HTiOH	1538.9	1107.7	699.7	697.3	1.39	3.79	2.40
HVOH	1583.0	1140.3	703.3	696.6	1.47	3.84	2.37
HCrOH	1639.9	1184.5	674.1	654.0	1.69	3.40	1.71
HMnOH	1663.4	1197.1	648.1	628.5	1.70	3.18	1.48
HFeOH	1731.9	1245.3	682.4	660.5	1.83	3.53	1.70
HCoOH	1790.4	1291.2	667.4	641.9	2.01	3.37	1.36
HNiOH	1901.0	–	707.0	681.2	2.25	3.80	1.55
	1837.3	1323.9	690.6	664.8	2.10	3.62	1.52
HCuOH	1910.8	1380.8	615.6	613.8	2.33	3.12	0.79

*$\Delta k = k(M–OH) - k(H–M)$.

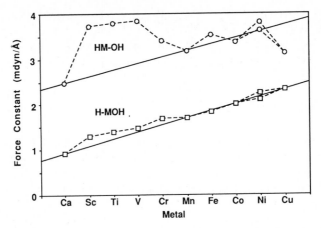

Fig. 1. Plot of metal hydride and metal hydroxide force constants versus atomic number of the first-row transition metals for the HMOH species.

rate of increase in the force constants for the two types of bonds is approximately the same. This suggests a common cause for the increase which is most likely a regular reduction in metal ionic radius with increasing atomic number across the series. One also notes a much larger deviation from the Ca–Mn line for the M–OH bond than the M–H bond. This deviation is thought to be caused by a considerable involvement of oxygen lone-pair electrons in the M–OH bond. The effect is largest for the early transition metals because of their largely vacant d-orbitals. Note also that for HCuOH, the bond force constant of the Cu–OH bond is approaching that of the Cu–H bond apparently as the result of an almost filled Cu d-shell which results in increased electron repulsion by the oxygen lone-pair electrons. The difference in magnitude for the respective Ca–Mn lines of the M–OH and M–H bonds is ~ 1.5 mdyn/Å. It may be taken as a relative measure of the difference in strength with the two respective σ-bonds. This assumes that Ca and Mn do not significantly interact with the lone-pair electrons of oxygen. The parallel behavior of the M–H and M–OH bonds for these elements strongly supports this assumption since hydrogen can only interact through its σ-orbital.

Bond-insertion reactions of the diatomic metals have been observed for Mn, Fe, and Co (Kauffman et al. 1985a,b), and Ni (Park et al. 1988). Three types of products were observed: two types of HM_2OH species, where the H and OH groups are found in either the bridging or end positions and a HMOMH species where the metal–metal bond and both oxygen–hydrogen bonds of water have been completely broken to form a C_{2v} species. The behavior of these metal diatomics suggests that the reaction of larger metal clusters with water may be quite complex.

Future reaction studies of water can be expected for the other transition metals. In addition to measurements of intensities as earlier mentioned, far-infrared measurements of the M_x–OH_2 stretching frequencies would be of interest since they directly reflect the metal–oxygen bond strength. The intensities of these modes are expected to be quite high due to the large induced dipole of the adduct.

2.2. Reactions with methane

Activation of the carbon–hydrogen bond of methane by metal atoms is of interest in part because it represents activation of the simplest, and most inert carbon–hydrogen bond of the alkane class. Only recently has bond activation of methane at a metal-atom center been shown to occur in solution (Janowicz and Bergman 1982, Hoyano and Graham 1982). In matrices it was first shown by Billups et al. (1980) and Ozin et al. (1981a,b) that the late first-row transition metals undergo oxidative insertion to form the $HMCH_3$ species when the metal atom is electronically excited. Ozin and McCaffrey (1982, 1983) also found that electronically excited $HFeCH_3$ reverts to the iron atom and methane. For copper, Ozin et al. (1981a) showed that the $HCuCH_3$ species also undergoes extensive fragmentation to CuH and $CuCH_3$. The ability to cycle the insertion reaction photolytically seems to occur quite frequently in photoactivated reactions of transition metal atoms. Often the insertion product can be electronically excited with longer wavelength light than is required to excite the metal atom–adduct complex. This allows cycling of the reaction by simply changing the wavelength of the photolysis source.

The manner in which methane interacts with a metal atom has been investigated both experimentally and theoretically. Kafafi et al. (1985a) suggested that iron is adducted to methane in a C_{3v} geometry, as evidenced by a splitting of the triply degenerate methyl deformation into two bands. Whether methane is bonded in a one-fold or three-fold manner could not be determined. Ozin et al. (1986) have argued that the presence of one-fold coordination provides an explanation of differences in product yields of Cu and Ag in methane matrices since this form of coordination would not lead to bond insertion. Theoretical calculations of Anderson and Baldwin (1987) suggest that one- and three-fold coordination are of similar stability and both more stable than two-fold coordination with iron.

Blomberg et al. (1984) calculated a barrier of 54 kcal for bond insertion into methane by the $Ni(^1D)$ d^9s^1 state as well as an H–Ni–C bond angle of 94°. Anderson and Baldwin (1987) suggest that the H–M–X bond for the transition metals will in general be nonlinear. Ozin et al. (1984) have suggested that $HFeCH_3$ and $HMnCH_3$ are bent since a splitting of the triply degenerate modes of the methyl group is observed. These modes would be degenerate if the molecule was linear. Kauffman et al. (1985a,b) have previously suggested that all of the similar HMOH first-row transition metal species are also nonlinear.

The oxidative insertion of the first-row transition metals into methane has required electronic excitation of the metal in all instances. Theoretical considerations suggest bond insertion occurs by donation of the excited p-orbital electron density to the antibonding orbital of the C–H bond while simultaneously the bonding orbital of the C–H bond donates electron density to the partially occupied metal 4s-orbital.

Billups et al. (1980) have found that all of the late first-row transition metals could be photo-inserted into methane with the exception of nickel. Nickel was later shown to also undergo photo-insertion (Chang et al. 1988a). The early first-row transition metals did not react. Similar observations were made by Ozin et al. (1984). Ozin et al. (1986) suggest that lack of reactivity is caused by the less stable d-orbitals of the early transition metals since they act as electron acceptors in the bond-insertion process. This suggestion implies that the excited states of the early transition metals provide a barrier to C–H insertion while for the late transition metals they do not. It is also possible that the species $HMCH_3$ is energetically and/or photolytically unstable for the elements lighter than manganese and that it spontaneously rearranges back to the metal atom and methane.

The measured vibrational frequencies for the known first-row transition metal $HMCH_3$ species are given in table 3. As was noted for the HMOH species the H–M stretching frequency increases with atomic number but not with the same regularity as for the HMOH species. The H–Ni value appears unusually high and the Cu–H value is low when compared to the respective HMOH values. The low Cu–H stretching frequency can be understood in terms of the high promotion energy of copper which is experienced to a larger extent by the M–H

Table 3

Vibrational frequencies (in cm^{-1}) for $HMCH_3$ species in argon.

Approximate vibrational mode	Metal						
	Mn	Fe	Co	Ni	Cu	Zn[a]	Zn[b]
CH_3, stretch	2924.0	2921[b]	2942.1	2950.5	–	–	–
	2895.0	2888[b]	2892.0	2861.5	–	–	–
M–H, stretch	1616.5	1683.6	1722.6	1945.1	1719.0	(1870)	1845.8
CH_3, deform	1148.6	1156.1	1168.5	1139.0	1012.0	(1070)	1065.9
CH_3, rock	543.3	544.0	573.2	642.7	–	(685)	689.1
	541.4	541.0	568.2				
M–C, stretch	506.2	523.5	530.8	554.9	–	(451)	447.1
HMC, bend	325[b]	300[b]	–	–	–	–	–
	316	293	–	–	–	–	–

Notes:

[a] Values were estimated by comparing known values in argon and methane matrices for other $HMCH_3$ species.

[b] Measured in a methane matrix.

bond in the $HMCH_3$ species due to the weaker $M-CH_3$ bond as compared to the $M-OH$ bond. The high value for the $Ni-H$ stretching frequency is more difficult to understand. It is possible that the frequency difference reflects a difference in the nature of the electronic ground states of $HNiCH_3$ and $HNiOH$ species. Finally, one notes a parallel behavior between the $M-H$ and $M-CH_3$ stretching frequencies which implies similar bonding for the $M-H$ and $M-CH_3$ bonds.

2.3. Reactions with diazomethane

Except for zinc, the first-row late transition metals react spontaneously with diazomethane in both argon and nitrogen matrices. Zinc forms an adduct with diazomethane which can be photolyzed to the respective metal methylene and dinitrogen. In solid argon, the products are both the ligated and the unligated metal methylenes. Iron methylene was identified by Chang et al. (1985), and was the first matrix-isolated metal methylene product to be characterized via FTIR spectroscopy. The presence of single iron and carbon atoms in this compound was deduced from ^{54}Fe and ^{13}C isotopic frequency measurements of the metal–carbon stretching mode. The measured frequency of 623.9 cm^{-1} was about 100 cm^{-1} higher in value than similar frequencies for molecules with single iron–carbon bonds. Double-bond character in the $Fe-C$ bond of $FeCH_2$ is thought to be responsible for the increased value of the stretching frequency. Figure 2 shows selected regions of the infrared spectra of $FeCH_2$, $Fe^{13}CH_2$ and $FeCD_2$. A comparison between all measured frequencies of the isotopomers of $FeCH_2$ and those calculated using normal coordinate analyses is given in table 4. Similar frequencies were measured and calculated for $CrCH_2$, $MnCH_2$, $CoCH_2$, $NiCH_2$, $CuCH_2$ and $ZnCH_2$ (see Chang 1987, Chang et al. 1987a,b, 1988a,b). Table 5 lists all of the frequencies measured for Cr, Mn, Fe, Co, Ni, Cu and Zn methylenes in solid argon. Of particular interest is the trend in the metal–carbon stretching force constants. This behavior is seen more clearly in fig. 3 where the variation of the metal–carbon single-, double- and triple-bond force constants for, respectively, $HMCH_3$, MCH_2 and $HMCH$ is displayed. For the metal–carbon single-bond force constant a monotonic increase is observed in going from Mn to Ni with a big drop for Zn. The frequencies of $Cr-C$ and $Cu-C$ single bonds have not been measured. The effect of periodicity is much more pronounced for the $M = C$ bond in MCH_2 where one may easily note half of the expected double hump behavior known for transition metals.

The bonding of the metal and carbon in MCH_2 is thought to be largely the result of interaction between the metal atom and the triplet $CH_2(^3B_1)$ state. This bonding can be viewed as simply a spin pair of the metal 4s- or 3d-orbital with the CH_2 σ-orbital, and the metal 3d-orbital with the CH_2 π-orbital. Theoretical calculations by Spangler et al. (1981), Rappe and Goddard (1977), and Brooks and Schaeffer (1977), indicated that the metal–carbon bond is quite covalent

R.H. Hauge et al.

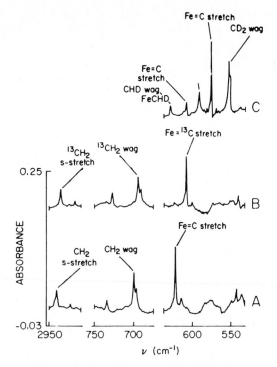

Fig. 2. Infrared spectra of iron methylene in an argon matrix from reactions of atomic iron with diazomethane: (A) $FeCH_2$; (B) $Fe^{13}CH_2$; (C) $FeCD_2$.

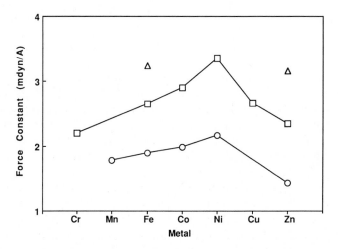

Fig. 3. Plot of metal–carbon force constants of $HM–CH_3$ ($\square$), $M = CH_2$ ($\bigcirc$) and $HM \equiv CH$ ($\triangle$) versus the atomic number of the first-row transition metals.

Table 4

Measured and calculated infrared frequencies (in cm^{-1}) for $FeCH_2$, $Fe^{13}CH_2$, $FeCHD$ and $FeCD_2$ in solid argon.[a]

Vibrational mode	$FeCH_2$		$Fe^{13}CH_2$		$FeCHD$		$FeCD_2$	
	Obs.	Cal.	Obs.	Cal.	Obs.	Cal.	Obs.	Cal.
CH_2 s-stretch	2941.6	2941.6	2936.1	2936.1	–	2168.3	2134.3	2134.3
CH_2 bend[b]	–	1319.2	–	1312.2	–	1175.2	–	1004.0
FeC stretch	623.9	624.0	607.7	608.1	608.2	608.9	575.2	573.6
CH_2 a-stretch	3011.5	3010.7	3002.0	3002.8	2975.9	2201.0	2201.0	2201.2
CH_2 rock	452.0	452.4	449.1	449.3	382.0	382.2	347.6	346.8
CH_2 wag	700.3	701.1	694.2	695.0	629.4	629.7	552.7	549.1
	697.4	–	691.4	–	628.0	–	550.9	–

Notes:

[a] The geometry and symmetry force constants used for the calculated frequencies can be found in the article by Chang et al. (1988b).

[b] The calculation of CH_2 bending frequencies was based on the measured values for $(N_2)_xFeCH_2$, etc., in nitrogen matrices (see Chang 1987).

Table 5

Infrared measured frequencies (in cm^{-1}) for MCH_2 with M = Cr, Mn, Fe, Co, Ni, Cu and Zn in solid argon.

Vibrational mode	$CrCH_2$	$MnCH_2$	$FeCH_2$	$CoCH_2$	$NiCH_2$	$CuCH_2$	$ZnCH_2$
CH_2 s-stretch	2907.4	2864.3	2941.6	2918.0	2917.1	2960.7	2956.1
CH_2 bend	–	–	–	1327.0	1328.5	1344.9	1341.5
M–C stretch	567.0	521.9	623.9	655.4	696.2	614.0	513.7
CH_2 a-stretch	2966.7	–	3011.5	2979.7	2973.2	3034.7	3047.2
CH_2 rock	450.3	–	452.0	587.5	586.7	–	543.8
CH_2 wag	687.7	–	700.3	757.4	791.9	526.0	524.8

with a small partial negative charge on the methylene carbon. The metal–carbon π-bond is considered to be quite weak due to the small overlap between the metal d-orbital and the methylene carbon π (p) orbital; thus, the metal–carbon σ-bond dominates the M = C bonding. The use of a metal 4s or 3d orbital to form the metal–carbon σ-bond varies from metal to metal as a function of the compensation of atomic promotion energy ($3d^n4s^2 \rightarrow 3d^{n+1}4s^1$, or $4s^2 \rightarrow 4s^14p^1$) and pairing energy.

A most interesting unimolecular photorearrangement was observed in the case of iron methylene and zinc methylene where broad band UV excitation led to a carbene → carbyne rearrangement. The first unligated hydrido iron and zinc methylidynes, HFeCH and HZnCH, were thus isolated and spectroscopically characterized (Chang et al. 1987b, 1988b). The iron and zinc metal–carbon frequencies were measured to be $674.2 \, cm^{-1}$ and $647.5 \, cm^{-1}$, respectively.

These high frequencies reflect a significant π-bonding in HMCH with a nominal triple $M \equiv C$ bond character. In the case of iron the unimolecular reaction is found to be photoreversible with UV excitation leading to oxidative-addition and formation of HFeCH, and visible photolysis resulting in reductive elimination with the regeneration of $FeCH_2$. Figure 4 shows this photoreversible process. Table 6 lists the measured and calculated frequencies for the isotopomers of HFeCH and HZnCH, respectively.

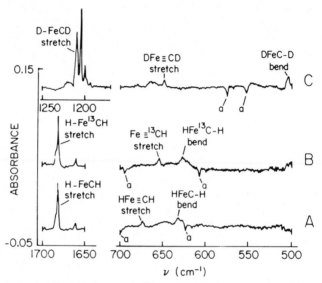

Fig. 4. Infrared spectra of hydrido iron methylidyne, HFeCH, in an argon matrix: (A) photolysis of $FeCH_2$ (a) with UV light $340 \geq \lambda \geq 280$ nm; (B) photolysis of $Fe^{13}CH_2$ (a); (C) photolysis of $FeCD_2$ (a). Note that these are difference spectra obtained from spectra measured before and after UV photolysis.

Table 6
Measured and calculated infrared frequencies (in cm^{-1}) for HMCH, $HM^{13}CH$, and DMCD with $M = Fe$, Zn in solid argon.*

		HMCH		$HM^{13}CH$		DMCD	
Vibrational mode		Obs.	Cal.	Obs.	Cal.	Obs.	Cal.
M–H stretch	Fe	1681.6	1681.8	1681.6	1681.5	1209.2	1209.0
	Zn	1924.4	1924.8	1924.4	1924.2	1386.8	1386.2
$M \equiv C$ stretch	Fe	674.2	675.2	655.0	655.4	648.3	646.8
	Zn	647.5	650.4	628.4	631.1	627.2	621.0
MC–H bend	Fe	632.1	635.4	627.6	630.9	503.7	489.6
	Zn	469.3	467.3	468.9	464.0	344.7	360.0

Note:

* The symmetry force constants used for the calculated frequencies can be found in the article by Chang et al. (1988b).

The ternary reaction between a transition metal atom, diazomethane and dihydrogen (with the exception of Zn) led to the formation of methane and the metal atom either directly as in the case of Cr or indirectly through the formation of the reaction intermediates, a methyl metal hydride and/or a methane metal-atom adduct. These intermediates have been previously isolated and characterized from the binary metal-atom/methane cryogenic reaction, and were discussed in section 2.2. Figure 5 depicts the $CoCH_2$ hydrogenation reaction which leads to the spontaneous formation of CH_4 and methylcobalt hydride. A photoinduced oxidative-addition/reductive elimination process is also shown to take place between the methane–cobalt complex and the methylcobalt hydride. Figure 6 gives the relative yield of methane formed as probed by the absorbance of the CH_4 deformation mode with respect to the transition metal methylene used in the hydrogenation reaction. It is interesting

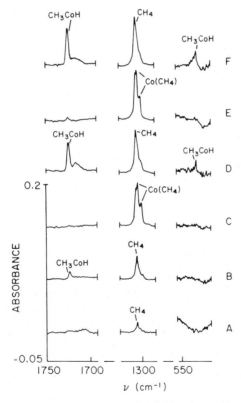

Fig. 5. Ternary argon matrix reaction of cobalt, diazomethane and hydrogen. FTIR spectra of selected regions of CH_4, $Co(CH_4)$, and CH_3CoH in argon matrices. (A) Without adding H_2, $Co:CH_2N_2:Ar = 1.0:0.9:100$; (B) Adding H_2 during deposition, $Co:CH_2N_2:H_2:Ar = 1.0:0.9:5:100$; (C) After 10 min photolysis of (B) with $\lambda \geq 400$ nm; (D) After 10 min photolysis of (C) with $360 \geq \lambda \geq 280$ nm; (E) After 10 min photolysis of (D) with $\lambda \geq 400$ nm; (F) After 10 min photolysis of (E) with $360 \geq \lambda \geq 280$ nm.

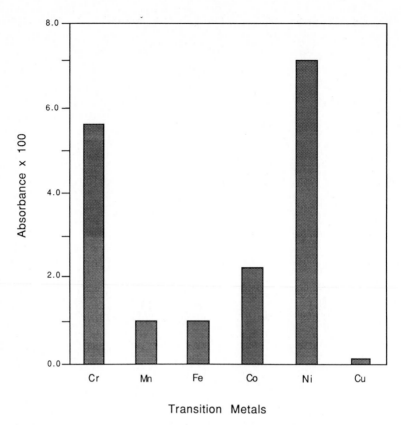

Fig. 6. The relative yield of methane formed in transition metal atoms ternary reactions with diazomethane and hydrogen on argon matrix surfaces. A = absorbance of the methyl deformation mode of methane.

to note that chromium and nickel seem to be the best catalysts for this hydrogenation process.

The ternary matrix reaction between a transition metal atom, diazomethane and water has also been investigated. This ternary reaction can provide basic information about the character of the methylene carbon. Although only $FeCH_2$ has been found, at this point, to react with H_2O spontaneously (Chang et al. 1988b), the reaction product methyliron hydroxide, CH_3FeOH, definitely indicates a nucleophilic methylene carbon. The reaction not only supports the theoretical finding of a negatively charged carbon in MCH_2 species, but also suggests that $FeCH_2$ is like a Schrock carbene. This reaction product has been previously isolated and characterized from the binary UV photoreaction between iron and methanol (Park 1988, Park et al. 1985).

A summary of the microsynthesis and overall chemistry of CrCH$_2$, FeCH$_2$, CoCH$_2$, NiCH$_2$, CuCH$_2$ and ZnCH$_2$ is given in fig. 7.

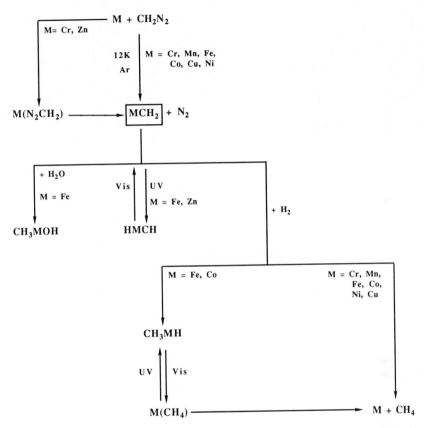

Fig. 7. Observed reaction pathways and photochemistry of unligated first-row transition metal methylenes.

2.4. Reactions with ethylene

The interaction between zero-valent first-row transition metal atoms and ethylene have been the subject of many experimental studies (Moskovits and Ozin 1976, Huber et al. 1976a,b, Ozin et al. 1977, 1978, Ozin 1977, Ozin and Power 1978, Hanlan et al. 1978, McIntosh et al. 1980, Kasai et al. 1980, Parker et al. 1983, Kafafi et al. 1985a, Ball et al. 1988a), and theoretical studies (Basch et al. 1978, Swope and Schaefer 1977, Widmark et al. 1985, Nicolas and Barthelat 1986, Barthelat et al. 1986, Anderson and Baldwin 1987). One of the most common modes of interaction for many transition metals follows the Dewar–Chatt–Duncanson model. In this formalism the ethylene molecule interacts with the metal atom both as a σ-electron donor through the filled (C–C) π-orbital and a π-electron acceptor through the empty (C–C) π^*-orbital. Recent theoretical calculations on the copper/ethylene complex, $Cu(C_2H_4)$, have shown that its $^2A_1$ ground state arises from the $3d^{10}4s^1$ atomic configuration. The weak bonding found in this complex has been described as a van der Waals interaction. Contrary to experimental results, theoretical studies on the $Ni(C_2H_4)$ complex have firmly established the ground state to be a singlet. These theoretical investigations suggested that the observed spectrum of $Ni(C_2H_4)$ is due to a weak van der Waals complex between a triplet nickel atom and ethylene. The singlet ground state of $Ni(C_2H_4)$ can only be reached from the nickel atom ground state via a barrier of approximately 10 kcal/mol.

A new mode of interaction between a transition metal atom and an ethylene molecule was recently observed by Kafafi et al. (1985b). Monatomic iron was found to hydrogen bond to an ethylene molecule in two distinct forms. One of these forms was shown to be a necessary precursor for the formation of the photo-insertion product, vinyliron hydride. An oxidative-addition/reductive-elimination reaction which is photoreversible, has been observed

$$Fe(C_2H_4) \underset{\lambda > 400\,nm}{\overset{280-360\,nm}{\rightleftharpoons}} HFeC_2H_3,$$

where interconversion of ethylene iron into vinyliron hydride and vice versa takes place. The iron/ethylene reversible photochemistry is displayed in fig. 8. The interaction of an iron atom with more than one ethylene molecule gives rise to a typical π-complex. An infrared spectrum characteristic of a π-complex results where activation and large frequency shifts of the $\nu(C=C)$ and the $\delta(CH_2)$ bending modes are observed. Near UV photolysis initially increases the yield of this complex. However, prolonged UV photolysis causes its partial loss with apparent disproportionation to methane and ethane.

Recent theoretical studies by Anderson and Baldwin (1987) have shown that ethylene undergoes π-coordination to $d^6s^2p^0$ Fe more weakly than σ coordination. The low stability of the π-complex is due to the closed-shell π^2–s^2 repulsion which prevents a close overlap between the π-orbital of ethylene and the σ-orbital of the metal atom. Similarly, theoretical studies by Swope and Schaefer (1977), examining the interaction between d^5s^2 Mn and ethylene, have predicted no π-bonding. On the other hand, atomic iron may σ-coordinate to one or two hydrogens of ethylene and form stable hydrogen-bonded complexes as shown in fig. 9. The most stable structure was found to be atomic iron interacting with two hydrogens of ethylene in the (1,2) two-fold coordination. When two ethylene molecules interact with atomic iron, a change in the electronic configuration of Fe takes place favoring π-coordination. Photo-excitation of the $Fe(C_2H_4)_2$ complex leads to its dissociation rather than C–H

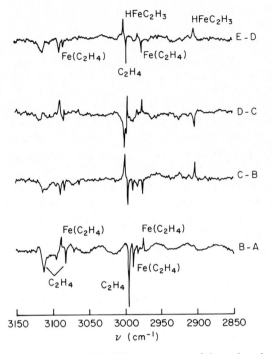

Fig. 8. A reversible photolysis study. FTIR difference spectra of the carbon–hydrogen stretching region of Fe/C_2H_4 in Ar: (A) no photolysis; (B) $\lambda > 400$ nm, 20 min; (C) 280 nm $< \lambda <$ 360 nm, 10 min; (D) $\lambda > 400$ nm, 20 min; (E) 280 nm $< \lambda <$ 360 nm, 10 min.

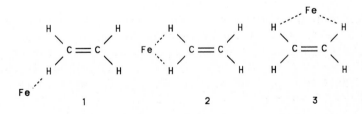

Fig. 9. Geometric isomers of hydrogen-bonded iron-ethylene adducts.

bond activation of ethylene. Photodissociation is favored since an electron from the excited s^1p^1 state of atomic iron is donated to the π^* orbital of ethylene which leads to a low barrier for dissociation of the excited π-complex.

2.5. Reactions with acetylene

The bonding patterns and reactivity between first-row transition metal atoms and acetylene are very similar to those discussed in the above section for the metal/ethylene systems. For instance, nickel and copper form π-complexes of the form $M(C_2H_2)_n$ where $n = 1,2$ as found by Kasai and McLeod (1978), Ozin et al. (1981b) and Chenier et al. (1983). Theoretical calculations by Cohen and Basch (1983) suggested that the matrix-isolated silver and possibly copper complexes arise from the excited p^1 state of the metal atom. For all the systems, the s^1 ground state of the metal atom is repulsive with the unpaired spin population residing almost completely on the s-orbital. ESR matrix-isolation studies by Kasai et al. (1980) support these theoretical results.

Most interesting is the photochemistry that has been recently observed for iron and nickel by Kline et al. (1985, 1987) in argon matrices. The nickel atom initially forms a strong π-complex with acetylene, where the triple bond of acetylene has been reduced to a double bond. Upon photolysis with visible violet light the $Ni(C_2H_2)$ adduct undergoes bond cleavage and rearrangement, and a metal vinylidene is formed. The mixed $^{12}C/^{13}C$ isotopic study clearly demonstrated the unequivalence of the two carbons. This photochemical rearrangement was found to be photoreversible

$$Ni(C_2H_2) \underset{280-360 \text{ nm}}{\overset{\lambda > 400 \text{ nm}}{\rightleftharpoons}} Ni = C = CH_2,$$

where violet light leads to the formation of nickel vinylidene and ultraviolet light causes it to revert back to the π-complex.

An iron atom prefers to hydrogen bond rather than π-bond to a molecule of acetylene again paralleling the Fe/C_2H_4 chemistry and substantiating the theoretical studies carried out on Fe/C_2H_4 and Mn/C_2H_2. These calculations have shown that the ground state of the metal atom is essentially repulsive towards π-complexation with the unsaturated hydrocarbon due to the doubly occupied s^2 electron pair of the metal. UV photolysis of the hydrogen-bonded complex leads to the activation of one of the C–H bonds of acetylene and the formation of ethynyliron hydride.

3. Concluding remarks

In the previous sections we have focused on the bond-insertion reactions of the first-row transition metals with small main-group molecules. As can be seen, a significant amount of information has been gathered which allows the comparison of similar reactions and products. It will be especially interesting in the future to compare the reactions of the HF, H_2O, NH_3, and CH_4 set since they are isoelectronic when one considers electron lone pairs as equivalent to hydrogen bonds. The reactions of NH_3 and HF have not been as extensively studied; however, reactions with iron have been studied by Kauffman et al. (1984). Parker et al. (1984) have also studied the reactions of the hydrogen halides with iron. Table 7 provides a listing of the H–Fe and Fe–X force constants for the HFeX species which form from insertion into HF, H_2O, NH_3 and CH_4. One sees that there is a small increase in the H–Fe force constant as the electronegativity of the associated anion increases. This can be explained as due to a slight reduction in size of the iron as it assumes greater positive charge which leads to a shorter and slightly stronger H–Fe bond. The variation in the Fe–X force constant is much more marked as the bonding X group increases in electronegativity. A significant increase is expected since a similar increase also occurs for the simple hydrides H–H, $H–CH_3$, $H–NH_2$, etc., as shown in the last column of table 7. This increase is due to the increased overlap of the σ-orbital of the respective X group with the hydrogen orbital. A similar effect is expected when these groups bond to the HFe σ-orbital. This appears to be the case when one compares the H–FeX bond to the HFe–X bond as shown by their respective ratios in table 7. Note however, that the NH_2, OH and F groups form stronger bonds than expected which suggests that the electron lone pairs of these groups are enhancing the bond strength with added π-electron donation to metal d-orbitals. A similar donation by the methyl group will be much reduced since the electron lone pairs are now present as hydrogen bonds.

Comparisons of metal single and double bonds are also of interest. By comparing the differences between the HM–OH bond and the M = O bond in table 8 one sees for the M = O bond a rapid drop in the π-bonding contribution to the M = O bond with increasing atomic number from Fe to Cu. This result is quite different from what is seen for the π-bonding contribution in $M = CH_2$ when one compares the $HM–CH_3$ and $M = CH_2$ bonds. Note that the additional π-bonding causes $NiCH_2$ to be the most strongly bound of the latter first-row transition metals. One might expect this to be true for NiO as well but

Table 7
Comparison of Fe–H force constants (in mdyn/Å) with Fe–X force constants.*

HFeX	$k(H–Fe)$	$k(Fe–X)$	$k(Fe–X)/k(Fe–H)$	$k(H–X)/k(H–H)$
HFeD	1.74	1.74	1.0	1.0
$HFeCH_3$	1.73	1.89	1.1	1.0
$HFeNH_2$	1.80	3.35	1.9	1.3
HFeOH	1.83	3.53	2.0	1.5
HFeF	1.88	3.55	2.0	1.8

Note:
* Calculated from a diatomic model.

Table 8
Force constant (in mdyn/Å) comparisons of metal–carbon and metal–oxygen bonds.

Bond	Metal					
	Mn	Fe	Co	Ni	Cu	Zn
$k(M = O)$	–	5.64	5.32	5.06	2.98	–
$k(HM–OH)$	–	3.58	3.47	3.80	3.12	–
$k(M = O) - k(HM–OH)$	–	2.06	1.85	1.26	–0.14	–
$k(M = CH_2)$	–	2.65	2.91	3.36	2.66	2.35
$k(HM–CH_3)$	1.78	1.89	1.99	2.17	–	1.43
$k(M = CH_2) - k(HM–CH_3)$	–	0.76	0.92	1.19	–	0.92
$k(HM \equiv CH)$	–	3.24	–	–	–	3.16
$k(HM \equiv CH) - k(M = CH_2)$	–	0.59	–	–	–	0.81

this clearly is not the case. Thus one concludes again that the electron lone pairs of oxygen are playing a significant role in the bonding as was earlier pointed out for the HMOH species. This may come in the form of additional bonding for FeO by π-electron donation into the less filled σ-orbitals of Fe and/or by increased electron repulsion between the oxygen lone pairs and the greater number of d-electrons for Ni and Cu. Note that CuO has essentially no π-bonding, presumably because it prefers a Cu electronic bonding configuration which is largely $d^{10}s^1$ in character.

In summary, evidence for involvement of oxygen and nitrogen electron lone pairs in their respective bonds with a metal is based on the following:

(1) A comparison of the H–MOH and HM–OH force constants across the transition metal series as shown in fig. 1 and discussed in a previous section.

(2) The abrupt rise in HM–NH$_2$ and HM–OH force constants when compared to those for HM–CH$_3$ and HM–H, as shown in table 7.

(3) The dependence of π-bonding contributions of the $M = O$ and $M = CH_2$ bonds on the number of metal d-electrons.

It is also interesting to note from table 8 that both Fe and Zn undergo increased π-bonding between the metal and carbon when $M = CH_2$ rearranges to HMCH if one assumes that the σ-bond is relatively unchanged. The formation of a nominal triple bond between carbon and a single metal center is novel and rather surprising.

Table 9 lists the additional frequency data which are available for HMX species that have not been discussed in detail in the previous sections. Reactions of metal atoms and small clusters with molecular hydrogen have been studied by a number of investigators: Cr by Van Zee et al. (1979), Mn and Fe by Ozin and co-workers (Ozin and McCaffrey 1984a,b, Ozin et al. 1984, 1986), Fe and Cu by

Hauge et al. (1987), and Fe by Rubinovitz and Nixon (1986). Atomic Cu has been extensively studied by Ruiz et al. (1986), Ozin and Gracie (1984) and Ozin et al. (1982). Nickel atom reactions have been very recently studied by Park (1988).

As Parker et al. (1983) have noted, the H–MX stretching frequency increases with increased electronegativity of the other group bound to the metal. Apparently because the metal becomes more positive thus shortening its bond with hydrogen and increasing the force constant. Very similar H–MX frequencies for the HMH and HMCH$_3$ species indicate the charge states of the metal are quite similar for the two types of species. A similar correspondence also exists between the H–MOH and H–MOCH$_3$ species (Park 1988).

Nickel appears to be an exception with respect to the relative ordering of the H–Ni stretching frequency versus electronegativity of the other group bound to the metal. Note that both the H–NiH and H–NiCH$_3$ stretching frequencies are higher than those for the H–NiOH and H–NiOCH$_3$ species. It is possible that an internal interaction of the nickel hydrogen with electron lone pairs on the oxygen is causing the lower H–Ni stretching frequency for the oxygen-containing species. This requires the H–Ni–O angle to be considerably less than 90°. It is also possible that a change in electronegativity of the group bound to the metal markedly changes the character of the hybrid nickel orbital that interacts with the hydrogen. For instance, greater involvement of the smaller 3d-orbital in the bonding should increase the H–NiX force constant. Theoretical studies of the preferred geometries and electronic bonding configurations for the HNiX species are clearly required for an improved understanding of this unusual behavior of nickel.

Table 9

Molecular stretching frequencies for hydrido metal compounds (in cm^{-1}).

Bond	Metal					
	Cr	Mn	Fe	Co[a]	Ni	Cu
H–MCH$_3$	1575	1617	1684	1723	1945	1719
H–MD	1575	1617	1691	1715	1967	–
HM–D	1129	1163	1215	1235	1421	–
H–MOH	–	–	1732	1790	1901[a]	1911
					1837	
H–MOCH$_3$	–	–	1741	1809	1869	–
H–MF	–	–	1753	–	–	–
H–MCl	–	–	1755	–	–	–
H–MNH$_2$	–	–	1717	–	1918[b]	1852[b]

Notes:

[a] Unpublished frequencies from work in progress at Rice University.

[b] Ball et al. (1988b).

Finally, it can be safely stated that the exploration of excited and ground-state bond activation reactions of metal atoms and clusters will continue to be a very active branch of matrix-isolation studies. This is especially true since all refractory metals can now be readily vaporized with the advent of relatively inexpensive and reliable pulsed lasers. Laser vaporization also provides a means of producing many novel molecular metal carbides, borides and oxides by vaporization of the appropriate refractory solid. The reactions of the above species with hydrogen can be used as an alternate means of producing the same intermediate species formed by bond insertion into the appropriate hydride. Such studies will allow one to explore reaction barriers from both the forward and reverse directions.

Acknowledgment

The authors would like to acknowledge the contributions of W.E. Billups to the work relating to metal/carbene and metal/methane chemistry. We would also especially like to acknowledge major contributions of J.W. Kauffman, S.C. Chang, M. Park, E. Kline, and D. Ball through their doctoral theses in which various aspects of bond activation via atomic metals and small clusters have been investigated. Financial support of this work has been provided by the Robert A. Welch Foundation and by the National Science Foundation.

References

Anderson, A.B., and S. Baldwin, 1987, Organometallics **6**, 1621.
Ball, D.W., 1987, Ph.D. Thesis (Rice University, Houston, TX).
Ball, D.W., Z.H. Kafafi, R.H. Hauge and J.L. Margrave, 1985, Inorg. Chem. **24**, 3708.
Ball, D.W., Z.H. Kafafi, R.H. Hauge and J.L. Margrave, 1986, J. Am. Chem. Soc. **108**, 6621.
Ball, D.W., Z.H. Kafafi, L. Fredin, R.H. Hauge and J.L. Margrave, 1988a, A Bibliography of Matrix Isolation Spectroscopy, 1950–1985 (Rice University Press, Houston, TX).
Ball, D.W., R.H. Hauge and J.L. Margrave, 1988b, High Temp. Sci., accepted for publication.
Barnes, A.J., A. Müller, W.J. Orville-Thomas and R. Gaufres, eds, 1981, Matrix Isolation Spectroscopy, Hingham, MA (Reidel, Dordrecht).
Barthelat, J.C., M. Hliwa, G. Nicolas, M. Pelissier and F. Spiegelmann, eds, 1986, Nato ASI Ser., Ser. C, Vol. 176.
Basch, H., M.D. Newton and J.W. Moskovits, 1978, J. Chem. Phys. **69**, 584.
Billups, W.E., M.M. Konarski, R.H. Hauge and J.L. Margrave, 1980, J. Am. Chem. Soc. **102**, 7393.
Blomberg, M.R.A., U.B. Brandemark and P.E.M. Siegbahn, 1984, J. Am. Chem. Soc. **106**, 2006.
Blomberg, M.R.A., U.B. Brandemark and P.E.M. Siegbahn, 1986, Chem. Phys. Lett. **126**, 317.
Brooks, B.R., and H.F. Schaefer III, 1977, Mol. Phys. **34**, 193.
Chang, S.C., 1987, Ph.D. Thesis (Rice University, Houston, TX).
Chang, S.C., Z.H. Kafafi, R.H. Hauge, W.E. Billups and J.L. Margrave, 1985, J. Am. Chem. Soc. **107**, 1447.
Chang, S.C., Z.H. Kafafi, R.H. Hauge, W.E. Billups and J.L. Margrave, 1987a, J. Am. Chem. Soc. **109**, 4508.

Chang, S.C., R.H. Hauge, Z.H. Kafafi, J.L. Margrave and W.E. Billups, 1987b, J. Chem. Soc., Chem. Commun., p. 1682.

Chang, S.C., R.H. Hauge, W.E. Billups, Z.H. Kafafi and J.L. Margrave, 1988a, Inorg. Chem. **27**, 205.

Chang, S.C., R.H. Hauge, Z.H. Kafafi, W.E. Billups and J.L. Margrave, 1988b, J. Am. Chem. Soc. **110**, 7975.

Chenier, J.H.B., J.A. Howard, B. Mile and R. Sutcliffe, 1983, J. Am. Chem. Soc. **105**, 788.

Cohen, D., and H. Basch, 1983, J. Am. Chem. Soc. **105**, 6980.

Cradock, S., and A.J. Hinchcliffe, 1975, Matrix Isolation (Cambridge University Press, Cambridge, UK).

Curtiss, L.A., and J.A. Pople, 1985, J. Chem. Phys. **82**, 4230.

Curtiss, L.A., E. Kraka, J. Gauss and D. Cremer, 1987, J. Phys. Chem. **91**, 1080.

Hallam, H.E., 1973, Vibrational Spectroscopy of Trapped Species (Wiley, New York).

Hanlan, A., J. Lee, G.A. Ozin and W.J. Power, 1978, Inorg. Chem. **17**, 3648.

Hauge, R.H., Z.H. Kafafi and J.L. Margrave, 1987, Reactions of Diatomic and Triatomic Metal Clusters of Iron and Copper with Dihydrogen in Inert Matrices, in: Physics and Chemistry of Small Clusters, eds P. Jena, B.K. Rao and S.N. Khanna (Plenum, New York) p. 787.

Howard, J.A., and B. Mile, 1987, Acc. Chem. Res. **20**, 173.

Hoyano, J.K., and W.A.G. Graham, 1982, J. Am. Chem. Soc. **104**, 3723.

Huber, H., G.A. Ozin and W.J. Power, 1976a, J. Am. Chem. Soc. **98**, 6508.

Huber, H., D.F. McIntosh and G.A. Ozin, 1976b, J. Organometal. Chem. **112**, C50.

Ismail, Z.K., R.H. Hauge, L. Fredin, J.W. Kauffman and J.L. Margrave, 1982, J. Chem. Phys. **77**, 1617.

Janowicz, A.H., and R.G. Bergman, 1982, J. Am. Chem. Soc. **104**, 352.

Kafafi, Z.H., R.H. Hauge and J.L. Margrave, 1985a, J. Am. Chem. Soc. **107**, 6134.

Kafafi, Z.H., R.H. Hauge and J.L. Margrave, 1985b, J. Am. Chem. Soc. **107**, 7550.

Kafafi, Z.H., W.E. Billups, R.H. Hauge and J.L. Margrave, 1987, J. Am. Chem. Soc. **109**, 4775.

Kasai, P.H., and D. McLeod Jr, 1978, J. Am. Chem. Soc. **100**, 625.

Kasai, P.H., D. McLeod Jr and T. Watanabe, 1980, J. Am. Chem. Soc. **102**, 179.

Kauffman, J.W., R.H. Hauge and J.L. Margrave, 1984, High Temp. Sci. – Brewer Issue – **17**, 237.

Kauffman, J.W., R.H. Hauge and J.L. Margrave, 1985a, J. Phys. Chem. **89**, 3541.

Kauffman, J.W., R.H. Hauge and J.L. Margrave, 1985b, J. Phys. Chem. **89**, 3547.

Klabunde, K.J., 1980, Chemistry of Free Atoms and Particles (Academic Press, New York).

Kline, E.S., 1987, Ph.D. Thesis (Rice University, Houston, TX).

Kline, E.S., Z.H. Kafafi, R.H. Hauge and J.L. Margrave, 1985, J. Am. Chem. Soc. **107**, 7559.

Kline, E.S., Z.H. Kafafi, R.H. Hauge and J.L. Margrave, 1987, J. Am. Chem. Soc. **109**, 2402.

Kline, E.S., R.H. Hauge, Z.H. Kafafi and J.L. Margrave, 1988, Organometallics **7**, 1512.

Manceron, L., A. Loutellier and J.P. Perchard, 1985, Chem. Phys. **92**, 75.

McIntosh, D.F., G.A. Ozin and R.P. Messmer, 1980, Inorg. Chem. **19**, 3321.

Morse, M.D., M.E. Geusic, J.R. Heath and R.E. Smalley, 1985, J. Chem. Phys. **83**, 2293.

Moskovits, M., 1986, Metal Clusters (Wiley, New York) p. 185.

Moskovits, M., and G.A. Ozin, 1975, Characterization of Products of Metal Atom–Molecule Cocondensation Reactions by Matrix Infrared and Raman Spectroscopy, in: Vibrational Spectra and Structure, Vib. Spectrosc. Struct. **4**, 187.

Moskovits, M., and G.A. Ozin, eds, 1976, Cryochemistry (Wiley-Interscience, New York).

Müller, R.P., and J.R. Huber, 1984, J. Phys. Chem. **88**, 1605.

Nicolas, G., and J.C. Barthelat, 1986, J. Phys. Chem. **90**, 2870.

Olsen, A.W., and K.J. Klabunde, 1987, Metal Particles and Cluster Compounds, Encyclopedia of Physical Science and Technology, Vol. 8 (Academic Press, New York) p. 109.

Ozin, G.A., 1977, Acc. Chem. Res. **10**, 21.

Ozin, G.A., and M.P. Andrews, 1986, Stud. Surf. Sci. & Catal. **29**, 265.

Ozin, G.A., and C. Gracie, 1984, J. Phys. Chem. **88**, 643.

Ozin, G.A., and J.G. McCaffrey, 1982, J. Am. Chem. Soc. **104**, 7351.

Ozin, G.A., and J.G. McCaffrey, 1983, Inorg. Chem. **22**, 1397.

Ozin, G.A., and J.G. McCaffrey, 1984a, J. Phys. Chem. **88**, 645.

Ozin, G.A., and J.G. McCaffrey, 1984b, J. Am. Chem. Soc. **106**, 807.

Ozin, G.A., and W.J. Power, 1978, Inorg. Chem. **17**, 2836.

Ozin, G.A., H. Huber and D.F. McIntosh, 1977, Inorg. Chem. **16**, 3070.

Ozin, G.A., W.J. Power, T.H. Upton and W.A. Goddard III, 1978, J. Am. Chem. Soc. **100**, 4750.

Ozin, G.A., D.F. McIntosh and S.A. Mitchell, 1981a, J. Am. Chem. Soc. **103**, 1574.

Ozin, G.A., D.F. McIntosh, W.J. Power and R.P. Messmer, 1981b, Inorg. Chem. **20**, 1782.

Ozin, G.A., S.A. Mitchell and J. Garcia-Prieto, 1982, Angew. Chem. **94**, 380.

Ozin, G.A., J.G. McCaffrey and D.F. McIntosh, 1984, Pure and Applied Chem. **56**, 111.

Ozin, G.A., J.G. McCaffrey and J.M. Parnis, 1986, Angew. Chem. **98**, 1076.

Park, M., 1988, Ph.D. Thesis (Rice University, Houston, TX).

Park, M., R.H. Hauge, Z.H. Kafafi and J.L. Margrave, 1985, J. Chem. Soc., Chem. Commun., p. 1570.

Park, M., R.H. Hauge and J.L. Margrave, 1988, High Temp. Sci., accepted for publication.

Parker, S.F., C.H.F. Peden, P.H. Barrett and R.G. Pearson, 1983, Inorg. Chem. **22**, 2813.

Parker, S.F., C.H.F. Peden, P.H. Barrett and R.G. Pearson, 1984, J. Am. Chem. Soc. **106**, 1304.

Perutz, R.N., 1985a, Chem. Rev. **85**, 77.

Perutz, R.N., 1985b, Ann. Rep. Proc. Chem. Sec. C **82**, 157.

Poliakoff, M., K.L. Lompa and S.D. Smith, eds, 1979, Infrared Laser Induced Photochemistry in the Solid State, in: Laser Induced Processes in Molecules (Springer, Berlin).

Rappe, A.K., and W.A. Goddard III, 1977, J. Am. Chem. Soc. **99**, 3966.

Ratajczak, H., and W.J. Orville-Thomas, eds, 1980, Vibrational Spectroscopy of Molecular Complexes in Low Temperature Matrices, in: Molecular Interactions (Wiley, New York) p. 273.

Richtsmeier, S.C., K.E. Parkas, K. Kiu, L.G. Pobo and S.J. Riley, 1985, J. Chem. Phys. **82**, 3659.

Rubinovitz, R.L., and E.R. Nixon, 1986, J. Phys. Chem. **90**, 1940.

Ruiz, M.E., J. Garcia-Prieto, E. Poulain, G.A. Ozin, A.R. Pairier, S.M. Mattar, I.G. Csizmadia, C. Gracie and O. Novaro, 1986, J. Phys. Chem. **90**, 279.

Spangler, D., J.J. Wendoloski, M. Dupis, M.M.L. Chen and H.F. Schaefer III, 1981, J. Am. Chem. Soc. **103**, 3985.

Swope, W.C., and H.F. Schaefer III, 1977, Mol. Phys. **34**, 1037.

Tachibana, A., M. Koizumi, H. Teramae and T. Yamabe, 1987, JACS **109**, 1383.

Van Zee, R.J., T.C. DeVore, W. Weltner Jr, 1979, J. Chem. Phys. **71**, 2051.

Weltner Jr, W., and R.J. Van Zee, 1984, Ann. Rev. Phys. Chem. **35**, 291.

Whetten, R.L., D.M. Cox, D.J. Trevor and A. Kaldor, 1985, Phys. Rev. Lett. **54**, 1494.

Widmark, P.-O., B.O. Roos and P.E.M. Siegbahn, 1985, J. Phys. Chem. **89**, 2180.

CHAPTER 11

Energy Transfer and Lifetime Studies in Matrix-Isolated Molecules

H. DUBOST and F. LEGAY

Laboratoire de Photophysique Moléculaire du CNRS*
Bâtiment 213
Université de Paris-Sud
91405 ORSAY Cedex
France

*Laboratoire associé à l'Université de Paris-Sud.

Chemistry and Physics of
Matrix-Isolated Species
L. Andrews and M. Moskovits, eds

Contents

1. Vibrational relaxation and excitation of diatomic molecules

Vibrational relaxation of matrix-isolated diatomic molecules has been extensively studied over the past fifteen years. A great deal of activity is still devoted to the subject and points now toward three major directions: non-radiative vibrational relaxation, vibrational stimulated emission (VSE) and strongly vibrationally excited molecules. The HCl–rare-gas systems constitute the most documented example of non-radiative vibrational relaxation into local phonons and provide a good test for theoretical models. New optical pumping schemes have allowed the production of strong vibrational population inversions in long-lived CO or NO species. Direct evidence for VSE has been obtained which opens the way to the realization of solid-state vibrational lasers. A strong vibrational excitation of these systems has been achieved, and in some cases the vibrational energy exceeds that of the lowest electronic states and is close to the first dissociation limit.

Besides thoroughly studied systems like HCl or CO in Van der Waals crystals that have become prototypic in the study of vibrational dynamics in the solid state, new systems have been found that exhibit a slow vibrational relaxation. As a matter of fact the vibrational dynamics of CN^- ions embedded in alkali halide crystals is quite similar to that of CO or NO in rare-gas crystals even though the crystal bonding strength is different. Long-lived laser-induced vibrational fluorescence has been observed in these systems and laser action on vibrational transitions has been obtained as well. The room-temperature stability of these systems should facilitate practical applications such as the realization of solid-state vibrational lasers.

1.1. Non-radiative vibrational relaxation

1.1.1. HCl in rare-gas matrices
The vibrational relaxation of HCl ($v = 1,2,3$) in Ar, Kr and Xe matrices was extensively studied by Young and Moore (1984). Optical pumping of HCl was achieved by means of an optical parametric oscillator ($0 \to 2$ excitation) or using the Raman-shifted output of a dye laser ($0 \to 3$ excitation). The relaxation rate has been measured as a function of host and temperature (9–42 K). The relaxation is found to occur stepwise non-radiatively in the sequence $3 \to 2$,

$2 \to 1$, $1 \to 0$. The corresponding rate $k_{v,v-1}$ increases in the order $k(\text{Ar}) < k(\text{Kr})$ $< k(\text{Xe})$. Such a behaviour is in contrast with that of most other systems. Only the $\text{NH}(A^3\Pi)$ has been found to exhibit the same trend in relaxation rate (Bondybey and Brus 1975a). In addition the v-dependence of the rate deviates considerably from linearity. This effect is more pronounced in Xe ($k_{32}:k_{21}:k_{10}$ $= 260:33:1$) at $T = 20$ K than in Ar ($11:4.3:1$). The study of vibrational relaxation of HCl ($v = 1,2,3$) in solid Xe has been extended up to the fusion point independently by Chesnoy (1985) and by Krueger et al. (1985). The rates increase more than linearly with v. However, the relative values become closer as the temperature is increased. For instance, $k_{32}:k_{21}:k_{10} = 31:7:1$ at 158 K. The rates display a temperature increase which is rather mild below 100 K and becomes stronger beyond this point. The most important observation is the continuity of the relaxation rate from the solid to the liquid state. The vibrational relaxation rate of HCl in liquid Xe extrapolated to infinite dilution is 9×10^4 s^{-1} at $T = 163$ K, whereas the rate extrapolated for the solid at the fusion temperature is 6.5×10^4 s^{-1}. A similar continuity of the relaxation behaviour across the liquid–solid phase transition has been observed for HD and D_2 in H_2 by Gale and Delalande (1978). The dependence of the relaxation rate upon the rare-gas host suggests that attractive forces play an important role in the vibrational relaxation of hydrides such as HCl or $\text{NH}(A^3\Pi)$. These attractive forces are expected to be strong for Xe, smaller for Kr and very weak for Ar. They account for the observed increase in rate from Ar to Xe. The strong non-linearity of the v-dependence of the rate can only be understood if it is assumed that the attractive part of the intermolecular potential is strongly non-linear in the vibrational coordinate Q. The relaxation behaviour of matrix-isolated HCl reflects the particularities of the rare-gas–HCl potential and should be more or less similar in any phase (Chesnoy 1985). This idea has received experimental confirmation from gas-phase experiments on HCl diluted in Kr and Xe (Schwahn and Schramm 1986).

The similar relaxation properties in gas, liquid and solid support the idea that collisional models should be applicable to any phase. Chesnoy (1985) used the isolated binary collision (IBC) model to calculate the vibrational relaxation rate of HCl in Xe. Using the interaction potential of Hutson and Howard (1982) and a collision frequency $v_c = 100$ cm^{-1}, i.e., twice the local phonon frequency, the experimental temperature dependence of the relaxation rate between 10 and 160 K is quite satisfactorily reproduced. Chesnoy (1985) suggested that the relaxation rate in dense phases can be deduced from the elementary de-excitation process resulting from an isolated collision by using an appropriate scaling law. This results from the fact that a strong interaction involving a large deviation from the intermolecular equilibrium position is necessary to relax a vibrational quantum.

1.1.2. CN⁻ in alkali halide matrices

Vibrational fluorescence from CN^-–F center complexes dispersed in a KCl matrix has been observed by Yang and Luty (1983). The IR emission of CN^- was induced by optical excitation of F centers followed by E–V transfer. The lifetime τ of the $1 \to 0$ emission was found to be temperature independent up to 70 K with $\tau \approx 6$ ms before decreasing at higher temperatures. A more extensive study has been reported by Koch et al. (1984). Vibrational emission of CN^- around 4.8 µm was studied at 80 K in twelve different alkali halide hosts. The $v = 2$ level of CN^- was optically pumped by the intracavity radiation of a KCl:Na color center laser. The fluorescence lifetime exhibits a strong dependence upon the matrix host, spreading over a broad range from 20 ms in KI (close to the radiative value) to 10 µs in NaCl. The fluorescence lifetime is shortened by host-dependent non-radiative relaxation. The data are consistent with an energy-gap law for either multi-phonon relaxation or vibration to libration transfer.

An important advantage of non-volatile solids such as alkali halides lies in the fact that it is possible to investigate the vibrational relaxation over an extremely wide temperature range. Gosnell et al. (1985b) measured the fluorescence lifetime of CN^- in several hosts from 1.7 up to 600 K. Surprisingly, the high-temperature lifetime is quite long, e.g., 100 µs for CN^- in CSI at $T = 600$ K. The functional form of the temperature dependence is nearly independent of the host lattice, contrary to the predictions of multi-phonon relaxation theories.

1.2. Vibrational stimulated emission (VSE)

1.2.1. VSE induced by fundamental pumping

It is well known that vibrational relaxation of matrix-isolated diatomic species like CO, NO, CN,..., in their ground electronic state is purely radiative (Legay 1977, Dubost 1984). In these systems the rate of vibration to vibration (V–V) transfer exceeds the rate of loss due to spontaneous emission. The energy deposited into the $v = 1$ level by optical pumping (fundamental pumping) is redistributed to higher levels of the vibrational ladder. At sufficiently high concentrations, excitation with 1–10 µJ, 200 ns pulses of a Q-switched frequency doubled CO_2 laser has been found to produce population inversions (Dubost and Charneau 1979). Population inversions have also been found to take place among the vibrational levels of $^{13}C^{18}O$ in pure solid CO at $T = 25$ K upon irradiation with a multi-line CW CO laser delivering 0.5–1.5 W (Legay-Sommaire and Legay 1980). However, no direct evidence for VSE has been obtained.

The first observation of VSE from a matrix-isolated molecular defect has been made in ionic crystals. Tkach et al. (1984) and Gosnell et al. (1985c) have irradiated samples of KBr doped with 0.5% CN^- immersed in superfluid helium ($T < 1.7$ K) with 300 µJ, 100 ns pulses of a frequency doubled TEA CO_2 laser resonant with the $0 \to 1$ vibrational transition of CN^-. In a way very similar to matrix-isolated CO, rapid V–V transfer to the $v = 2$ level, accompanied by the emission of 25 cm^{-1} phonons, creates a population inversion between $v = 2$ and $v = 1$. Laser oscillation has been obtained on the $2 \to 1$ transition from samples with coated end faces. VSE has been found to be restricted to very low temperatures ($T \leq 4$ K). This is due to the fact that the population inversion is destroyed by thermally activated back transfer.

1.2.2. VSE induced by overtone pumping

Optical pumping on the weak $0 \to 2$ vibrational overtone transition (overtone pumping) produces a strong population inversion between $v = 2$ and $v = 1$ without any need for V–V transfer. CW laser action has been obtained on the $2 \to 1$ transition of CN$^-$ using a tunable $(F_2^+)_A$ color center laser as the excitation source (Gosnell et al. 1985a). However, the temperature range for laser action was limited to $T \leq 4$ K. Galaup et al. (1985a,d) have found direct evidence for VSE on the $2 \to 1$ transition of NO and CO in solid N_2. The energy-level configuration is shown in fig. 1. Overtone pumping is achieved with a modulated KCl:Li F_A(II) color center laser delivering 10–20 mW. At sufficiently low concentrations ($\leq 2 \times 10^{-3}$) and temperatures ($T \leq 7$ K), the $2 \to 1$ emission of CO or NO consists of intense spikes superimposed on the regular fluorescence background. The pulsations are in general irregular and chaotic, although regular pulsing is sometimes observed, and they are always undamped. Their intensity can be 10–100 times larger than that of the spontaneous fluorescence and their period, ranging typically from 1 to 100 μs, is orders of magnitude shorter than the radiative lifetime (30–50 ms). Such pulsations are very similar to the well-known spiking which occurs in most solid-state lasers. This spiking

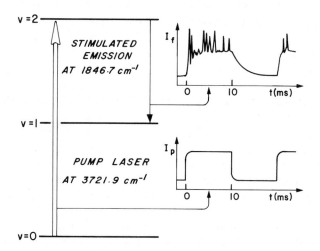

Fig. 1. Schematic diagram showing the lowest vibrational levels of NO. Optical pumping through the $0 \to 2$ overtone results in stimulated emission on the $2 \to 1$ transition.

has been interpreted as the evidence for VSE. Surprisingly, the pulsations have never been observed from the NO or CO systems when a rare gas is used as a matrix under conditions identical to that met for N_2 matrices.

Clearly, a model based on the Einstein rate equations for populations and photon density in the medium is unable to account for the occurrence of undamped pulsations. Such a description, which neglects the phase information carried by the polarization of the medium, is no longer justified for the $2 \rightarrow 1$ transition of CO in solid N_2. Indeed previous transient hole-burning experiments by Harig et al. (1982) have shown that the low-temperature value of the T_2 time associated with the $2 \rightarrow 1$ transition was larger than 10^{-7} s, i.e., much longer than the transit time of light across a 1 mm thick sample (10^{-11} s). These considerations have led Zondy (1986) and Zondy et al. (1987) to use a model derived by Casperson (1978) to describe instabilities in lasers. This model is based on a semi-classical description of the interaction between a monochromatic field (the stimulated emission at the $2 \rightarrow 1$ band center) and the inhomogeneously broadened medium ($\Delta\nu_{\text{hom}} \ll \Delta\nu_{\text{inhom}}$). Numerical integration of the equations showed that for reasonable values of the molecular parameters undamped pulsations appear in N_2 but not in rare-gas matrices. Physically, each pulse can be viewed as a coherent emission from in-phase dipoles and in that sense the phenomenon is close to superradiance or self-induced optical nutation.

1.2.3. VSE induced by optical pumping of an electronic state

The severe restriction on the tunability of solid-state vibrational lasers optically pumped through $0 \rightarrow 1$ or $0 \rightarrow 2$ transitions (less than 1 cm^{-1}) would make such devices not very attractive in practice. The temperature quenching of the gain can be avoided by using as active centers complexes that can be electronically excited and, subsequently, convert their electronic excitation into vibrational energy (this is the E–V process described in section 3.2). There are two known examples of VSE resulting from direct electronic excitation of a atom–diatom complex. The first one is the $F_H(CN^-)$ center in alkali halide crystals. Such a complex is constituted by the association of an F center to CN^-. Optical excitation of the F-center produces several IR emission lines around 4.8 µm which correspond to $\Delta\nu = 1$ vibrational transitions of CN^- originating from $\nu = 5$ to 1 (Yang and Luty 1985).

More recently, Yang et al. (1985) showed that the electronic energy is transferred to the $\nu = 4$ level of CN^-. Laser oscillation has been obtained at temperatures up to 77 K on the $4 \rightarrow 3$, $3 \rightarrow 2$ and $2 \rightarrow 1$ vibrational transitions of CN^- using the 647 nm line of a Kr^+ ion laser as the pump source (Gellermann et al. 1986).

The second example is provided by matrix-isolated Hg–CO Van der Waals complexes. Crépin et al. (1986) used the 249 nm line of a KrF excimer laser to electronically excite Hg–CO complexes trapped in solid N_2 to the $^3P_1$ state of mercury. Strongly non-resonant V–E transfer populates levels between $\nu = 11$

and $v = 5$ and creates a strong population inversion between $v = 5$ and $v = 4$. At temperatures $T \geq 17$ K, a short VSE pulse is emitted on the $5 \rightarrow 4$ transition.

1.3. Vibrational excitation beyond 5 eV

1.3.1. CO

Vibrational fluorescence of $^{13}C^{18}O$ in solid argon induced with a CW chopped KCl : Li color center laser was recently studied by Zondy et al. (1985) and Zondy (1986). Despite the weakness of the $0 \rightarrow 2$ overtone transition, efficient optical pumping of the $v = 2$ level can be achieved with ≈ 10 mW of tunable IR radiation at $\lambda = 2.47$ μm. The distributions of vibrational populations resulting from V–V transfer were deduced from the $\Delta v = 1$, 2 or 3 infrared emission spectra. These distributions are characterized by an extreme broadness. Figure 2 shows a typical population distribution obtained in solid argon doped with 1%

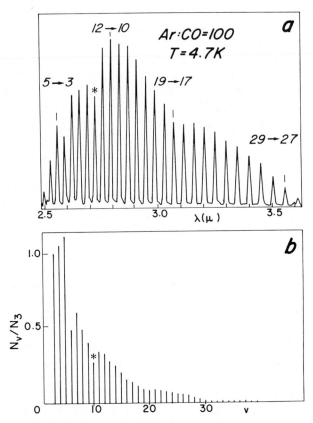

Fig. 2. (a) First overtone emission spectrum of $^{13}C^{18}O$ in solid argon. The intensity of starred lines is perturbed by atmospheric water absorption. (b) Corresponding relative vibrational populations
$$N_v/N_3 = (I_{v \rightarrow v-2}/I_{3 \rightarrow 1})(A_{3 \rightarrow 1}/A_{v \rightarrow v-2})(\nu_{3 \rightarrow 1}/\nu_{v \rightarrow v-2}).$$

$^{13}C^{18}O$. A succession of population inversions is observed between the lowest levels of the vibrational ladder. The vibrational energy deposited into the $v = 2$ level is redistributed over 42 levels. The energy of the topmost detected state is fairly high (≈ 8 eV) and is approximately 75% of the CO dissociation energy. Compared to excitation up to $v = 12$ resulting from irradiation of a 1% $^{12}C^{16}O$/argon sample with a pulsed frequency doubled CO_2 laser (Dubost and Charneau 1979), overtone pumping appears to be much more efficient than fundamental pumping. The modulated CW laser is well-suited for optical pumping of long-lived species. The energy deposited into the CO vibration is two orders of magnitude larger than that delivered by the pulsed system. A detailed time and frequency resolved study of the $\Delta v = 2$ IR emission of $^{13}C^{18}O$ in solid argon have allowed Zondy et al. (1985) and Zondy (1986) to get some insight on the energy transfer mechanisms.

Are there processes which limit the up conversion of vibrational excitation and thus prevent the CO molecules to be excited up to dissociation? The fluorescence decay of the $41 \rightarrow 39$ emission is in the ms range, indicating that vibrational relaxation from the highest detected states is radiative. Vibrational energy transfer can still be competitive with populations decay. Since the energy of vibrational levels with $v \geq 29$ in the ground electronic state is similar to that of several excited electronic states, an ultimate limit to up pumping could be set by intramolecular V–E transfer. The fact that the highest detected vibrational state is nearly isoenergetic with the $A^1\Pi$ state suggests that $A \rightarrow X$ internal conversion could limit the V–V pumping of CO in the ground electronic state. Actually, IR pumping does not induce any UV emission. In particular the Cameron bands which are easily induced in matrix-isolated CO by excitation of any of the electronic states between 6 and 8 eV (Fournier et al. 1980, Bahrdt et al. 1987) are not observed at all. Therefore, anharmonic V–V pumping should bring a small fraction of the CO molecules further up, leading ultimately to dissociation.

1.3.2. NO

IR luminescence experiments are more difficult to carry out on NO than on CO. $(NO)_2$ dimers are easily formed in matrices and quench efficiently the monomer emission. In order to get a good fluorescence quantum yield, the samples should be deposited from the gas phase at very low temperature (4 K) and the NO concentration should not exceed 2×10^{-3}. Great care should also be taken to remove the impurities from gaseous NO. Under such conditions, Dubost et al. (1989) have observed a strong IR fluorescence from NO in solid argon. Vibrational levels up to $v = 20$ have been detected by their $\Delta v = 2$ emission. As in the case of CO, vibrational relaxation from the highest detected levels is radiative. Actually, the distribution of vibrational populations extends well beyond $v = 20$, as shown by the following experimental observations. In addition to IR emission, an ultraviolet fluorescence originating from the lowest valence states $a^4\Pi$ and $B^2\Pi$ is also induced by IR excitation. These electronic states are populated by intramolecular V–E transfer (see section 3.1.2) and they constitute a very sensitive probe for the high lying vibrational levels of NO $(X^2\Pi)$. Moreover, the intensity of the UV emissions coming from a freshly irradiated sample decays on a 10 min time scale to a small fraction (5–30%, depending on the NO concentration) of its initial value. Evidence for the NO dissociation is obtained from the observation of thermoluminescence. Upon warming of the sample up to 13 K, phosphorescence signals are observed at the wavelengths of the $a^4\Pi \rightarrow X^2\Pi$ transition in absence of any optical excitation. Such an emission results from the recombination of N and O atoms produced by IR-induced dissociation of the NO molecules. The various processes possibly occurring in matrix-isolated NO are shown in fig. 3.

Infrared multi-photon dissociation of polyatomic molecules, such as SF_6 or N_2F_4, has been observed in the gas phase but is suppressed in low-temperature matrices (Crocombe et al. 1978). As discussed by Tasumi et al. (1979) the absence of dissociation in matrices can be due either to the quenching of the

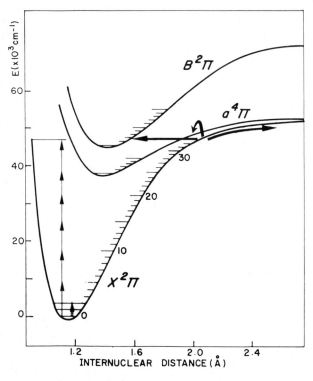

Fig. 3. Potential energy curves of the ground and lowest valence states of NO. The various processes taking place in the molecule (laser excitation of $X^2\Pi$ $v = 2$, V–V pumping, $X^2\Pi \rightarrow B^2\Pi$ internal conversion, dissociation) are indicated by arrows.

multi-photon excitation by radiationless relaxation or to a fast recombination of the products in the matrix cage. The IR laser-induced dissociation of NO shows that atomic fragments may not recombine and gives additional support to the idea that multi-photon events do not occur in matrices.

1.3.3. Mechanism of the anharmonic V–V pumping

The population distribution among the lowest vibrational levels of CO or NO is reasonably well described by the model derived by Manz (1977), slightly modified in order to take into account stimulated emission which occurs between inverted levels, in particular between $v = 2$ and $v = 1$. Accordingly, anharmonic V–V pumping occurs in four steps:

   (i) optical excitation of $v = 2$ by the laser;
   (ii) fast population transfer from $v = 2$ to $v = 1$ by stimulated emission;
   (iii) migration of the $v = 1$ excitation; and
   (iv) repetitive fusion of $v = 1$ excitations resulting from phonon assisted V–V transfer:

$$AB(v = 1) + AB(v \geq 2) \rightarrow AB(v = 0) + AB(v + 1) + \Delta E_v.$$

For large values of v, the fraction of the donor energy which must be converted into phonons becomes as large as the vibrational energy received by acceptor. The rate of such processes becomes vanishingly small. In order to account for the excitation of high-lying levels, exchange of energy among two molecules both vibrationally excited need to be considered. The probability of processes such as:

$$AB(v' \approx v) + AB(v) \rightarrow AB(v' - 1) + AB(v + 1) + \Delta E_{vv'},$$

becomes significant for large v values.

   The localization of high-v excitations which is much stronger than that of $v = 1$ is counterbalanced by the significant increase in the strength of dipole–dipole interaction and by the important decrease in the energy defect. In addition V–V exchange processes involving multi-quantum transitions ($\Delta v = 2, 3$) should also be considered.

2. Vibrational relaxation in polyatomic molecules

A number of polyatomic molecules, ions and radicals have been studied the last few years. Vibrational levels of small molecules (HCN, NH_3, CH_3F, SF_6) in their ground electronic state have been excited by optical pumping, either by a CO_2 laser when coincidences exist between laser lines and molecular absorption, or by tunable infrared lasers. In the first case, fundamental vibrational modes near $1000 \, cm^{-1}$ can be excited, in the second case, an optical parametric oscillator with a spectral range limited to the near-infrared has allowed the excitation of overtone or combination modes. The number of molecules which can be excited by pumping in the mid-infrared is severely limited by the lack of tunable lasers available in this spectral region. On the other hand near-UV and visible are easily accessible to dye lasers. This allows optical excitation of vibrational levels in electronic states, but with the restriction to the molecular species with electronic absorption in this region, i.e., radicals, ions or some large molecules. This technique has been extensively used by Bondybey and co-workers in their studies of triatomic radicals (CF_2, CNN, ClCF, CCl_2, NCO, CBrCl), of ions (CS_2^+, halogenated benzene cation) and some large molecules (naphthazarine, 9-hydroxyphenalenone).

Two approaches were used for the theoretical interpretation of the experimental results:

(i) The polyatomic molecule is assimilated to a diatomic one, and treated by the theories based on the interactions of the internal vibrational mode with the phonon bath via the local modes and possibly the rotational levels.

(ii) Or a collision theory very similar to the gas-phase case is used, the partners of collision being the molecule and the cage in which it is enclosed. Nevertheless because of the complexity of the problem, theoretical effort has been relatively small. There is, so far, no model elaborate enough to be tested quantitatively from the experiment and there is no convincing arguments in favor of a definite mechanism.

2.1. Vibrational relaxation in ground electronic state

2.1.1. HCN

HCN, $HC^{15}N$ and DCN have been extensively studied by Abbate and Moore (1985a–c) in rare-gas matrices, particularly in Xe. Their work is especially interesting in relation with the controversial problem of the dominant channels in the relaxation process. The infrared spectra of these three molecules in Ar, Kr and Xe, show the presence of two trapping sites in Kr and Xe. One of these sites in Xe induces a splitting of the degeneracy of the bending vibration. A single unsplit site is observed in Ar. The molecules do not seem to undergo rotation in these matrices. Relaxation measurements were carried out by a laser-induced fluorescence technique. The v_3 vibrational levels of the three isotopes were excited with 8 ns pulses from a YAG-laser-pumped optical parametric oscillator (OPO). v_3 corresponds to the C–H stretch in HCN at ≈ 3300 cm^{-1} (at ≈ 2600 cm^{-1} for DCN). The fluorescence spectra were time resolved with a fast Cu–Ge infrared detector at 4 K and coarsely frequency resolved with variable interference filters. The relevant levels are shown in fig. 4. Results in Xe are given in table 1.

Site effect. The v_3 relaxation rates differ in the two sites for HCN in xenon. The site with the faster relaxation rate (6.0×10^5 s^{-1} versus 6.4×10^5 s^{-1}) is also the one that induces degeneracy splitting and blue shift of the transition, in this case

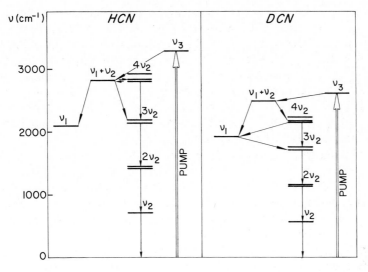

Fig. 4. Energy level diagram showing the relaxation paths in HCN and DCN.

Table 1
Relaxation rates (in s⁻¹) and, in parentheses, transition wave-numbers (in cm⁻¹).

Transition	$HC^{14}N$	$HC^{15}N$	DCN
$(0,1,0) \rightarrow (0,0,0)$	1.0×10^5	8.0×10^4	2.1×10^5
	(716.7)	(715.7)	(573.2)
$(0,0,1) \rightarrow (1,1,0)$	6.3×10^5	4.7×10^5	3×10^8
	(476.6)	(508.4)	(120.5)
$(1,0,0) \rightarrow (0,2,0)$	8.0×10^3	1.4×10^4	
	(671.6)	(641.2)	
$(1,0,0) \rightarrow (0,3,0)$			8.8×10^6
			(~ 200)

a stronger guest–host interaction is probably the common cause of these effects.

Concentration effect. The relaxation of the $(0,0^0,1)$ and $(1,1^1,0)$ levels in HCN is non-exponential. The relaxation rate decreases with the dilution to become constant at very high dilution. These effects have been successfully interpreted by the Förster (1948) theory of energy diffusion by dipolar interactions. The diffusion among identical molecules (donors) allows the energy to be trapped by a species relaxing to the ground state (acceptors). If the concentration of donors is very small and/or dipolar transitions are very weak, the diffusion is negligible among the donors and each donor relaxes independently with a rate depending on the distances to the acceptors (non-diffusion limit). The net result is a non-exponential fluorescence. On the other hand, if the diffusion is very fast, the energy is averaged among the donors, the relaxation becomes exponential (rapid-diffusion limit). In the HCN case, the vibrational energy is trapped by the dimers which relax very rapidly due to the presence of low-frequency modes. Increasing the concentration, increases both the diffusion rate and the dimers concentration. In addition, the $v_1 + v_2$ level which shows a very strong concentration effect is involved in a quasi-resonant intermolecular process:

$$(1,1^1,0) + (0,0^0,0) \rightleftarrows (1,0^0,0) + (0,1^1,0) - 3 \text{ cm}^{-1}.$$

The $(1,1^1,0)$ excitation diffuses very slowly because the $(1,1^1,0) \rightarrow (0,0^0,0)$ transition is very weak, thus the negligible diffusion limit should apply for the direct process. The opposite is true for the reverse process, the v_2 transition is strong and therefore the fast-diffusion limit is a better assumption.

Temperature effect. The temperature dependence is rather small for the relaxation of levels not in resonance with other levels. Thus, from 9 to 30 K, the relaxation rates of v_3 for HCN and of $4v_2$ for DCN increase by factors of 1.5 and 1.9, respectively. The results can be fitted by the formula for the exchange of one phonon,

$$k \propto 1 + [\exp(\hbar\omega_\text{p}/kT) - 1]^{-1},$$

where ω_p is the phonon wavenumber (25 cm^{-1} for HCN and 15 cm^{-1} for DCN). In contrast, the rate for the v_1 level, in near resonance with the $3v_2$ levels, increases by a factor of 15 for HC^{14}N and a factor of 5.7 for HC^{15}N, from 9 to 30 K. This behavior is due to the endothermic process of transfer to the $3v_2$ levels, slightly higher than v_1.

Host effect. The relaxation rate increases noticeably from Xe to Ar. For HC^{15}N it increases by a factor of 3.4 for v_3 and of 41 for v_1. The logarithm of the relaxation rates are almost perfectly linearly correlated with the matrix shift of the corresponding level: a blue shift corresponding to an increase of the relaxation rate. A similar behavior is observed in ammonia and methyl fluoride as we shall see later. This suggests that the relaxation mechanism is probably dominated by the repulsive interactions between guest and matrix, in contrast with the HCl case where the opposite is true (section 1.1.1).

Mechanism of the relaxation. The authors have built a model of collisions with the matrix cage that is an adaptation of the Schwartz et al. (1952) theory of vibrational relaxation in the gas phase. HCN vibrates in a Xe cage where the entire matrix is considered as the collision partner. The probability of transfer by collision is related to the variation of linear momentum by

$$k \propto \Gamma(\theta_0^2 - \theta_\text{f}^2)^2 \exp(\theta_0 - \theta_\text{f}), \tag{2.1}$$

where Γ is the collision frequency and $\theta_0 = 4\pi^2 \mu v_0 / \alpha h$, with a similar expression for θ_f with v_f, v_0 is the initial relative velocity due to the phonon zero-point energy, v_f is the final relative velocity, and μ is the reduced mass of the collision pair. The interaction is assumed purely repulsive of the type $\exp(-\alpha R)$. Equation (2.1) gives a good fit of the experimental data with $\alpha = 6$ Å^{-1} and an initial translation energy of 100 cm^{-1}. The value of α is close to the value in gas phase for similar molecules (≈ 5 Å^{-1}) and the initial translation energy is in reasonable agreement with the local phonon frequency.

On the other hand, vibrational transfer to rotation can be estimated in the framework of this collision theory using v_f as the rotational velocity. The agreement with experiment is not good. This strongly suggests that V–R transfer is not the dominant relaxation mechanism.

2.1.2. NH_3

NH_3 was the first polyatomic molecule whose vibrational relaxation was measured in the ground electronic state (Abouaf-Marguin et al. 1973). Its lower vibrational level is relatively high ($v_2 \approx 950$ cm^{-1}), thus one can expect a fairly long vibrational lifetime. Furthermore, the absorption band is in the CO_2 laser region, allowing an efficient optical pumping. Spectroscopic studies have led to a good understanding of the dynamics of NH_3 in matrix. In rare-gas matrices, NH_3 can rotate almost freely (Abouaf-Marguin et al. 1977b, Abouaf-Marguin 1973). The rotational barrier, estimated with the theory of Devonshire (1936) is very small: from 4 cm^{-1} in Ne to 51.5 cm^{-1} in Xe, and the inversion barrier is only slightly increased in matrix relatively to the gas phase. The two quantum numbers of the symmetrical top, J and K, remain good quantum numbers in matrix and the levels belong to two symmetry types: A_2 for $K = 0, 3, 6, \ldots$, and E for $K = 1, 2, 4, 5, \ldots$, with different nuclear spin states. The levels for the v_2 transition are shown in fig. 5 for the gas phase. At temperatures below 10 K, the nuclear spin conversion takes hours and thus one can consider that the two ammonia species, A_2 and E, are co-existing in the matrix. In contrast, in nitrogen matrix (Nelander 1984, Girardet and Lakhlifi 1986), the rotation of the molecule around the axis perpendicular to the C_3 symmetry axis is blocked by a barrier of 750 cm^{-1}; however, the molecule undergoes a hindered rotation around the C_3 axis, through a rotational barrier of ≈ 90 cm^{-1}. This corresponds to a rotation with $K = J$. The inversion doubling is much smaller than in the gas phase giving an apparent increase of the inversion barrier due to the strong coupling between the intrinsic inversion and the translational space inversion (Girardet et al. 1984a,b).

Abouaf-Marguin et al. (1973), Abouaf-Marguin (1973), and, more recently, Boissel (1985) and Boissel et al. (1985) carried out relaxation measurements by a double-resonance method, using a Q-switch CO_2 laser as pump and a CW CO_2 laser as probe. This technique is very sensitive compared to the laser-induced fluorescence; the two lasers can operate at the same frequency, since the pump-

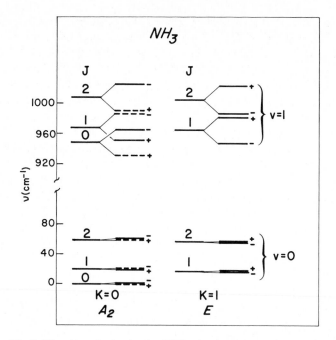

Fig. 5. Vibration–rotation levels of NH_3 showing the inversion splitting.

laser scattered light can be eliminated from the probe channel by crossing the polarisation of the two beams. In rare-gas matrices, the lines are rotationally broadened and at least two consecutive CO_2 laser lines can be found in the absorption range. In nitrogen matrices, since v_2 appears as a quadruplet under high resolution, with narrow components (Girardet et al. 1984b) there is no good coincidence with the laser lines.

The results are very dependent upon the matrix (Boissel 1985, Boissel et al. 1985):

Rare-gas matrices
In rare-gas matrices with pump and probe laser at the same frequency an exponential relaxation is observed with an increase of the relaxation rate with concentration noticeable under $M/R < 2500$. As usual this effect is due to a transfer to the dimers or polymers. At low concentration, the host effect is important: ≈ 1 ns in Ne, 90 ns in Ar, 580 ns in Kr and 2600 ns in Xe. A relatively small temperature effect is observed, the relaxation rate does not change for $T < 20$ K. From $T = 20$–50 K, in Xe and Kr, the relaxation rate increases by a factor of ≈ 7. The experimental curve can be interpreted, using the expression derived from the theory of Berkowitz and Gerber (1979),

$$k \propto (\langle n_L \rangle + 1)^N (\langle n_S \rangle + 1),$$

where $\langle n_L \rangle$ is the occupation number of N local phonons and $\langle n_S \rangle$ the occupation number of a bulk phonon. A direct conversion of energy into phonons implies 12 phonons of 79 cm^{-1} to match the experiment. A transfer to the nearest rotational level of the ground vibrational state at 8.5 cm^{-1} below the v_2 level is ruled out. However, with the $J = 8$, $K = 0$ rotational level as an acceptor, the remaining energy of 255 cm^{-1} can be released into 6 phonons of 42.5 cm^{-1}.

The two fits obtained with or without rotation, are equally good. Thus it is not possible to discriminate between the two mechanisms on this experimental basis only. However, a recent theoretical work by Lakhlifi and Girardet (1987) shows that the transfer to high rotational levels is the dominant relaxation channel, faster than multi-phonon processes by at least a factor 10^3. The calculated lifetimes, obtained without any adjustable parameter, are in good agreement with the experimental measurements of Boissel et al. (1985) at $T = 10$ and 40 K.

Nitrogen matrices
In nitrogen matrices, two types of studies have been performed:

(i) Experiments carried out with pump and probe laser at the same frequency in the center of the v_2 quadruplet yield non-exponential decays, in the 10 μs range, even at low concentrations. The best fit of the decay curves is obtained with a law of the Förster type, $\exp[-kt - (\alpha t)^{1/2}]$ with $k \ll \alpha$. A very strong temperature dependence is observed: the signal amplitude decreases dramatically with increasing T, and vanishes above 11 K, whereas its duration is decreased by a factor ≈ 30 between 6.5 and 10 K.

(ii) New experimental results have been obtained, using a diode laser tuned on the v_4 mode of NH$_3$ (Boissel et al. 1989). The width of the R$_0$ and P$_1$ lines of the $v = 0 \to v_4 = 1$ transition is 0.07 cm^{-1} at a concentration of 1:10 000 ($T = 8$ K), and their profile is highly temperature dependent. The double-resonance signals obtained by pumping v_2 and probing v_4 do not show the decay in the μs range, previously observed when probing v_2. They are merely of thermal origin, and their risetime suggests that the vibrational energy relaxation is fast (≈ 50 ns). The signals observed on v_2 should then come from a relaxation process occurring inside the vibrational ground state. This could be related to the inversion splitting: the two components of the transition originating from the ground state sublevels, well-separated below 10 K on v_2, are completely overlapping on v_4.

The observation of a vibrational decay time in the tens of nanoseconds region for NH$_3$ in N$_2$ is satisfactory since the influence of the matrix can be described by a mass effect: a logarithmic plot of the experimental relaxation rates, from neon to xenon including N$_2$, shows a linear correlation with the variation of momentum $(2\mu E_v)^{1/2}$, a basic parameter in binary collision models.

2.1.3. CH_3F, CD_3F

Methyl fluoride has been the subject of detailed studies by several groups. Its spectroscopy in condensed media is now well known, although the theoretical interpretation is still controversial. CH_3F has three fundamental bands of type A_1: v_1-v_3 and three of type E: v_4-v_6. The lowest vibrational mode is the stretching mode $v_3 \simeq 1040$ cm^{-1} for CH_3F but the bending mode $v_6 \simeq 910$ cm^{-1} for CD_3F. The levels are represented in fig. 6. The linewidths vary over a wide range depending both on the mode and on the matrix: from 0.22 cm^{-1} (v_3) to 6.5 cm^{-1} (v_4) for CH_3F in Ar (Jones and Swanson 1982a), down to 0.07 cm^{-1} in N_2 (v_3 of CH_3F and v_6 of CD_3F) and <0.03 cm^{-1} in Ne (v_3 of CH_3F) (Abouaf-Marguin et al. 1985). Using dilute, but very thick, samples, Gauthier-Roy et al. (1981) observed, on the high-frequency side of the v_3 line, a structure which they attributed to the residual absorption of a weakly hindered rotor. Double-resonance experiments confirm that this structure is due to the same molecules as the v_3 line (Gauthier-Roy 1980).

On the other hand, Jones and Swanson (1982a,b) in a study at high resolution of CH_3F in Ar and Kr did not find any evidence of rotational structure and concluded free or hindered rotation to be absent in these systems. Apkarian and Weitz (1982) built an electrostatic model for symmetric top molecules trapped in an octahedral field. They found that the spectra are compatible with a minimum barrier to tumbling (rotation around the B-axis perpendicular to the symmetry axis) of ≈ 80 cm^{-1} and with free spinning motion (around the symmetry axis). Lakhlifi and Girardet (1984) using a pairwise potential between CH_3X

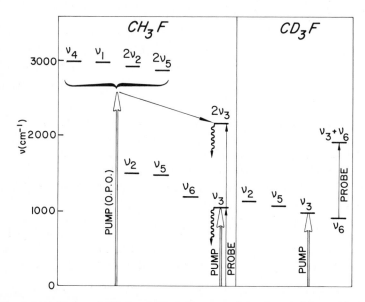

Fig. 6. Energy level diagram of CH_3F and CD_3F. The wavy arrows show the observed fluorescence.

and a nitrogen or a rare-gas molecule calculated upper and lower limits of the barrier for the tumbling motion of CH_3F in rare-gas matrices: 220–430 cm^{-1} in Ar, 183–354 cm^{-1} in Kr and 58–122 cm^{-1} in Xe. They also found a value of ≈ 300 cm^{-1} in N_2. Thus these calculations confirm the results of Jones and Swanson (1982a) as far as the tumbling motion is concerned.

CH_3F and CD_3F are appropriate systems for study by CO_2-laser pumping technique. The v_3 line can be pumped by a selected CO_2 laser line in all the rare-gas matrices. Abouaf-Marguin et al. (1977b), Abouaf-Marguin and Gauthier-Roy (1980), Gauthier-Roy (1980) and Gauthier-Roy et al. (1980) made an extensive study of these systems, using the double-resonance method as in the NH_3 case. Since v_3 of CH_3F is the lowest mode of the isolated molecule, it decays either by relaxation to the phonon bath, possibly via librational or rotational levels, or to some other species with lower frequencies, e.g. dimers or polymers. On the other hand, v_3 of CD_3F is ≈ 150 cm^{-1} above v_6 and an exothermic intramolecular transfer faster than the apparatus response time (< 500 ns) takes place between these levels, which is consistent with linewidths arguments (Abouaf-Marguin et al. 1985). The probe-laser line monitors the population of the v_6 level by the transition $(v_3 = 0, v_6 = 1) \rightarrow (v_3 = 1, v_6 = 1)$.

Two other groups used a laser-induced fluorescence technique to observe the decay of the v_3 mode:

(i) Apkarian and co-workers (Apkarian and Weitz 1980, Janiesch et al. 1982, Apkarian et al. 1986), with a CO_2 laser as pump, studied the decay of the $2v_3$ level populated from the v_3 level by the process:

$$CH_3F(v_3 = 1) + CH_3F(v_3 = 1) \rightarrow CH_3F(v_3 = 0) + CH_3F(v_3 = 2). \qquad (2.2)$$

(ii) Young and Moore (1982) pumped higher levels, around 3000 cm^{-1}, with an optical parametric oscillator, as in the HCN studies, and observed fluorescence from $2v_3$ and v_3. The $2v_3$ level is populated by fast V–V transfer from the excited levels (see fig. 5). The results of relaxation measurements of v_3 or v_6 at high dilution and low temperature (< 9 K) are given in table 2.

Temperature effect. The temperature dependence of the relaxation rate is small for CH_3F in Kr, this rate is increased by a factor of about 1.9 between 9 and 60 K (Abouaf-Marguin et al. 1977a, Brueck et al. 1977), and its temperature dependence is fitted quite well by a model where the dominant process is the emission of only one bulk phonon (Abouaf-Marguin and Gauthier-Roy 1980). A similar temperature dependence is observed in the other rare-gas matrices, as well as for $2v_3$ (Young and Moore 1982). The temperature dependence is larger for v_6 of CD_3F. In krypton the relaxation rate increases by a factor of 3 between 7 and 60 K. A two-phonon process is required in this case to fit the curve $k = f(T)$ (Abouaf-Marguin and Gauthier-Roy 1980, Gauthier-Roy et al. 1980).

Concentration effect. As usual, quasi-resonant transfer to dimers or polymers causes an increase of the relaxation rate with concentration (Abouaf-Marguin

Table 2
Methyl fluorides relaxation rates ($\times 10^{-3}$ s^{-1}).

	Ne	Ar	Kr	Xe	N$_2$
$\nu_3(CH_3F)$	250 ± 25 [1]	330 ± 30 [1]	87 ± 8 [2]	17 ± 2 [1]	
		320 ± 40 [3]	88 ± 10 [3]	13 ± 2 [3]	
$\nu_6(^{12}CD_3F)$	25 ± 3 [4]	83 ± 8 [4]	6.9 ± 0.6 [4]	0.5 ± 0.05 [4]	333 ± 30 [4]
$\nu_6(^{13}CD_3F)$	20 ± 2 [1]	77 ± 8 [1]	5 ± 0.5 [1]	0.17 ± 0.04 [1]	
$\nu_6(CD_3F-CD_3F)$	120 [4]	225 ± 30 [4]	83 ± 13 [4]	4.5 ± 0.05 [4]	

References:
  [1] Abouaf-Marguin and Gauthier-Roy (1980).
  [2] Abouaf-Marguin et al. (1977a).
  [3] Young and Moore (1982).
  [4] Gauthier-Roy et al. (1980).

et al. 1977a, Gauthier-Roy et al. 1980, Young and Moore 1982). Assuming that resonant transfer between monomers is fast enough to ensure averaging of the vibrational energy, a global transfer rate to the dimers can be defined. Transfer rates to CH$_3$F and CD$_3$F dimers are given in table 3. A careful study of the population of $2\nu_3$ at different concentrations led Apkarian et al. (1986) to identify resonant and non-resonant intermolecular energy transfer processes and to determine the probability of the phonon assisted V–V process (3.8×10^{-35} and 6.7×10^{-35} cm^6 s^{-1} in Xe at 10 and 20 K, respectively).

Table 3
Methyl fluorides monomers → dimers transfer rates ($\times 10^{16}$ s^{-1} dimer^{-1} cm^3).

	Ne	Ar	Kr	Xe
$\nu_3(CH_3F)$ [1]			400	
$\nu_6(CD_3F)$ [2]	20 ± 10	70 ± 20	12 ± 3	3 ± 1

References:
  [1] Abouaf-Marguin et al. (1977a).
  [2] Gauthier-Roy et al. (1980).

Host effect. As can be seen in table 2, the intrinsic relaxation rate decreases in the order of increasing mass: Ar, Kr, Xe, with the exception of Ne where the relaxation rate is smaller than in Ar. Abouaf-Marguin and Gauthier-Roy (1980) have found a very good linear correlation of log k with the matrix shift of ν_6 for all the methyl fluoride species, while correlation with the matrix shift of ν_3 is good for Ar, Kr and Xe but not for Ne. A similar correlation with ν_6 is found for the monomer to dimer transfer rate. But the intermolecular V–V transfer rate constants increase going from Ar to Kr (Apkarian et al. 1986).

Mixed-matrices effect. Gauthier-Roy et al. (1980) have studied CD_3F isolated in a mixture of argon and krypton. The decrease of k from pure argon to pure krypton is not at all linear: 8% of Kr gives a decrease of $\approx 40\%$. This proves that the interactions governing the relaxation process are not simply pairwise additive: the dominant effect seems to be due to the interaction of CD_3F with a sole Kr atom. Similar observations have been made by Janiesch et al. (1982) in CH_3F isolated in mixed Kr/Xe matrices.

Irradiation effect. Janiesch et al. (1982) have observed an increase of the relaxation rate with the duration of irradiation by the CO_2 pump laser. After blocking the laser, the relaxation rate returns exponentially, on a minute timescale, to its lower value with a rate increasing with temperature. The authors attribute this effect to a slow molecule–cage reorientation process.

Mechanism of relaxation. Abouaf-Marguin and Gauthier-Roy (1980) proposed a model for the relaxation mechanism based on the theory of Berkowitz and Gerber (1979) elaborated for diatomic molecules. The minimum rotational quantum J to match the lower vibrational level is $J = K = 14$ in CH_3F and 18 in CD_3F. The remaining energy mismatch is filled by one or two phonons. Thus the deuterated molecule decays more slowly as the relaxation process requires a larger variation of J. The decrease of the relaxation for the ^{13}C species is explained by a mass asymmetry effect which increases the anisotropy of the impurity–crystal coupling. The host effect may be due to the variation in the dynamical contribution of the repulsive interactions, dominant in the relaxation mechanism, from Ar to Kr. Ne is, again, the exception as the neon lattice has a very low rigidity.

The above model has been questioned by Jones and Swanson (1982a) who argued that the rotational barrier is much too high to allow a deactivation by a rotational channel. They pointed out that a direct comparison between CH_3F and CD_3F is questionable as it is not the same vibrational mode which is involved in the two cases (v_3 and v_6). They propose a model where the relaxation channel is the molecular libration. The libration around the C_3 axis would be more effective in the relaxation process than the libration around the perpendicular axis. The stretching mode v_3 would be more coupled to the libration around the C_3, while the bending mode v_6 would be preferentially coupled to the other libration. Nevertheless, if one notes that the vibrational energy is much higher than the rotational barrier, one may wonder whether a quasi-free rotation is not possible on the high J levels. Moreover, the observed temperature dependence described above fits quantitatively with the model involving rotation (Abouaf-Marguin and Gauthier-Roy 1980).

Young and Moore (1982) used a binary collision model where the short-range repulsive forces are dominant. Although this model gives a qualitative under-standing of the relaxation dependence on host lattice, an estimation from the experimental data of the relaxation probabilities per collision gives different values for the solid and liquid phases at the same temperature, contrary to the requirement of a consistent model. In fact, the relaxation rate of CH_3F in solid argon at 20 K (3.4×10^5 s^{-1}) (Gauthier-Roy 1980) is fairly close to its value in liquid argon at 77 K (7.8×19^5 s^{-1}) (Brueck et al. 1977).

2.1.4. SF_6

Relaxation of SF_6 in rare-gas and nitrogen matrices has been studied by Abouaf-Marguin et al. (1981), Boissel (1985) and Boissel et al. (1988) using the double resonance technique with CO_2 lasers as pump and probe. The spec-troscopy of SF_6 in a matrix is fairly complicated due to the presence of multiple sites and of site symmetry splitting (Jones and Swanson 1984). The v_3 mode absorbs in the 930 cm^{-1} region where several CO_2 laser lines are available. Measurements in a Xe matrix with different pump and probe lines gives several relaxation rates which have received the following interpretation: v_3 (931 cm^{-1}), directly pumped, decays to v_5 (523 cm^{-1}) with 1 ns $< \tau <$ 10 ns. Then v_5 decays to the lowest level v_6 (347 cm^{-1}) with 30 ns $< \tau <$ 80 ns and finally v_6 decays slowly to the phonon bath with $\tau = 12.7$ or 18.5 µs, according to the site where the pumped molecule is located. The longest time corresponds to a site of high symmetry, which is the most stable one (Swanson et al. 1986). The results for v_6 in all the studied matrices are given in table 4.

Table 4
SF_6 relaxation rates of v_6 (in s^{-1}) (Boissel 1985, Boissel et al. 1988).

Level	Ne	Ar	Kr	Xe	N_2
v_6	1.4×10^8	5×10^6	1.5×10^5	8.2×10^4	$> 10^8$
				5.4×10^4	

The relaxation rate of v_6 in Xe has a moderate dependence on temperature with an increase by a factor of 3.3 between 7 and 50 K. The variation can be fitted well by a process involving 6 phonons of 59 cm^{-1} which match the v_6 level. As in the case of ammonia, there is a good linear correlation between log k and the variation of momentum $(2\mu E_v)^{1/2}$ for all the rare-gas matrices, confirming the validity of a collisional model (Boissel 1985, Boissel et al. 1988).

2.2. Vibrational relaxation in excited electronic states

Bondybey and co-workers have made extensive investigations of excited electronic states of several molecular species trapped in rare-gas matrices.

With excitation by a tunable dye laser, they have often observed vibrationally unrelaxed emissions from electronic states. By direct time-resolved emission spectroscopy or by relative intensity measurements, they have determined relaxation times of vibrational levels in these excited electronic states, in addition to the electronic lifetime.

2.2.1. Triatomic radicals

Vibrational relaxation has been studied by time-resolved spectroscopy in the first excited electronic state of six triatomic radicals: CF_2 (Bondybey 1976a), CNN (Bondybey and English 1977a), ClCF (Bondybey and English 1977b, Bondybey 1977b), CCl_2 (Bondybey 1977a), NCO (Bondybey and English 1977c) and CBrCl (Bondybey and English 1980a). The results in an argon matrix are given in table 5. The electronic lifetime is purely radiative. There is no obvious correlation between the vibrational energy gap and the vibronic lifetime. Evidently, the nature of the radical plays an important role in the relaxation rate of the v_2 level and the nature of the matrix can also be important.

Table 5
Relaxation in triatomic radicals in an argon matrix.

Species	Electronic lifetime	Vibrational level	Wavenumber (cm^{-1})	Vibrational lifetime
$NCO(A^2\Sigma^+)$	180 ns	(010)	529.5	$> 20\ \mu s$
CBrCl	5.6 μs	(001)	468	$\gg 5.6\ \mu s$
		(010)	186	$< 1\ \mu s$
ClCF	340 ns	(010)	394	< 2 ns
		(001)	712	15 ns
$CNN(A^3\Pi)$	220 ns	(010)	525	52 ns
CF_2	27 ns	(010)	668	9.5 ns
		(020)	1336	5 ns

The rate of the transfer between different modes is very sensitive to the intramolecular anharmonic coupling. For ClCF, and CF_2 the transfer is inefficient between the v_3 and v_2 modes, but a fast transfer is observed between modes with levels coupled by Fermi resonance, as in NCO and CNN. Such an effect has also been observed in CS_2^+ (Bondybey and English 1980a).

2.2.2. Polyatomic cations. Large polyatomics

The large number of vibrational modes of these species implies the presence of low-frequency vibrational modes and a high density of states. Relaxation proceeds via low-order phonon processes and, therefore, is expected to be much faster than in small molecules, i.e., in the ps rather than in the ns range. Measurements were carried out either by ps spectroscopy (Bondybey et al.

1983a, 1984a), or by intensity measurements comparing the intensities of the non-relaxed spectrum with these of the relaxed one (Bondybey et al. 1980, 1983a, 1984a, Rentzepis and Bondybey 1984).

Halogenated benzene radical cations. The following species have been extensively studied in neon matrices near 4 K (Bondybey et al. 1980, 1983a): $C_6F_6^+$, sym-$C_6H_3F_3^+$, 1,2,3,5-$C_6H_2F_4^+$, sym-$C_6H_3Cl_3^+$, 1,2,4,5-$C_6H_2F_4^+$, 1,2,4-$C_6H_3F_3^+$. The electronic lifetimes are in the 30–50 ns range while for low vibrational levels, lying between 200 cm^{-1} and 1250 cm^{-1} above the vibrationless level, the vibrational lifetimes are in the range of few ps to 120 ps. The principal conclusions stressed by the authors are the following:

(i) Within a given species there is a trend towards increasing relaxation rates with increasing vibrational energy. However, this trend is not systematic.

(ii) Molecules of higher symmetry tend to relax more slowly than less symmetric molecules.

(iii) The absence or presence of a dipole moment does not seem to play a significant role in the relaxation mechanism, as well as the overall density of states.

Naphthazarin. Spectroscopic studies (Bondybey et al. 1983b, Rentzepis and Bondybey 1984) provided strong evidence that naphthazarin (5,8-dihydroxy-1,4-naphthoquinone) has a C_{2v} symmetry (1,4-quinoid structure with the two OH residing on the same ring). The vibrational lifetimes are in the few ps to 320 ps ranges, similarly to the case of the halobenzene cations, but with a much shorter electronic lifetime, 350 ps. Rentzepis and Bondybey (1984), Huppert et al. (1985) and Huppert and Rentzepis (1987) have measured the lifetimes in the S_1 state for vibrational levels below 1600 cm^{-1}. The longest lifetimes (100 and 150 ps) were found for the 311 and 386 cm^{-1} levels. Higher levels relax with shorter lifetimes. In general, overtone and combination bands relax faster than fundamental modes near the same energy. The most important conclusion is that relaxation proceeds via specific pathways. The nature of the matrix induces a negligible change in the relaxation rates and does not affect the overall pathways and general trends. The temperature effect appears to be relatively small.

9-hydroxyphenalenone and 5-methyl derivative. 9-HPLN is structurally and spectroscopically related to naphthazarin (Bondybey et al. 1984a,b). It contains a strong hydrogen bond between the phenolic hydrogen and the carboxylic hydrogen, characterized by a symmetric double-minimum potential. The lifetimes are similar to those of naphthazarin (the longest-lived, low vibrational levels of 9-HPLN are of the order of 100 ps in neon) and the relaxation also proceeds via specific pathways. On the other hand, as the 5-M-9-HPLN shows no evidence of vibrationally unrelaxed fluorescence, the vibrational relaxation is at least one order of magnitude faster in this compound than in the unsubstituted molecule. The insertion of the methyl group which undergoes an almost free rotation, introduces low-frequency modes, increasing the efficiency of the relaxation process.

3. Energy transfer between vibrational and electronic degrees of freedom

In electronically excited diatomic molecules, the interstate transitions induced by intermolecular interactions are in some cases much faster than intrastate vibrational relaxation (Brus and Bondybey 1980). The $CN(A^2\Pi)$ in solid neon is a particularly explicit example of interstate cascading. Vibrational relaxation of the $A^2\Pi$ vibronic states proceeds through the high vibrational levels of the ground electronic state, giving rise to successive $E \rightarrow V$ and $V \rightarrow E$ crossings (Bondybey 1977c).

Interstate cascading has also been observed in the collisional quenching of gas-phase molecules such as CO^+ (Bondybey and Miller 1978), CN (Katayama et al. 1979, Furio et al. 1986, Katayama 1984, 1985). This similarity between gas phase and solid phase suggests that the energy-transfer mechanism is typically the same in the two phases. Hence a collisional model may be used to describe matrix-isolated molecules.

Electronic to vibrational energy transfer induced by collisions between gas-phase atoms and molecules was discovered a long time ago. For instance, it has been known for many years that the resonance radiation of sodium atoms is efficiently quenched in the presence of molecular gases such as H_2, D_2 and N_2 (Jenkins 1966). The high quenching efficiency for these molecules, in comparison to the rather inefficient quenching by atomic gases, has been attributed to efficient energy transfer involving molecular vibrational states. A similar behavior has also been reported for Hg vapor mixed with CO and NO gases (Karl et al. 1967). In both cases, only 30–40% of the electronic energy of the atoms is transferred to vibration of the colliding molecules. Recent theoretical and experimental studies of $E \rightarrow V$ exchange from $Hg(6^3P)$ (Horiguchi and Tsuchiya 1979) or $Na(3^2P)$ (Reiland et al. 1982) atoms colliding with CO molecules have demonstrated that energy transfer proceeds via an intermediate collision complex.

Energy transfer from electronically excited halogen atoms $Br(^2P_{1/2})$ and $I(^2P_{1/2})$ to various molecules including HX, H_2, CO and NO has been investigated by IR fluorescence (Houston 1981). Energy transfer is found to be nearly resonant and a curve-crossing mechanism seems to be unlikely in this case.

Direct evidence for collisional $E \rightarrow V$ transfer between singlet oxygen and NO, CO and H_2O molecules has been obtained through the detection of the IR emission from the quencher species (Ogryzlo and Thrush 1973, 1974). Vibrational excitation of several molecules induced by collision with $O_2(a^1\Delta_g)$ and $O_2(b^1\Sigma_g^+)$ has been reported (Ogryzlo 1979, Singh et al. 1985). The E–V transfer rate constants for $O_2(a)$ are several orders of magnitude smaller than those for $O_2(b)$. Such a large difference should be attributed to the fact that energy transfer from $O_2(a^1\Delta_g)$ is a spin-forbidden process whereas the relaxation of $b^1\Sigma_g^+$ occurs to the $a^1\Delta_g$ and thus is spin allowed. In addition, the vibrational product states do not necessarily support an energy resonance mechanism.

Examples of $V \to E$ transfer are much more scarce, probably because of the difficulty of preparing molecules in vibrational levels high enough to match that of atomic or molecular states. The evidence that the D-line radiation of sodium could be induced by energy transfer from vibrationally excited molecules such as H_2, D_2 and N_2 has been obtained long ago in experiments carried out on gas mixtures in shock tubes (Clouston et al. 1958). More recently crossed-beam experiments have shown that most of the Na excitation comes from internal vibrational energy, but not exclusively (Krause 1975).

Electronic emission from $Br(^2P_{1/2})$ has been found to be induced by selective IR laser excitation of $HF(v = 1)$ (Quigley and Wolga 1975). Intermolecular V–E transfer from $N_2(v)$ to $CN(A^2\Pi$ and $B^2\Sigma^+)$ was discovered in the CO-N_2 laser medium produced by an RF discharge (Taïeb and Legay 1970). A similar $V \to E$ conversion from $CO(v)$ to $CN(B^2\Sigma^+)$ and $C_2(A^3\Pi_g)$ has been observed in IR laser pumped CO–N_2–Ar mixtures (Rich and Bergman 1979).

In the past, little attention has been paid to intermolecular E–V or V–E processes in matrices (see, however, section 1.2.3), although several molecular electronic states exhibit vibrationally unrelaxed fluorescence (see section 2.2) and that energy exchange among electronic states has been the subject of many studies (see, e.g., Meyer 1971). Very recently, the systems Hg–CO, Hg–N_2 and CO–O_2 have been extensively studied, using laser excitation and UV–visible and infrared fluorescence detection.

3.1. Intramolecular $E \to V$ and $V \to E$ transfer

3.1.1. Electronically excited systems

Interstate cascading has been observed in several matrix-isolated species: C_2^- (Bondybey and Brus 1975b), C_2 (Bondybey 1976b), CN (Bondybey 1977c, Bondybey and Nitzan 1977), O_2 (Rossetti and Brus 1979). The rate constants for several individual E–V and V–E processes have been measured. They are listed in table 6 together with the associated energy gaps and Franck–Condon factors. Additional examples were found recently (see below).

(i) $N_2(W^3\Delta_u \to A^3\Sigma_u^+$ or $B^3\Pi_g)$. Zumofen et al. (1984) have used electron bombardment to excite high lying electronic states (≈ 10 eV) of matrix-isolated N_2. They observed an IR emission corresponding to transitions between excited triplet states $W^3\Delta_u$, $v' = 4 \to A^3\Sigma_u^+$ ($v'' = 1$, 2). Interstate cascading causes relaxation of higher vibronic states down to the $W^3\Delta_u$, $v' = 4$ level. The large energy gap $\Delta E \simeq 1000$ cm^{-1} to the next lower levels then acts as a bottleneck for radiationless relaxation. In this case the radiative channel dominates over the radiationless channel. The upper bound to the rate for the two possible internal conversion processes for N_2 in Kr is also given in table 6. This rate is at least 100 times smaller than those of the $A \rightarrow X$ processes in CN/Ar with comparative energy gaps. The authors relate this difference in the rates to the matrix dependence of the efficiency of multi-phonon relaxation which is known to increase as the mass of the host atoms decreases. The results of Zumofen et al. (1984) have been confirmed recently by Kühle et al. (1986) who used mono-chromatic synchrotron radiation as a selective excitation source. Several bottlenecks for radiationless relaxation which allow slow radiative transitions to compete with non-radiative processes have been identified in $A^3\Sigma_u^+$, $B^3\Pi_g$, $a^1\Pi_g$ and $w^1\Delta_u$ states of N_2 in solid Kr. In particular, the existence of a bottleneck for $B'^3\Sigma_u^-$ ($v = 6$) confirms that the energy gap is not a criterion sufficient to decide whether internal conversion should occur or not. In this case the energy gap to the next lower level $W^3\Delta_u$, $v = 11$ is only 300 cm^{-1}. This would suggest that symmetry selection rules may also be important in determining the magnitude of the coupling strength between electronic states.

(ii) $ND(a^1\Delta \rightarrow X^3\Sigma^-)$. Intersystem crossing from $A^3\Pi_i$ to $b^1\Sigma^+$ and from $a^1\Delta$ to $X^3\Sigma^-$ has been suggested by Bondybey and English (1980c) in order to

Table 6
Intramolecular $E \rightarrow V$ and $V \rightarrow E$ transfer in matrix-isolated diatomic molecules.

System	Process	ΔE (cm^{-1})	FC factor	Rate constant (s^{-1})	Ref.*
CN/Ne	X7 → A2	1043	1.3×10^{-3}	2.0×10^5	[1]
	X8 → A3	1168	3.1×10^{-3}	1.3×10^5	[1]
	A10 → X13	1295	1.9×10^{-1}	7.7×10^6	[1]
	A9 → X12	1469	0.23	2.6×10^6	[1]
N_2/Kr	W4 → A11	980	3.3×10^{-2}	$<1.7 \times 10^4$	[2]
	W4 → B3	1350		$<1.7 \times 10^4$	[2]
NH/Ne, Ar, Kr	b0 → a2	2300	$\sim 10^{-7}$	$\geq 10^3$	[3]
ND/Ar	a0 → X6	69	2×10^{-9}	24	[3]
NI/Ar	b0 → a7	300–400	$<10^{-6}$	$<10^4$	[4]
C_2^-/Ar	a0 → B0	730	10^{-7}	1.4×10^5	[5]
C_2/Ar	a0 → X0	620	0.099	4×10^4	[6]
O_2/Ar, Kr	c2 → C0	20	~ 1	$\sim 10^5$	[7]

*References
[1] Bondybey (1977c).
[2] Zumofen et al. (1984).
[3] Ramsthaler-Sommer et al. (1986).
[4] Becker et al. (1986).
[5] Bondybey and Brus (1975b).
[6] Bondybey (1976b).
[7] Rossetti and Brus (1979).

explain the temperature dependence of the $X^3\Sigma^-$, $v = 1$ lifetime of ND in Kr. In their optical double-resonance experiment, the $A^3\Pi_i$ ($v = 0$) state is excited with a pulsed laser and the X, $v = 1$ level is probed by a second laser. The abrupt lengthening of the $v = 1$ lifetime from 330 ms at 4 K to >1 s at 8 K is attributed to the feeding of the ground-state vibrational manifold from the long-lived singlet states.

The relaxation behavior of NH/ND ($a^1\Delta$) in rare-gas hosts has been studied in detail by Ramsthaler-Sommer et al. (1986). The lifetime of the $a^1\Delta$, $v = 0$ level of NH in Ne, Ar and Kr matrices is of the order of 1 s at $T = 4$ K, i.e. very close to the expected value of the radiative lifetime, and is only weakly temperature dependent. Indeed, radiationless crossing to the next lower level,

$$NH(a^1\Delta, v = 0) \rightarrow NH(X^3\Sigma^-, v = 4), \quad \Delta E \simeq 1000 \text{ cm}^{-1}$$

is expected to be quite inefficient. In contrast to NH, the decay rate of ND($a^1\Delta$, $v = 0$) exhibits a strong increase with temperature. The energy gap to the next lower level $v = 5$ of $X^3\Sigma^-$ is quite large ($\Delta E = 2300 \text{ cm}^{-1}$) and makes exothermic transfer very unlikely. The $v = 6$ level lies above the $v = 0$ level of $a^1\Delta$ but the energy difference is small. Thus, the observed temperature dependence of the decay is attributed to the onset of a near resonant endothermic process,

$$ND(a^1\Delta, v = 0) \rightarrow ND(X^3\Sigma^-, v = 6), \quad \Delta E \simeq 70 \text{ cm}^{-1}.$$

The rate constant for the corresponding exothermic process, which can be deduced from the experimental lifetimes, is 24 s^{-1}. Such a small rate should be related to the forbidden character of the a–X transition and to the smallness of the FC factors (table 6).

(iii)   *NH, NI, NBr* ($b^1\Sigma^+ \rightarrow a^1\Delta$). Ramsthaler-Sommer et al. (1986) have been able to prepare matrix-isolated NH and ND radicals in the $b^1\Sigma^+$, $v = 0$ state using a pulsed tunable dye laser. Subsequent emission was found to occur mostly from $a^1\Delta$ while the fluorescence from the $b^1\Sigma^+$ was extremely weak. This observation indicates that the internal conversion b → a is very efficient in comparison to the radiative relaxation ($\tau_{\text{rad}} = 50–100$ ms in the gas phase), despite the large energy gap to the next-lower level (cf. table 6).

In similar experiments, Becker et al. (1986) have found that a different situation prevails in the case of the isoelectronic species NI. Five vibronic species of the $b^1\Sigma^+$ state were excited with the laser, giving rise to vibrationally unrelaxed fluorescence from $b^1\Sigma^+$, v'. However, no emission from $a^1\Delta$ could be detected, indicating that in an Ar matrix the internal conversion b → a does not occur for NI. As shown in table 6, the small FC factor balances the reduced energy gap. In a recent account, Becker et al. (1987) reported a study of NBr in solid Ar. Excitation of $b^1\Sigma^+$, v' results in fluorescence from both a and b singlet states. The NBr radicals are trapped in several sites and the b → a crossing was found to exhibit a strong site dependence.

(iv) *NO*. Chergui (1986) has observed, by pumping matrix-isolated NO in the first Rydberg state, $A^2\Sigma^+$ fluorescence from this state and from the two first valence states $a^4\Pi$ and $B^2\Pi$, except in Xe where only the Rydberg emission was detected. He has attributed this effect to a large spin–orbit coupling in the Xe matrix, causing a fast E–V transfer to the ground state $X^2\Pi$ probably via the $a^4\Pi$ state. This assumption was recently confirmed by the observation of infrared emission with a risetime $<1\,\mu$s from the vibrational levels $v = 22$ to $v = 15$ of $X^2\Pi$ in NO/Xe, pumped at 193 nm by an excimer ArF laser (Legay et al. 1988).

(v) *CO*. Bahrdt and Schwentner (1988) have excited the vibrational levels $v = 0$–8 of the $A^1\Pi$ state of CO in solid neon by monochromatized synchrotron radiation. By monitoring the various singlet and triplet emissions, they have observed several processes of intersystem crossing and internal conversion in the triplet manifold. The measured rate constants are explained by an intramolecular mixing of electronic states and electron–phonon coupling with the matrix.

3.1.2. Vibrationally excited systems

Matrix-isolated diatomic molecules, such as CO or NO, can be pumped to extremely high vibrational levels (see section 1.3). Intramolecular energy transfer from the high lying vibrational levels of the ground state to excited electronic states becomes energetically possible. Such a process does not seem to occur in matrix-isolated CO in spite of the fact that excitation to vibrational levels of the $X^1\Sigma^+$ state higher than the $A^1\Pi$ singlet state have been detected in the infrared (Zondy 1986). From the previous studies of E–V transfer, the probability of the spin forbidden process,

$$X^1\Sigma^+,\ v \approx 30 \to a^3\Pi,\ v \approx 0,$$

is expected to be very small in view of the extremely small Franck–Condon factors associated with a $\Delta v = 30$ transition. To a lesser extent the same argument should also hold for the spin-allowed exchange

$$X^1\Sigma^+,\ v \approx 40 \to A^1\Pi,\ v \approx 0.$$

However, it should be noted that this process has been found to be quite efficient in highly vibrationally excited gaseous CO (Deleon and Rich 1986).

In the case of matrix-isolated NO the situation is different. In solid argon, UV–visible emission from the $a^4\Pi$ and $B^2\Pi$ valence states of NO has been found to be induced by vibrational up-pumping in the ground electronic state (Dubost et al. 1989). Intramolecular V–E transfer is thought to take place near the first dissociation limit from high-lying v-levels of $X^2\Pi$ to the $a^4\Pi$ state. Such a phenomenon is quite analogous to the chemiluminescence resulting from the recombination of N and O atoms in the gas phase. The $B^2\Pi$ state is then fed by a $\to$ B intersystem crossing which is known to occur from $v > 0$ of the $B^2\Pi$ state (Goodman and Brus 1978).

3.2. Intermolecular $E \rightarrow V$ and $V \rightarrow E$ transfer

3.2.1. Electronic to vibrational energy transfer

(i) *Hg–CO system*. Many years ago, the Hg–CO system was studied in gas phase by Karl et al. (1967), and, more recently, by Horiguchi and Tsuchiya (1979). After excitation of the 6^3P_1 level of mercury, the collisions induce a non-resonant transfer to the vibrational levels of the ground electronic state of CO. The process takes place in a (CO–Hg)* excimer. The excess of energy, representing about half the 6^3P_1 energy, is dissipated to kinetic energy. The potential surfaces of the Hg–CO system were calculated by Kato et al. (1983).

Recently, this system has been studied in a pure CO matrix by Legay and Legay-Sommaire (1984) and in N_2 and Kr matrices by Crépin et al. (1986, 1987a,b). The mercury was excited with a 15 ns pulse of a KrF excimer laser to the 6^3P_1 level, taking advantage of coincidences at 249 nm between the laser line and the $^3P_1 - ^1S_0$ absorption line, strongly broadened in the matrix. The subsequent processes are

$$Hg(^3P_1) \rightarrow Hg(^3P_0); \tag{3.1}$$

$$Hg(^3P_0) + CO \rightarrow Hg(^1S_0) + CO(v). \tag{3.2}$$

The relevant levels are shown in fig. 7. Process (3.1) is induced by the presence of CO or N_2 and process (3.2) populated the vibrational levels of CO, observed by IR fluorescence. The process is not resonant and takes place in a Hg–CO complex, as in the gas phase. The curve $K = f(v)$, giving the transfer rate to each vibrational level, shows a strong maximum around $v = 6$ and negligible value above $v = 12$. This curve is almost the same for the solid and gas phase. In pure

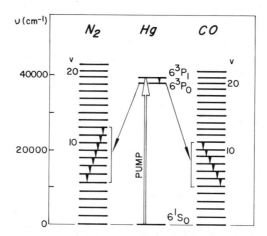

Fig. 7. Relaxation paths in the HgCO and HgN$_2$ systems.

CO, the lifetime of levels with $v > 7$ is purely radiative, while lower levels decay by transfer to the vibron of the CO matrix,

$$CO(v) + CO(0) \rightarrow CO(v-1) + CO(1).$$

The anharmonicity of the CO oscillator makes this process strongly endo-thermic for the high v and therefore very slow compared to the radiative lifetimes.

The transfer time, measured by the risetime of the IR fluorescence is less than 200 ns in all the matrices at 12 K. It was estimated to have a value of $\tau \simeq 25$ ns at $T = 22$ K in N_2 (Crépin et al. 1986).

(ii) *Hg–N_2 system*. Crépin et al. (1987a,b) studied this system in mercury-doped nitrogen as well as in $Hg:N_2:Kr$ mixed matrices. In nitrogen a strong, broad-band fluorescence was observed in the near-UV region and was assigned to the $(Hg–N_2)^*$ exciplex. Its decay time varies from 850 to 100 µs for T going from 12 to 25 K. Mercury isolated in pure krypton gives rise to a strong band at 39 270 cm^{-1}, corresponding to the $^3P_1 \rightarrow {}^1S_0$ transition with a purely radiative lifetime of 60 ns and a weak emission at 38 300 cm^{-1} corresponding to the $^3P_0 \rightarrow {}^1S_0$ transition with a 600 ± 100 ms decay time. This means that the transfer from the $^3P_1$ to the $^3P_0$ state is very inefficient in the pure krypton matrix and probably induced by impurities or defects. Upon addition of N_2, the exciplex fluorescence becomes apparent and the $^3P_0$ emission is strongly enhanced. Then nitrogen induces very efficiently a transfer $^3P_1 \rightarrow {}^3P_0$. This effect has been previously observed in gas-phase collisions (Horiguchi and Tsuchiya 1977), and in the dissociation of Van der Waals complexes (Jouvet and Soep 1984). With increasing temperature the $^3P_0$ emission is reduced while the exciplex fluorescence is strongly enhanced.

At low N_2 concentration and assuming a statistical repartition of the species, the largest fraction of the sites where Hg is close to N_2 is of the type $Hg(N_2)Kr_{11}$, and the experimental results are conveniently explained by a model where Hg in $^3P_0$ state and the $(Hg–N_2)^*$ exciplex are formed both at $Hg(N_2)Kr_{11}$ sites. The $^3P_1$ state decays rapidly to the $^3P_0$ state which, in turn, decays by three channels:

(i) $^3P_0$ emission;
(ii) rearrangement to the exciplex configuration, and
(iii) electronic-to-vibrational energy transfer.

At low temperature, two types of $Hg(N_2)Kr_{11}$ sites are assumed: channel (i) is favoured at site I and channel (ii) at site II.

Infrared emission was observed in mercury-doped nitrogen matrix, or in $Hg:Kr:N_2$ matrices with a high proportion of N_2 ($> 50\%$). In the 2000 cm^{-1} region, six emissions lines were observed and attributed to the $\Delta v = -1, v' = 6$ to 11 vibrational transitions of $N_2(X^1\Sigma_g^+)$. Although in a centrosymmetric N_2 crystal, the N_2 internal vibrational mode is infrared inactive, a weak dipole moment is induced when the site symmetry is modified by the presence of a Hg atom. The resulting transition moment is very small; nevertheless, the emission was easily observable through a grating monochromator equipped with a conventional phase detection system, as the very long lifetime of the IR fluorescence makes the emission practically continuous even with an excitation rate of only a few Hz. The $E \to V$ process takes place from the $^3P_0$ state populated by process (3.1), as for CO (see fig. 7),

$$Hg(^3P_0) + N_2(0) \to Hg(^1S_0) + N_2(v).$$

The decay time of the high levels with $v > 9$ was extremely slow (≈ 7 s) and probably mostly radiative. As for the CO case, the transfer to N_2 ($v = 1$) vibron causes the decay time to decrease with v. Thus the $v = 6$ level has a decay time of ≈ 1 s and the lower levels decay too fast to be observable. The curve $k = f(v)$, giving the $E \to V$ transfer rate as a function of the vibrational level quantum number, has a shape similar to the shape in the CO case, but with a maximum shifted to $v = 9$.

A rather surprising temperature effect was observed: the intensity of the emission lines increases strongly with temperature in the 14–20 K range. Thus the intensity of the $v = 10 \to 9$ line increases by a factor of 3. On the other hand, the transfer rate to the vibron is strongly enhanced by the temperature increase, leading to a decrease of the vibrational lifetime.

Due to the low transition moment in N_2, the rise time of the IR fluorescence was not observed. Nevertheless, measurements in $Hg:N_2:CO$ mixed matrices of the rise time of the CO ($v = 1$) fluorescence has given an upper bound of the $E \to V$ transfer time in N_2, i.e. $\tau < 5$ µs. The CO concentration was kept sufficiently low to have a vibration-to-vibration transfer from N_2 to CO dominant relatively to the Hg to CO $E \to V$ transfer. Then CO was used as a probe of the vibrational excitation of N_2.

(iii) *Mechanism of the $E \to V$ transfer.* It would seem reasonable to assume a similar mechanism for the electronic to vibrational energy transfer in the Hg–CO and Hg–N_2 systems, but in fact it was very difficult to find, in the Hg–N_2 case, a model able to explain both UV and IR results.

Transfer in the Hg–CO system was conveniently described in a semi-qualitative way by using the potential surfaces of Kato et al. (1983). The bottom of the crossing between ground and excited surfaces is lower than the $^3P_0$ state in the Hg–CO linear geometry. A fast transfer to the ground surface by passing through the barrier or by a tunnelling effect leads to a vibrational distribution similar to the distribution in gas phase, in agreement with experiment. The fact that no UV emission from a Hg–CO excimer has been detected can be due either to a fast relaxation by $E \to V$ transfer or by an emission in a not yet investigated spectral region.

In the Hg–N_2 case, calculations of potential surfaces by Crépin and Millié (1989) give a much shallower minimum for the first excited state of $(Hg-N_2)^*$, in agreement with the UV observations. But the exciplex fluorescence has a much longer lifetime (850 µs) than the E → V transfer time (<5 µs). This seems to exclude the excimer as an intermediate in the transfer process. Then, an other possibility is the presence of two deactivating channels:

(i) direct transfer to vibration, and

(ii) excimer formation by trapping in the excited potential surface minimum, followed by a slow radiative deactivation.

Nevertheless, such a model cannot easily explain the simultaneous intensity increase of both the IR and UV emissions with temperature.

3.2.2. *Vibrational to electronic energy transfer*

(i) *CO–O_2 system.* Galaup et al. (1985b,c) observed UV-visible fluorescence from solid argon doped with 1% of $^{13}C^{18}O$ upon irradiation by a color-center laser whose frequency is coincident with the first overtone absorption band of CO. These emissions have been identified with $C^3\Delta_u \to X^3\Sigma_g^-$ and $b^1\Sigma_g^+ \to X^3\Sigma_g^-$ transitions of molecular oxygen which was present in the samples as an unintentional impurity. More recently, an extensive study of V–E transfer between CO and O_2 in double-doped Ar, Kr and N_2 matrices has been done by Dubost et al. (1987). As usual, the vibrational CO populations were probed through the $\Delta v = 2$ overtone emission. Electronic excitation of O_2 was detected by monitoring the various emissions in the near-IR, visible and near-UV spectral regions. In addition to the UV-visible bands, the $b^1\Sigma_g^+ \to a^1\Delta_g$ and $a^1\Delta_g \to X^3\Sigma_g^-$ transitions have been observed in the near-IR. Excitation of the O_2 luminescence bands is into the CO vibrational overtone absorption, indicating that the vibrational energy of CO is transferred to the $a^1\Delta_g$, $b^1\Sigma_g^+$ and $C^3\Delta_u$ states of O_2.

The intensity ratio of the singlet and triplet emission bands is strongly sensitive to isotope substitution of O_2. In the case of $^{16}O_2$, the intensity of the b–X and C–X emissions is quite comparable. When $^{18}O_2$ is used instead of $^{16}O_2$, the intensity of the triplet emission is decreased by three orders of magnitude and is almost not observed. On the contrary, the singlet bands of $^{18}O_2$ are ≈ 10 times stronger than that of $^{16}O_2$. In solid argon, the molecular oxygen bands have been observed for a large range of O_2 (10 to 0.001%) and CO (2 to 0.02%) concentrations. The intensity ratio of the b–X and C–X emissions is sensitive to the experimental parameters which affect the distribution of vibrational populations in CO. For instance, in argon doped with 1% $^{16}O_2$ and $^{13}C^{18}O$, the C–X intensity of $^{16}O_2$ becomes one order of magnitude weaker than that of b–X as the CO concentration is decreased from 1 to 0.02%. At 0.01% CO, the triplet emission is no longer observed whereas the singlet bands are still present. In the krypton matrix, the O_2 electronic emissions behave quite similarly compared to solid Ar. On the contrary, in the N_2 matrix they are no longer observed.

The time development of the various emissions upon excitation by a square laser pulse has given two important pieces of information. The population of O_2 electronic states evolves on two different time scales:

  (i) an ms one which is determined by the 20 ms lifetime of the CO vibration and of the $b^1\Sigma_g^+$ of O_2 (Becker et al. 1988), and
  (ii) a minute time scale which reflects the 80 s lifetime of $O_2(a^1\Delta_g)$ (Crone and Kügler 1985).

When the sample is irradiated after being left 2–3 min in the dark in order to allow for complete relaxation of the $a^1\Delta_g$ state, the intensity of the b emission induced by the first pulse is very weak, then grows steadily upon the next excitation pulses and reaches a steady-state value after irradiation for 1 min. This behavior, which is very similar to that of the $a^1\Delta_g$ emission, indicates that the b state is populated mostly from the a state and not from the ground state. On the contrary, the intensity of the C–X signal is independent of the excitation regime and keeps a constant value. Moreover, the 1 ms delay of the CX emission with respect to the laser pulse and the 2–3 ms decay time suggests that the C state of O_2 is fed from very high lying levels of CO ($v > 42$).

The experimental results demonstrate the following properties of V–E transfer between CO and O_2:

  (1) Singlet and triplet states are populated via two independent channels;
  (2) The b state of O_2 is fed from the long-lived a state which acts as an energy reservoir;
  (3) The C state receives its energy from CO vibrational levels with $v > 42$ which are higher than those feeding the singlet states; and
  (4) The O_2 vibration is involved in the transfer process.

(ii) *CO–NH, CO–NX (X = I, Br or Cl) and CO–CN systems.* Since many open-shell species have low lying electronic states, one could expect V–E transfer to occur from vibrationally excited CO or NO to many other systems. In particular, radicals such as NH or NX have the same electronic configuration as O_2. Therefore, they all have two low-lying singlet states likely to be sensitized by V–E transfer. A systematic search for V–E transfer from vibrationally excited CO to diatomic radicals has been undertaken by Becker et al. (1989). The NH and CN species were prepared by in situ photolysis of argon samples doped with CO and NH_3 or HCN with the Lyman α-line of H_2 emitted by a RF discharge lamp. The NX radicals were obtained by co-deposition of a gaseous $N_2 : X_2 : Ar$ mixture through a RF discharge together with CO/Ar. The vibrational excitation of CO did not induce any electronic fluorescence from NH or NX, in spite of the fact that an intense singlet emission resulted from direct laser excitation of the $b^1\Sigma^+$ state and that the quenching of the CO vibrational excitation by the precursors was moderate. On the contrary, a strong emission from the $A^2\Pi$, $v = 0$ state of CN resulted from the excitation of CO with the IR laser.

(iii) *Mechanism of V–E transfer.* As discussed by Galaup et al. (1985b,c) there are three possible transfer mechanisms:

(1) Direct intermolecular V–E transfer involving the conversion of several vibrational quanta of CO into a single quantum of electronic energy in the acceptor. The excitation of the C state of O_2 by such a process, which would require the transfer *en bloc* of at least 19 vibrational quanta of CO, is very unlikely. The number of vibrational quanta of the donor necessary to match the energy of the singlet states of the O_2 acceptor is much smaller. Therefore, processes such as

$$^{13}C^{18}O(v=4) + O_2(X^3\Sigma_g^-) \rightarrow {}^{13}C^{18}O(v=0) + O_2(a^1\Delta_g), \quad \Delta E = 130 \text{ cm}^{-1},$$

could be efficient on the minute time scale of the $a^1\Delta_g$ state lifetime. However, they do not account for the observed isotopic dependence in O_2. In addition, they should occur in the NH and NX species as well.

(2) Intramolecular V–E crossing to $a^3\Pi$ or more likely to $A^1\Pi$ in CO followed by intermolecular transfer of electronic excitation. There is direct experimental evidence against this process (Bahrdt and Schwentner 1986). In solid argon doped with 1% CO and O_2, the direct excitation of the $a^3\Pi$, $a'^3\Sigma^+, d^3\Delta, e^3\Sigma^-$ or $A^1\Pi$ states of CO does not induce any electronic emission from O_2.

(3) Intermolecular V–V transfer from CO to O_2 followed by intersystem crossing or internal conversion in O_2. As a matter of fact vibrational energy transfer from $^{13}C^{16}O$ $(v=1)$ to $^{16}O_2$ $(v=1)$ is known to occur in solid argon. However, the probability of this strongly non-resonant process ($\Delta E = 537$ cm^{-1}) is five orders of magnitude smaller than that of the near-resonant first steps in the CO up pumping (Dubost 1984). The exchange of vibrational energy between strongly vibrationally excited CO and unexcited O_2 is much more probable because the energy of the $\Delta v = 1$ transitions in CO and O_2 becomes quite similar. For instance, the process

$$^{13}C^{18}O(v=21) + {}^{16}O_2(v=0) \rightarrow {}^{13}C^{18}O(v=20) + {}^{16}O_2(v=1), \quad \Delta E = 17 \text{ cm}^{-1},$$

may have a probability quite comparable to that of further up pumping in CO. Since the vibrational anharmonicity of CO and O_2 are similar, successive near-resonances should result in the simultaneous vibrational excitation of the two species as soon as CO levels above $v = 20$ are populated. The strong isotope effect can be explained according to the following model. In matrix-isolated O_2, intersystem crossing is known to take place between nearly degenerate singlet and triplet vibronic levels (Rossetti and Brus 1979). In $^{18}O_2$, the $X^3\Sigma_g^-$, $v = 15$ and the $a^1\Delta_g$, $v = 9$ levels are nearly resonant ($\Delta E = 37$ cm^{-1}). Therefore, an efficient X $\rightarrow$ a crossing is expected to by-pass the ground-state vibrational excitation. The a and b singlet manifolds which are loosely coupled (Becker et al. 1988) are then both strongly populated while the $C^3\Delta_u$ state receives almost no energy. On the contrary, near resonance between a and X in $^{16}O_2$ occurs at a much higher energy,

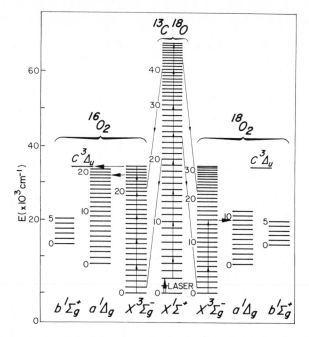

Fig. 8. Energy diagram showing the vibrational levels of $^{13}C^{18}O$, $^{16}O_2$, $^{18}O_2$ and the vibronic levels of the lowest excited singlet and triplet states of O_2. The arrows show the strong vibrational excitation in CO and O_2 resulting from intermolecular V–V transfer and the subsequent O_2 electronic excitation.

thus allowing for the excitation of high-lying vibrational levels in the ground state. Like in the case of NO (section 3.1.2), energy transfer can occur near the dissociation limit from $X^3\Sigma_g^-$, v to the excited electronic states which correlate with ground-state N and O atoms, and in particular to the $C^3\Delta_u$ state. In solid nitrogen, V–E transfer between CO and O_2 is not observed, probably because the excitation of the high-lying vibrational levels of CO and consequently those of O_2 is prevented by near-resonant exchange of two CO vibrational quanta against a single N_2 quantum. Finally, V–E transfer takes place from CO to CN but not to NH or NX, because only the CN species can gain a sufficient vibrational excitation from CO.

Acknowledgments

We are very grateful to our colleagues L. Abouaf-Marguin, P. Boissel and B. Gauthier-Roy for providing us with unpublished material and for their invaluable help in the preparation of the NH_3 and CH_3F sections of this chapter.

References

Abbate, A.D., and C.B. Moore, 1985a, J. Chem. Phys. **82**, 1255.
Abbate, A.D., and C.B. Moore, 1985b, J. Chem. Phys. **82**, 1263.
Abbate, A.D., and C.B. Moore, 1985c, J. Chem. Phys. **83**, 975.
Abouaf-Marguin, L., 1973, Thesis (University of Paris VI).
Abouaf-Marguin, L., and B. Gauthier-Roy, 1980, Chem. Phys. **51**, 213.
Abouaf-Marguin, L., H. Dubost and F. Legay, 1973, Chem. Phys. Lett. **22**, 603.
Abouaf-Marguin, L., B. Gauthier-Roy and F. Legay, 1977a, Chem. Phys. **23**, 443.
Abouaf-Marguin, L., M.E. Jacox and D.E. Milligan, 1977b, J. Mol. Spectrosc. **67**, 34.
Abouaf-Marguin, L., P. Boissel and B. Gauthier-Roy, 1981, J. Chem. Phys. **75**, 495.
Abouaf-Marguin, L., B. Gauthier-Roy, J. Dupré and C. Meyer, 1985, J. Mol. Spectrosc. **110**, 347.
Apkarian, V.A., and E. Weitz, 1980, Chem. Phys. Lett. **76**, 68.
Apkarian, V.A., and E. Weitz, 1982, J. Chem. Phys. **76**, 5796.
Apkarian, V.A., L. Wiedman, W. Janiesch and E. Weitz, 1986, J. Chem. Phys. **85**, 5593.
Bahrdt, J., and N. Schwentner, 1986, J. Chem. Phys. **85**, 6229.
Bahrdt, J., and N. Schwentner, 1988, J. Chem. Phys. **88**, 2869.
Bahrdt, J., P. Gürtler and N. Schwentner, 1987, J. Chem. Phys. **86**, 6108.
Becker, A.C., J. Langen, H.M. Oberhoffer and U. Schurath, 1986, J. Chem. Phys. **84**, 2907.
Becker, A.C., K.P. Lodemann and U. Schurath, 1987, J. Chem. Phys. **87**, 6266.
Becker, A.C., U. Schurath, J.P. Galaup and H. Dubost, 1988, Chem. Phys. **125**, 321.
Becker, A.C., R. Charneau, H. Dubost and U. Schurath, 1989, to be published.
Berkowitz, M., and R.B. Gerber, 1979, Chem. Phys. **37**, 369.
Boissel, P., 1985, Thesis (University Paris-Sud, Orsay).
Boissel, P., B. Gauthier-Roy and L. Abouaf-Marguin, 1985, J. Chem. Phys. **82**, 1056.
Boissel, P., B. Gauthier-Roy and L. Abouaf-Marguin, 1988, Chem. Phys. **119**, 419.
Boissel, P., B. Gauthier-Roy and L. Abouaf-Marguin, 1989, to be published.
Bondybey, V.E., 1976a, J. Mol. Spectrosc. **63**, 164.
Bondybey, V.E., 1976b, J. Chem. Phys. **65**, 2296.
Bondybey, V.E., 1977a, J. Mol. Spectrosc. **64**, 180.
Bondybey, V.E., 1977b, J. Chem. Phys. **66**, 4237.
Bondybey, V.E., 1977c, J. Chem. Phys. **66**, 995.
Bondybey, V.E., and L.E. Brus, 1975a, J. Chem. Phys. **63**, 794.
Bondybey, V.E., and L.E. Brus, 1975b, J. Chem. Phys. **63**, 2223.
Bondybey, V.E., and J.H. English, 1977a, J. Chem. Phys. **67**, 664.
Bondybey, V.E., and J.H. English, 1977b, J. Mol. Spectrosc. **68**, 89.
Bondybey, V.E., and J.H. English, 1977c, J. Chem. Phys. **67**, 2868.
Bondybey, V.E., and J.H. English, 1980a, J. Mol. Spectrosc. **79**, 416.
Bondybey, V.E., and J.H. English, 1980b, J. Chem. Phys. **73**, 3098.
Bondybey, V.E., and J.H. English, 1980c, J. Chem. Phys. **73**, 87.
Bondybey, V.E., and T.A. Miller, 1978, J. Chem. Phys. **69**, 3597.
Bondybey, V.E., and A. Nitzan, 1977, Phys. Rev. Lett. **38**, 889.
Bondybey, V.E., T.A. Miller and J.H. English, 1980, Phys. Rev. Lett. **44**, 1344.
Bondybey, V.E., J.H. English and T.A. Miller, 1983a, J. Phys. Chem. **87**, 1300; and references therein.
Bondybey, V.E., S.V. Milton, J.H. English and P.M. Rentzepis, 1983b, Chem. Phys. Lett. **97**, 130.
Bondybey, V.E., R.C. Haddon and P.M. Rentzepis, 1984a, J. Am. Chem. Soc. **106**, 5969.
Bondybey, V.E., R.C. Haddon and J.H. English, 1984b, J. Chem. Phys. **80**, 5432.
Brueck, S.R.J., T.F. Deutsch and R.M. Osgood, 1977, Chem. Phys. Lett. **51**, 339.
Brus, L.E., and V.E. Bondybey, 1980, Spectroscopic and Time Resolved Studies of Small Molecules Relaxation in the Condensed Phase, in: Radiationless Transitions, ed. S.H. Lin (Academic Press, New York) ch. 6, p. 259.

Casperson, L.W., 1978, IEEE J. Quantum Electron. **QE-14**, 756.

Chergui, M., 1986, Thesis (University Paris-Nord).

Chesnoy, J., 1985, J. Chem. Phys. **83**, 2214.

Clouston, J.C., A.G. Gaydon and I.I. Glass, 1958, Proc. R. Soc. A **248**, 429.

Crépin, C., and P. Millié, 1989, to be published.

Crépin, C., F. Legay, N. Legay-Sommaire and A. Tramer, 1986, Opt. Commun. **58**, 100.

Crépin, C., F. Legay, N. Legay-Sommaire and A. Tramer, 1987a, Chem. Phys. **111**, 169.

Crépin, C., F. Legay, N. Legay-Sommaire and A. Tramer, 1987b, Chem. Phys. **111**, 183.

Crépin, C., F. Legay, N. Legay-Sommaire and A. Tramer, 1989, to be published.

Crocombe, R.A., N.R. Smyrl and G. Mamantov, 1978, J. Am. Chem. Soc. **100**, 6526.

Crone, K.P., and K.J. Kügler, 1985, Chem. Phys. **99**, 293.

Deleon, R.L., and J.W. Rich, 1986, Chem. Phys. **107**, 283.

Devonshire, A.F., 1936, Proc. R. Soc. A **153**, 601.

Dubost, H., 1984, Spectroscopy of Vibrational and Rotational Levels of Diatomic Molecules in Rare Gas Crystals, in: Inert Gases - Potentials, Dynamics and Energy Transfer in Doped Crystals, ed. M.L. Klein (Springer, Berlin) p. 145–257.

Dubost, H., and R. Charneau, 1979, Chem. Phys. **41**, 329.

Dubost, H., J.P. Galaup and R. Charneau, 1987, J. Lumin. **38**, 147.

Dubost, H., I. Hadj-Bachir and R. Charneau, 1989, to be published.

Förster, Th., 1948, Ann. Phys. **2**, 55.

Fournier, J., H.H. Mohammed, J. Deson, C. Vermeil and J. Schamps, 1980, J. Chem. Phys. **73**, 6039.

Fredin, L., and B. Nelander, 1981, Chem. Phys. **60**, 181.

Furio, N., A. Ali and P.J. Dagdigian, 1986, J. Chem. Phys. **85**, 3860.

Galaup, J.P., J.Y. Harbec, R. Charneau and H. Dubost, 1985a, Chem. Phys. Lett. **118**, 252.

Galaup, J.P., J.Y. Harbec, R. Charneau and H. Dubost, 1985b, Chem. Phys. Lett. **120**, 188.

Galaup, J.P., J.Y. Harbec, R. Charneau and H. Dubost, 1985c, J. Phys. C **7**, 293.

Galaup, J.P., J.Y. Harbec, J.J. Zondy, R. Charneau and H. Dubost, 1985d, J. Phys. C **7**, 299.

Gale, G., and C. Delalande, 1978, Chem. Phys. **34**, 205.

Gauthier-Roy, B., 1980, Thesis (University Paris-Sud, Orsay).

Gauthier-Roy, B., L. Abouaf-Marguin and F. Legay, 1980, Chem. Phys. **46**, 31.

Gauthier-Roy, B., C. Alamichel, A. Lecuyer and L. Abouaf-Marguin, 1981, J. Mol. Spectrosc. **88**, 72.

Gellerman, W., Y. Yang and F. Luty, 1986, Opt. Commun. **57**, 196.

Girardet, C., and A. Lakhlifi, 1986, Chem. Phys. **110**, 447.

Girardet, C., L. Abouaf-Marguin, B. Gauthier-Roy and D. Maillard, 1984a, Chem. Phys. **89**, 415.

Girardet, C., L. Abouaf-Marguin, B. Gauthier-Roy and D. Maillard, 1984b, Chem. Phys. **89**, 431.

Goodman, J., and L.E. Brus, 1978, J. Chem. Phys. **69**, 1853.

Gosnell, T.R., A.J. Sievers and C.R. Pollock, 1985a, Opt. Lett. **10**, 125.

Gosnell, T.R., R.W. Tkach and A.J. Sievers, 1985b, Solid State Commun. **53**, 419.

Gosnell, T.R., R.W. Tkach and A.J. Sievers, 1985c, Infrared Phys. **25**, 35.

Harig, M., R. Charneau and H. Dubost, 1982, Phys. Rev. Lett. **49**, 715.

Horiguchi, H., and S. Tsuchiya, 1977, Bull. Chem. Soc. Jpn. **50**, 1661.

Horiguchi, H., and S. Tsuchiya, 1979, J. Chem. Phys. **70**, 762.

Houston, P.L., 1981, Electronic to Vibrational Energy Transfer from Excited Halogen Atoms, in: Photoselective Chemistry, eds J. Jortner, R.D. Levine and S.A. Rice (Wiley, New York).

Huppert, D., and P.M. Rentzepis, 1987, J. Chem. Phys. **86**, 603.

Huppert, D., V.E. Bondybey and P.M. Rentzepis, 1985, J. Chem. Phys. **89**, 581.

Hutson, J.M., and B.J. Howard, 1982, Mol. Phys. **45**, 769.

Janiesch, W., V.A. Apkarian and E. Weitz, 1982, Chem. Phys. Lett. **85**, 505.

Jenkins, D.R., 1966, Proc. R. Soc. London Ser. A **293**, 493.

Jones, L.H., and B.I. Swanson, 1982a, J. Chem. Phys. **76**, 1634.

Jones, L.H., and B.I. Swanson, 1982b, J. Chem. Phys. **77**, 6338.

Jones, L.H., and B.I. Swanson, 1984, J. Chem. Phys. **80**, 2980.

Jouvet, C., and B. Soep, 1984, J. Chem. Phys. **80**, 2229.

Karl, G., P. Kraus and J.C. Polanyi, 1967, J. Chem. Phys. **46**, 224.

Katayama, D.H., 1984, J. Chem. Phys. **81**, 3495.

Katayama, D.H., 1985, Phys. Rev. Lett. **54**, 657.

Katayama, D.H., T.A. Miller and V.E. Bondybey, 1979, J. Chem. Phys. **71**, 1662.

Kato, S., R.L. Jaffe, A. Komornicki and K. Morokuma, 1983, J. Chem. Phys. **78**, 4567.

Koch, K.P., Y. Yang and F. Luty, 1984, Phys. Rev. B **29**, 5840.

Krause, L., 1975, Adv. Chem. Phys. **28**, 267.

Krueger, H., J.T. Knudtson, Y.P. Vlahoyannis and E. Weitz, 1985, Chem. Phys. Lett. **119**, 298.

Kühle, H., R. Fröhling, J. Bahrdt and N. Schwentner, 1986, J. Chem. Phys. **84**, 666.

Lakhlifi, A., and C. Girardet, 1984, J. Mol. Struct. **19**, 73.

Lakhlifi, A., and C. Girardet, 1987, J. Chem. Phys. **87**, 4559.

Legay, F., 1977, Vibrational Relaxation in Matrices, in: Chemical and Biochemical Applications of Lasers, Vol. II, ed. C.B. Moore (Academic Press, New York) p. 43–86.

Legay, F., and N. Legay-Sommaire, 1984, Chem. Phys. **89**, 151.

Legay, F., N. Legay-Sommaire, A. Tramer, M. Chergui and N. Schwentner, 1988, J. Phys. Chem. **92**, 261.

Legay-Sommaire, N., and F. Legay, 1980, IEEE J. Quantum Electron. **QE-16**, 308.

Manz, J., 1977, Chem. Phys. **24**, 51.

Meyer, B., 1971, Low Temperature Spectroscopy (Elsevier, New York).

Nelander, B., 1984, Chem. Phys. **87**, 283.

Ogryzlo, E.A., 1979, Gaseous Singlet Oxygen, in: Singlet Oxygen, eds H.H. Wasserman and R.W. Murray (Academic Press, New York) ch. 2, p. 35.

Ogryzlo, E.A., and B.A. Thrush, 1973, Chem. Phys. Lett. **23**, 34.

Ogryzlo, E.A., and B.A. Thrush, 1974, Chem. Phys. Lett. **24**, 314.

Quigley, G.P., and G.J. Wolga, 1975, J. Chem. Phys. **62**, 4560.

Ramsthaler-Sommer, A., K. Eberhardt and U. Schurath, 1986, J. Chem. Phys. **85**, 3760.

Reiland, W., H.U. Tittes, I.V. Hertel, V. Bonacic-Koutecky and M. Persico, 1982, J. Chem. Phys. **77**, 1908.

Rentzepis, P.M., and V.E. Bondybey, 1984, J. Chem. Phys. **80**, 4727.

Rich, J.W., and R.C. Bergman, 1979, Chem. Phys. **44**, 53.

Rossetti, R., and L.E. Brus, 1979, J. Chem. Phys. **71**, 3963.

Schwahn, H., and B. Schramm, 1986, Chem. Phys. Lett. **130**, 29.

Schwartz, R.N., Z.I. Slawsky and K.F. Herzfeld, 1952, J. Chem. Phys. **20**, 1591.

Singh, J.P., J. Bachar, D.W. Setser and S. Rosenwaks, 1985, J. Phys. Chem. **89**, 5347.

Swanson, B.I., L.H. Jones, S.A. Ekberg and H.A. Fry, 1986, Chem. Phys. Lett. **126**, 455.

Taïeb, G., and F. Legay, 1970, Can. J. Phys. **48**, 1956.

Tasumi, M., H. Takeuchi and H. Nakano, 1979, Chem. Phys. Lett. **68**, 44.

Tkach, R.W., T.R. Gosnell and A.J. Sievers, 1984, Opt. Lett. **10**, 122.

Yang, Y., and F. Luty, 1983, Phys. Rev. Lett. **51**, 419.

Yang, Y., and F. Luty, 1985, J. Phys. C **7**, 287.

Yang, Y., W. Von der Osten and F. Luty, 1985, Phys. Rev. B **32**, 2724.

Young, L., and C.B. Moore, 1982, J. Chem. Phys. **76**, 5869.

Young, L., and C.B. Moore, 1984, J. Chem. Phys. **81**, 3137.

Zondy, J.J., 1986, Thesis (University Paris-Sud, Orsay).

Zondy, J.J., J.Y. Harbec, J.P. Galaup, R. Charneau and H. Dubost, 1985, J. Phys. C **7**, 305.

Zondy, J.J., J.P. Galaup and H. Dubost, 1987, J. Lumin. **38**, 255.

Zumofen, G., J. Sedlacek, R. Tautenberger, S.L. Pan, O. Oehler and K. Dressler, 1984, J. Chem. Phys. **81**, 2305.

CHAPTER 12

Solid-State Aspects of Matrices

H.J. JODL

Fachbereich Physik
Universität Kaiserslautern
6750 Kaiserslautern
Fed. Rep. Germany

Chemistry and Physics of
Matrix-Isolated Species
L. Andrews and M. Moskovits, eds

Contents

1. Introduction

1.1. Apologia and scope

Over 4000 articles, 100 reviews and 10 books have been written dealing with matrix isolation. These are collected in a bibliography (Ball et al. 1987). Because of this and the ease with which the literature is searched on-line only selected examples of matrix-isolated species need be discussed in detail.

Parallel to this increase in publications is an increase in the complexity of the species and the matrix materials that are being investigated. This chapter will be restricted to systems comprising diatomic molecules in rare-gas, nitrogen or oxygen (anisotropic) matrices. More complicated molecules were recently reviewed by Knözinger and Schrems (1987). From the physical point of view, this chapter will stress solid-state aspects such as crystal-field splitting or anharmonic phonon coupling, i.e. the matrix material in the defect/matrix interaction will be treated as a crystal with discrete structure, lattice dynamics and so on; in contrast to the traditional approach wherein the action of the matrix on the defect was described in such a way that the properties of the free molecule were compared to the molecules in the matrix-isolated case; i.e., in general, the matrix was introduced more or less as a dielectric continuum.

Spectroscopically, mostly Raman scattering and IR absorption data will be considered. This means excitations of internal (vibrations) and external (rotations, translations) modes within the electronic ground state only. Electronic excitations in molecular crystals are well reviewed by Rössler (1976) with respect to pure rare-gas solids and by Jortner (1974) discussing impurity states in insulators. Although this relaxation channel is more complicated, there are more data available as VUV-spectroscopy of matrix-isolated species is a few years older. This is natural because the only allowed matrix bulk excitations are in the VUV. In rare-gas solids, first-order Raman scattering and IR absorption are not allowed; excitations of the electronic ground state are only known from inelastic neutron scattering experiments and very weak second-order Raman scattering. Because of recent spectroscopic improvements (e.g., FTIR), the weak defect-induced phonon spectra in O_2 and N_2, used as matrices, are nowadays measurable.

One of my aims is to fill the gap in our knowledge of the condensed-matter

345

aspects of matrix investigations. Of course, there already exists some work in this area (Swanson and Jones 1983). In general, however, these solid-state effects were observed either only for specific species, although not always, or were interpreted globally as due to sites or phonon sidebands and were not studied on purpose in detail. Since the development of the matrix-isolation technique in 1954, knowledge regarding molecular crystals has increased enormously and detailed reviews are available (Lascombe 1987).

A second aim is to revise the concept of matrix shifts as a descriptive parameter for the matrix-isolated species. The change of this value as a function of matrix material, concentration of impurities, sample temperature, and pressure will be discussed in the framework of solid-state physics; the solid shift as a special case will be included. By increasing the impurity concentration we can develop a new point of view of the matrix-isolated case, namely as the low-concentration limit of mixed molecular crystals.

Third, it is undoubtfully true that sample quality (amorph, polycrystal, single crystal) affects strongly those physical values that describe the solid state of the system, such as inhomogeneous line broadening or crystal-field splitting. Unfortunately, the spectra of glassy films obscure those fine details which characterize the solid state. Hence, the classical method for sample generation – spraying on thin films – although very effective, as regards time and material consumption, is not necessarily adequate for these type of studies. Other crystal growing techniques must be considered.

Fourth, the claim of matrix isolation is that the matrix hardly influences the defect. However, if one looks more closely, one finds that an anisotropic matrix (such as N_2, O_2) interacts differently with the defect than isotropic matrices such as rare-gas solids. In turn, the defect also influences the matrix.

Aim number five is to consider anharmonicities also when discussing matrix-isolation data. The theoretical concept of harmonic, quasi- and anharmonic contributions to the lattice dynamics of pure molecular crystals is now well established (see, e.g. Califano et al. 1981), and it is a challenge to transfer the idea to doped molecular crystals. In addition, spectroscopic techniques are now sensitive enough to measure the consequences of anharmonicity, such as small line broadening or frequency shifts. Moreover, by applying ultra high pressure at varying low temperatures one is able to distinguish between different sources of anharmonicity, e.g., volume expansion against phonon–phonon interaction. On top of that the combination of high pressure and matrix-isolation spectros-copy allows an effective direct test of theoretically developed inter- and intramolecular potentials. Last but not least, pressure-induced chemistry and physics of matrix-isolated systems and mixed molecular crystals will therefore open a further, new research area.

A glance at table 1 (column 4) indicates the scope of the solid-state physics that can be studied in matrix-isolated systems. The other columns of the table contain information about the guest and the host material (a slash means

Table 1
Summary of selected examples.

Sample	Method*	Reference(s)	Indication to solid-state effects
N_2 in Ar	BS, HRRS	Kiefte et al. (1982)	Single crystal/concentration-
O_2 in Ar		Ahmad et al. (1982)	dependent measurements/alloy,
N_2/CO	RS	Witters and Cahill (1977)	mixture, cluster, matrix
CO_2/N_2O			isolated/hcp or fcc Ar phase-
			stabilized by impurities/
			rotation–translation-mode
			coupling/one- and two-mode
			behaviour
Kr in Ar		Malzfeldt (1985)	Local distortion of matrix around
Xe in Ar			impurity or cluster/change in
Sm in RGS	EXAFS	Niemann et al. (1987)	Franck–Condon factors – change
$(Ag)_x$ in Ar		Montano et al. (1984)	in bond length/local distortion of
Br_2 in Ar	LIF	Langen et al. (1987)	impurity
K in Ar	OS	Stöckmann (1984)	Light-induced site
	MCD		modifications/line shape
	ESR	Steinmetz et al. (1987)	analysis/shift and splitting of
Na, K in RGS	OS	Forstmann et al. (1977)	electronic levels due to static crystal
Cu, Ag, Au	Th	Weinert et al. (1983)	field or dynamic Jahn–Teller
in RGS			effect/stable trapping sites, matrix
			cage/temperature-dependent
			measurements
SF_6 in RGS	FTIR	Swanson and Jones (1981)	Tetrahedral molecule rotating in an
SeF_6 in RGS	FTIR	Jones and Swanson (1981)	octahedral field – experiment and
CH_4 in RGS	FTIR	Jones et al. (1986)	group theory/local-site phonon
CH_4 in RGS	RS	Regitz et al. (1984)	mode in resonance with phonon
CH_4 in RGS	Th	Kobashi et al. (1975)	bath/multiple trapping sites and
CH_4 in RGS	FTIR	Nanba et al. (1980)	crystal field splitting/dynamic of
			matrix/molecule interaction/
			rotational and translational modes
			in large single crystals
NH_3 in RGS	FTIR	Girardet et al. (1984a)	Quadruplet structure of v_2,
	Th	Girardet et al. (1984b)	inversion doubling with
NH_3 in N_2	FTIR	Girardet and Lakhlifi (1985)	rotations/matrix is not a matrix or
	Th	Girardet and Lakhlifi (1986)	different equilibrium eccentricity/in
			a very sophisticated model the
			theoretically generated IR
			spectrum is in excellent agreement
			with experiment/isotopes used and
			different temperatures
		Galaup et al. (1985)	Phonon-assisted vibrational energy
CO in RGS	HB	Dubost et al. (1982)	transfer/vibron, libron, phonon
	FTIR	Dubost and Charneau (1979)	coupling and relaxation/

Table 1 (continued)

Sample	Method*	Reference(s)	Indication to solid-state effects
CO in N_2		Dubost and Charneau (1976)	zero-phonon line and phonon sidebands/linewidth and lineshape
	Th	Allavena and Seiti (1986)	analysis/temperature-,
	VUV	Bahrdt et al. (1987)	concentration- and isotope dependence/mode coupling for theoretical explanation of vibrational dephasing
N_2/RGS	RS	Löwen et al. (1987)	Zero-phonon line of impurity and
N_2 in Xe	VUV	Kühle et al. (1985)	phonon sidebands of matrix/Stokes
Eu in Ar	OS	Jakob et al. (1977)	and anti-Stokes sideband as a
CO in RGS	IR	Dubost (1976)	function of concentration and temperature/electron- and vibron–phonon coupling discussed in terms of relaxation processes/ anharmonicities and symmetries
RG in RGS	Th	Cohen and Klein (1974)	Defect-induced phonon spectra/
RG in RGS	IR	Keeler and Batchelder (1972)	local, resonant and gap modes/ modified phonon density of states
HCl in Ar	Th	Mannheim (1972)	by mass and force constant
H_2 in Kr	IR	De Remigis and Welsh (1970)	changes in a linear chain
RG in RGS	Th	Mannheim (1968)	

*Methods used:

RS is Raman scattering, HRRS high resolution RS, BS Brillouin scattering, EXAFS extended X-ray absorption fine structure, LIF laser-induced fluorescence, OS optical spectroscopy, MCD magnetic circular dichroism, ESR electron-spin resonance, Th theory, FTIR Fourier transform infrared spectroscopy, HB hole burning, VUV vacuum-ultraviolet spectroscopy and IR infrared absorption.

mixtures), the method used and the reference concerned. There is no one-to-one correlation within the lines because publications were formed into groups around a solid-state topic – such as crystal structure and distortions; molecule symmetry, site and crystal symmetry; static and dynamic interaction between guest and host; different kinds and ways of mode coupling; zero-phonon line and phonon sideband; and defect-induced spectra.

1.2. Background

In former days, the matrix-isolated molecules were described by the frozen-gas model. Most spectroscopic measurements were performed with disordered specimens (Ball et al. 1987). The material under investigation was formed by condensing the molecular and matrix gases on a cold substrate at low temperature (<30 K). The spectroscopic properties of the molecule stand out, but the dynamics of the matrix–defect system are not observed. Small spectroscopic features – such as asymmetric lines due to crystal-field splitting, or broad phonon wings instead of structured phonon sidebands – were sufficiently described by the amorphous or disordered state of the molecular crystals. The applicability and the limits of the frozen-gas model have now been reached, and the smooth transition from the matrix-isolated case via clusters towards microcrystals is under theoretical development. Current spectroscopic techniques, such as FTIR and Raman scattering, are capable of resolving small features (10^{-3} a.u.) next to a fundamental mode (intensity 1 a.u.).

Molecular crystals are built out of molecules that retain in the crystal almost the same internal structure that they have in the gas phase. Internal vibrations of the molecules making up the lattice are, in general, separable from external or lattice vibrations due to the considerably greater strength of intramolecular forces (frequencies of the order of 1000 cm^{-1}) as compared to the intermolecular forces (lattice modes near 100 cm^{-1} or below). In dealing with lattice vibrations, we shall consider the molecules to be rigid bodies, ignoring the internal degrees of freedom. This has the advantage of reducing to six the number of degrees of freedom of each non-linear molecule. With this separation of the types of motion we can use a type of adiabatic approximation. One can average the potential energy over the vibrational wavefunctions of the molecules and obtain the effective potential for the lattice vibrations. The geometry of molecules in crystals is usually not very different from the free-molecule geometry, but this is an assumption.

Let the total Hamiltonian of the system consisting of vibrons (normal coordinate q) and phonons (Q) be written as

$$H = T_Q + T_q + V(q, Q), \tag{1.1}$$

where the T are kinetic energy operators and V the interaction potential. Then the solution of the Schrödinger equation in the Born–Oppenheimer approximation is

$$\psi = \phi(q, Q) \ominus (Q). \tag{1.2}$$

The adiabatic separability into vibron and phonon motions is determined by the magnitude $[\omega(Q)/\omega(q)]^{1/2}$. The situation here is similar to that of separating the motion of molecular systems into nuclear and electronic motion.

The equilibrium position of the atoms or molecules form a set of points called a lattice with clearly defined symmetry properties. The solution of the dynamical matrix can be classified according to the irreducible representations of the space group.

If the vibrations of the particles have only small amplitudes, then they can be calculated quite accurately. The potential energy of the crystal can be expanded in a Taylor series in the displacement coordinates, and for small amplitudes be truncated after the quadratic term. This is the so-called harmonic approximation, a standard method for making lattice dynamics calculations. The terms that are neglected in the harmonic approximation, the so-called anharmonic terms, are responsible for a large number of principal properties such as thermal expansions, conductivity of heat, phonon–phonon interaction, etc. The quasi-harmonic approximation acts as a link between harmonic and anharmonic methods by assuming a pure volume dependence in the phonon modes.

As molecules are not spherically symmetric, they can perform two types of motions and, in general, they are trapped in relatively deep potential wells. They can perform small translational oscillations around well-defined equilibrium positions in which the orientation does not change and small orientational (rotational) oscillations – so-called librations – around well-defined equilibrium orientations in which the position of the center of gravity does not change. The translational vibrations have been studied for many years and are well understood. In most cases, the harmonic approximation gives a good description of these vibrations or at least a good starting point for including anharmonic corrections by perturbation theory. Librations are still poorly understood, the main reason being that for librations the harmonic approximation and its extensions yield bad results. The librations often have a large amplitude (e.g. ~ 10–$15°$ in solid α-N_2) and many terms in the Taylor series of the potential energy should be included in the calculations.

At least two complications are known:

(1) It is not unusual for there to be no preferred orientation for a molecule and for the molecules to rotate in the pure crystal. Such a crystal is called plastic or orientally disordered. In that case, the harmonic approximation cannot be used at all. An N_2 molecule matrix-isolated in a rare-gas crystal is an example of this situation. It is, therefore, unrealistic to describe this system as a regular harmonic molecular crystal and to assume the defect to be in a substitutional site, and not rotating. Neither the one nor the other requirement is fulfilled.

(2) A strict separation of center-of-mass motion from librational motion is also unjustified. Most molecular crystals are rather closely packed, which forces molecules to separate as they change their orientations. As a result, there will be considerable coupling between the translational vibrations of the molecules and their rotational motions; causing translational phonons and librons to mix. In the matrix-isolated case, a further complication of this situation is present since in addition to host–host interactions one has host–guest interactions. Moreover, the center of mass of the guest may not occupy a lattice site. Consequently, all types of coupling must be considered: vibrational–translational, librational–translational, vibrational–librational–translational.

The harmonic approximation for translations consists of expanding the potential up to second order in the atomic or molecular displacement around some local minimum and the diagonalizing of the quadratic Hamiltonian. In the case of a pure intramolecular potential the appropriate displacement is, of course, the intramolecular coordinate – i.e. the change in bond length – whereas for a pure intermolecular potential the appropriate displacement is the change in nearest-neighbour distance, and in the case of an interaction potential $V(q, Q)$ between the vibron and phonon system a suitable physical parameter must be a displacement in the combined motion, e.g. vibrational–translational motion. For librations the rotational part of the kinetic energy, expressed in Euler angles, must be approximated too. The expansion of the *specific* potential V (either intermolecular, or intramolecular) in a Taylor series in the appropriate displacement coordinates x (x for translation, orientations, combined motion) possess the well-known form

$$V(x) = V_0 + \sum_i \left(\frac{\partial V}{\partial x_i} \right)_0 x_i + \frac{1}{2!} \sum_i \sum_j \left(\frac{\partial^2 V}{\partial x_i \, \partial x_j} \right)_0 x_i x_j$$

$$+ \frac{1}{3!} \sum_i \sum_j \sum_k \left(\frac{\partial^3 V}{\partial x_i \, \partial x_j \, \partial x_k} \right)_0 x_i x_j x_k + \text{higher-order terms.} \quad (1.3)$$

The index 0 indicates that the quantity must be evaluated at the equilibrium position, the summation and indices refer to components of vectors pointing from a crystal-fixed origin to the instantaneous position of an atom belonging to the molecule in the unit cell with the periodicity of the crystal lattice. The first term V_0 is the potential energy of the static crystal and affects the zero of energy only. Likewise, for a crystal at equilibrium the second term equals zero. Thus, in the harmonic approximation the series reduces to the third term only. The remainder of the expansion forms the anharmonic contributions.

As a consequence of this method, the complicated motion of a molecular crystal is separated into a set of independent particles whose motions, when quantized, behave as quasi-particles such as vibrons, librons or phonons.

If light interacts with matter, the dipole moment d is changed due to the electric field vector E of the electromagnetic wave

$$d = \mu + \alpha E + \beta E^2 + \text{higher-order terms,} \quad (1.4)$$

where μ is the static dipole moment, α the polarizability and β the hyper-polarizability. Each of these quantities is expandable in a Taylor series in the appropriate displacement x (q normal coordinate of internal and Q of external motion) for small motion about the equilibrium positions of the particles in the crystal:

$$\mu = \mu_0 + \sum_i \left(\frac{\partial \mu}{\partial x_i} \right)_0 x_i + \frac{1}{2!} \sum_i \sum_j \left(\frac{\partial^2 \mu}{\partial x_i \, \partial x_j} \right)_0 x_i x_j + \cdots, \quad (1.5)$$

$$\alpha = \alpha_0 + \sum_i \left(\frac{\partial \alpha}{\partial x_i} \right)_0 x_i + \frac{1}{2!} \sum_i \sum_j \left(\frac{\partial^2 \alpha}{\partial x_i \, \partial x_j} \right)_0 x_i x_j + \cdots. \quad (1.6)$$

The second term on the right-hand side of eq. (1.5) is responsible for an ordinary IR absorption and the one of eq. (1.6) for Raman scattering between two levels $|i\rangle$ and $|f\rangle$, whereas the third terms contribute to anharmonic phenomena because of mode coupling. In the simplest case, overtone excitation by light absorption is describable by $\partial^2 \mu / \partial q^2$.

The intensity I is proportional to

$$I \sim |\langle f|\mu|i\rangle|^2 \quad \text{for IR absorption,}$$

$$\sim |\langle f|\alpha|i\rangle|^2 \quad \text{for Raman scattering,}$$

$$\sim |\langle f|\beta|i\rangle|^2 \quad \text{for hyper Raman scattering.}$$

In the literature it is common to refer to the higher-order terms in the potential expansion as mechanical anharmonicity and to the higher-order terms in the dipole moment as electrical anharmonicity.

The inclusion of anharmonic terms has the consequence that normal modes are no longer independent, and phonon–phonon interaction may occur. Two-phonon bands can be classified as

(1) vibron/vibron combination band due to $\partial^2 A / \partial q \, \partial q$;
(2) vibron/phonon combination band due to $\partial^2 A / \partial Q \, \partial q$;
(3) phonon/phonon combination band due to $\partial^2 A / \partial Q \, \partial Q$;

where A stands for either potential or dipole moment.

As an example, the well-known Fermi resonance in solid CO_2 is due to the coupling of the normal modes v_1 and $2v_2$ via mechanical anharmonicity (Bogani and Salvi 1984); the vibron–phonon sideband intensity in α-N_2 by Raman scattering is explainable only via electric anharmonicity (Zumofen and Dressler 1976). The coupling constant between different modes is now produced by the second or third derivative of the appropriate physical parameter A (i.e. force constant, oscillator strength or change in polarizability) and is directly related to the measurable combination band intensity,

$$I_{i \to f} \sim \left[\frac{\partial^2 A}{\partial x(k) \, \partial x(k')} \right] |\langle f|x(k) \, x(k')|i\rangle|^2 \, [g(w(k) + w(k'))] \quad \delta(k + k').$$

| anharmonicity | symmetry | two-phonon density of states | momentum conservation |

(1.7)

Spectroscopically, the vibron–vibron bands appear as sharp lines because dispersion is negligible due to the fact that the intramolecular bonds are affected only slightly by intermolecular forces. A vibron–phonon transition consists of a broad band on both sides of the zero-phonon line (sum and difference bands) which are called phonon sidebands and are attributable to the one-phonon density of states, weighted to reflect the fact that different phonon modes can couple to varying degrees with the vibrons. In the third case – phonon/phonon combination bands – the whole two-phonon density of states has to be considered, and appears in the spectra as a complicated structured broad band.

In most cases the crystal potential is not known a priori. In the case of internal modes, it is straightforward to use the potential of the free molecule modified by minor corrections, such as static relaxation or changes in potential depth. For the external modes, the usual procedure is to introduce some model potential containing several parameters which are subsequently found by fitting the calculated crystal properties to the observed data available. This procedure has the drawback that the empirical potential thus obtained includes the effects of the approximations made in the lattice dynamics model, which is, in most cases, the harmonic model. It is useful to have independent and detailed information about the potential from quantum mechanical ab initio calculations.

A further approximation which is applied in almost all practical treatments is to write the intermolecular potential of the crystal as a sum over molecular pair potentials. For Van der Waals solids – like our matrix materials – this is fairly well justified: crystal symmetry in the first, second and other shells of nearest-neighbours is appropriately considered in summing over the two-body inter-actions over all molecules in the crystal. Three-body and higher contributions, which are smaller than about 10% of the pair potential in molecular crystals consisting of molecules with small dipole moments, are very rarely taken into account via perturbation techniques (Bobetic and Barker 1970, Bobetic et al. 1972).

The harmonic approximation is undoubtedly of extreme importance in the interpretation of the main features of the infrared, Raman and neutron scattering spectra of molecular crystals at low temperatures when displacement from equilibrium is small. This approximation is, however, insufficient to account for many physical properties of these crystals, especially for those which depend on temperature. The inclusion of terms of the crystal potential higher than the quadratic produces phonon–phonon coupling and results in:

- Phonon frequency shifts with respect to the harmonic values;
- An increase in line width that can be related to the phonon lifetime;
- The temperature and volume dependence of phonon frequencies and line widths;
- The appearance of multi-phonon bands due to the simultaneous excitation of two or more interacting phonons.

When the anharmonicity is small, standard perturbation theory can be used to work out the anharmonic corrections. A convenient parameter for classifying the order of the perturbation term in these treatments is the ratio ε between the mean square amplitude of vibration and the average next-neighbour distance. We can write the crystal Hamiltonian in orders of magnitude

$$H = H_0 + H_1 + H_2 + H_3 + \cdots, \tag{1.8}$$

where H_0 is the Hamiltonian for the harmonic problem and H_1 – the cubic term

– is of the order of εH_0, and H_2 – the quartic term – is proportional to $\varepsilon^2 H_0$. Usually, $\varepsilon \ll 1$ and thus the higher terms are small with respect to H_0 and can be treated as perturbations to the harmonic Hamiltonian.

When anharmonicity is large, other methods have been described in the literature to treat anharmonicity: renormalization of the crystal Hamiltonian (Wallace 1972); the temperature Green function (Wallace 1972), and the self-consistent phonon method (Werthamer 1970).

If a phonon is absorbed or scattered, two vibrational quanta may be excited on different molecules. These processes differ from combination and overtone transitions in a single molecule mainly because they are affected by the intermolecular interaction. Contrariwise, the study of two-phonon spectra through, e.g. bandshapes, absolute intensities in IR or Raman scattering, provides information concerning phonon density of states and intermolecular forces.

If one of the two interacting modes is an internal one, having a very small dispersion in comparison to external ones, and the other mode is an external vibration then the one-phonon density of states is the appropriate quantity to be compared with the observed shape of the phonon sideband.

Some concluding remarks regarding the use of *lattice dynamics*. Lattice dynamics can be applied only if a well-defined equilibrium configuration of the crystal exists (plastic crystals are therefore excluded) and if the full translation symmetry of the lattice can be utilized (the disordered matrix-isolated case is also excluded). Only if these conditions are satisfied, the crystal normal coordinates can be defined – assuming cyclic boundary conditions – and dynamical matrix and eigen-value equation be solved. In spite of complex and sophisticated intermolecular potentials, this method is rather simple and actual calculations require little computer time. The most important limitation of lattice dynamics is that, in the harmonic approximation, it does not account for temperature effects. Actually, temperature dependent anharmonic corrections can be included by means of perturbative methods as already outlined. Such calculations are, however, very complicated and time-consuming and represent the "frontier" of lattice dynamics.

Very recently a complementary method, *molecular dynamics*, was developed and applied to simple molecular crystals. Newton's equations of motion for a system of N molecules in a volume V are solved by numerical integration. Methods have been developed to calculate, e.g., the structure of the unit cell at finite temperature and its change with temperature, the density operator and dynamical structure factor, from the velocity–time correlation function the phonon density of states, etc. It must be distinctly pointed out that the anharmonic contributions to the structural and dynamical properties of the crystal are implicitly accounted for in this type of simulation. While for ordered phases, both lattice dynamics and molecular dynamics can be utilized, molecular dynamics is the only choice available for disordered systems. Through access to larger and ultrafast computers one is now able to perform calculations with approximately 5000 molecules.

Righini (1985) compared both simulation methods on solid pure CH_4, CS_2, CO_2 and N_2, whereas Cohen and Klein (1984) applied molecular dynamics to several matrix-isolated systems (rare-gas atoms and small molecules in rare-gas solids).

Let us now consider qualitatively some spectral features which may be attributed to solid-state phenomena as well as some recent results. Figure 1 shows a schematic comparison between a gas-phase and a crystal spectrum. The former contains internal modes and their overtones, the rotational fine structure has been omitted for simplicity; the crystal spectrum shows additional new bands in the low-frequency region – external modes – that are clearly separated from internal modes. Comparing one line of the gas spectrum to one of the crystal spectrum, one sees, in general, a frequency shift, extra splitting, line broading, modified lineshape and additional broad weak bands that result from a variety of solid-state effects. These will be discussed below.

(1) Frequency shift: By convention the frequency difference $\Delta\omega_s = (\omega_{solid} - \omega_{gas})$ defines the crystal shift and $\Delta\omega_M = (\omega_{Matrix} - \omega_{gas})$ the matrix shift which is a measure of the raising or lowering of the energy levels of the molecule due to environment. Typically, the relative frequency shift $\Delta\omega/\omega$ is less than or equal to 1%. Obviously, one is interested in the way that this quantity changes as a function of matrix material, temperature and pressure. Theoretical models estimate a frequency shift between harmonic and anharmonic contributions. For instance: internal modes are red-shifted by $\lesssim 1\%$, for N_2, $\omega_e = 2359$ cm^{-1} and $\omega_e x_e = 14$ cm^{-1}, whereas theoretical models for external modes predict an anharmonic shift of about $\pm 10\%$ uncertain in size and sign for α-N_2, $\omega(E_g) = 32$ cm^{-1} and $\Delta\omega = +3$ cm^{-1}. The difference in relative size of this shift

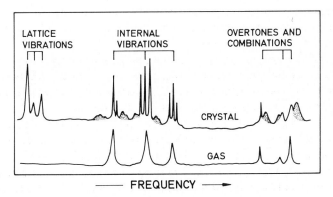

Fig. 1. Schematic drawing of vibrational (Raman-)spectrum of a molecular crystal compared to the spectrum of the gas phase (shaded area indicate sidebands, see text).

(1% and 10%) is understandable for molecular crystals; as outlined before, external modes are generated by the interaction of particles and anharmonicity comes along with this intermolecular interaction, whereas the internal modes are described in the free molecule by the intramolecular potential.

Doping by isotopes also produces a measurable frequency shift of several cm^{-1}; one can estimate how the free-molecule data must be modified when the molecules are matrix-isolated or embedded in a solid via the well-known normal coordinate analysis method. Crude models to calculate the frequency shift $\Delta\omega_M$ or $\Delta\omega_s$ (see section 2) are based on inter- and intramolecular potentials which

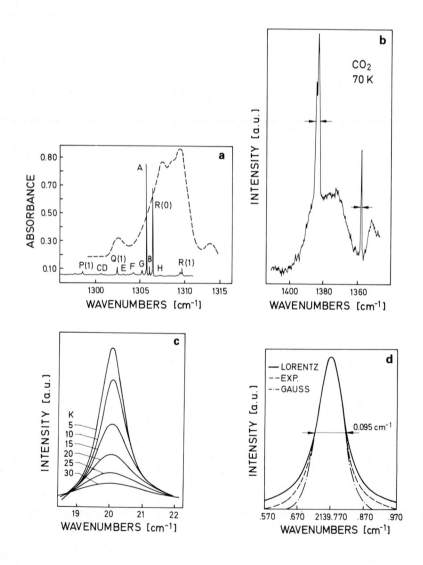

contain distances between molecules or bond length between atoms. To adjust the theoretical data to experimental data, one allows local distortions or static relaxation of the defect environment. Due to the high power of distance in the potential, this is a very sensitive parameter.

(2) Splitting. If the symmetry of the molecule and the crystal structure is known, then one is able to determine the symmetry of an optical transition using standard methods, such as correlation method or factor group analysis combining the molecule with the site, and with the crystal symmetry (Fateley et al. 1972). For example: two neighbouring N_2 molecules can vibrate in-phase and out-of-phase resulting in an A_g–T_g crystal-field splitting of the fundamental. Another example is the site splitting of the v_4-mode of matrix-isolated CH_4 (fig. 2a). Crystal-field splitting and site splitting are typically a few cm^{-1}, and the intensity ratio corresponds to the number of defects on one site or the other.

The group theoretical analysis of frequency splitting is, in principle, a very powerful technique; unfortunately, these methods have only been applied to pure and not to doped crystals. Miller and Decius (1973) showed in a very general manner the applicability to the matrix-isolated case, provided that the

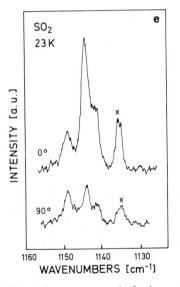
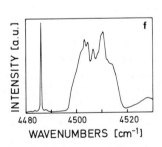

Fig. 2. Solid-state features next to the fundamental mode. (a) Site splitting of the v_4 absorption of CH_4 in Ar measured by conventional IR (dashed line; Cabana et al. 1963) and by modern FTIR (solid line; Jones et al. 1986). (b) Different linewidth of v^+ absorption of CO_2 and of matrix-isolated isotope. (c) Calculated bandshape of the $R(0)$ line of HCl in Ar at various temperatures (Allavena et al. 1982). (d) Different lineshapes of the fundamental IR absorption of CO in N_2 (Dubost et al. 1982). (e) Polarised Raman intensity of the v_1–SO_2 mode; × denotes the plasma line (Conrad 1983). (f) IR combination band and broad phonon sideband of solid para-H_2 (Van Kranendonk and Karl 1968).

specific site of the impurity molecule is known. For example, rare gases condense in space group O_h^5 containing the following possible site groups: $2O_h(1)$; $T_d(2)$; $D_{2h}(6)$; $C_{4v}(6)$; $C_{3v}(8)$; $3C_{2v}(12)$; $2C_s(24)$; $C_1(48)$. In most of the literature, isolation is assumed to occur in substitutional sites. This is not necessarily so. Second, the number of molecules per unit cell, obtained by structure analysis, is modified by the impurity/matrix doping ratio. A third problem exists if the impurity molecule rotates; then the site group cannot be correlated to the molecular point group. The shift found by assuming a certain position and orientation of the impurity molecule on a substitutional site and expanding the interaction potential is a very very special case specific to what is measured spectroscopically. Another way out might be to consider the matrix-isolated case – e.g. 100 Ar atoms and 1 N_2 molecule – as a supermolecule to describe the full symmetry of this unit and to produce the real crystal by translation operation.

A complication arises when within the same frequency interval, similar lines next to the fundamental mode occur which can be definitely assigned to aggregates. These can be distinguished from solid-state effects by their concentration dependence. The different components of crystal-field split lines can be investigated by polarized light. To study site splitting one needs single crystals and intentional site-selective changes by temperature or light.

(3) Line broadening. This phenomenon has several origins. The first contribution to the inhomogeneous linewidth Γ is due to crystal quality; amorphous and polycrystal state as well as head-to-tail arrangement, (CO–OC, for example). This static effect is caused by summing over several frequency-shifted components with different statistical weight: such as one N_2 molecule surrounded by twelve nearest neighbours, one with 11 and so on. Kikas and Rätsep (1982) developed a statistical theory to explain this line broadening by an elastic interaction of impurities with point defects.

A second contribution to line broadening is related to the fact that a sharp energy level of a free molecule becomes always broad in a solid consisting of N molecules. Figure 2b shows the crystal-field splitting of v^+ IR absorption of solid CO_2 and, shifted to lower frequencies, the equivalent fundamental of the matrix-isolated isotope $^{12}C^{18}O^{16}O$. Dubost et al. (1982) analyzed very carefully the increase in linewidth with increasing concentration of CO in solid N_2. Ortho/para hydrogen would be another good matrix-isolated system to study line broadening, due to the transition from the matrix-isolated case, to clusters, alloys, or mixtures.

A third contribution is the line broadening due to an increase in temperature that has its anharmonic origin either in phonon–phonon interaction and/or in a volume dependence; i.e. the lattice constant increases with temperature. This strong effect is well-known either phenomenologically or quantitatively; for example, Allavena et al. (1982) modelled the bandshape of the rotation–vibrational line $R(0)$ of HCl in Ar at various temperatures (fig. 2c).

A fourth contribution to line broadening is related to relaxation processes such as dephasing or energy relaxation from internal modes to the phonon bath. This complex is extensively studied in the chapter by Dubost and Legay (ch. 11).

(4) *Lineshape:* The detailed form of a line, e.g. the bandcontours shown in fig. 2d (Dubost et al. 1982), or its asymmetry (Weinert et al. 1982) can be studied by the well-known moment analysis. Henry and Slichter (1968) developed and applied this analysis to the lineshape of F-center absorption in alkali halides. This lineshape analysis of a molecule trapped in a matrix at low temperatures can yield valuable information about energy dissipation processes as a consequence of the molecule–lattice-vibration interaction.

In addition, polarization experiments provide information about the orientation of the matrix-isolated impurity and about the symmetry of tentatively assigned modes. Figure 2e shows polarized Raman scattering on solid SO_2 by which we solved the assignment puzzle known in literature: the high-energy triplet consists of the crystal-field split v_1-SO_2 and an $(SO_2)_x$ aggregate component. We measured the depolarization ratio ρ of the fundamental vibration of almost all small molecules and found that for the gas phase $\rho < 0.01$, for the liquid $\rho \sim 0.2$, for the matrix-isolated case $\rho \sim 0.3$ and for the solid $\rho \lesssim 0.1$ (Bier 1986). The procedure to determine the Raman intensity components $I(\text{parallel})$ and $I(\text{vertical})$, is well described by Bogani and Salvi (1984) who applied it to solid CO_2: First, they measured the polarizability tensor components of an isolated molecule, then on a site in the unit cell, and afterwards embedded in the crystal.

(5) Sidebands: As outlined, phonon side bands occur due to different mode coupling near the fundamental zero-phonon line. Figure 2f shows a sharp zero-phonon combination, rotation–vibration $S_1(0)$ in the IR spectrum of solid para-H_2, and a broad feature at higher energy: according to Van Kranendonk and Karl (1968), this sideband is due to a vibration $Q_1(0)$ in one molecule and a rotation $S_0(0)$ in another one. The vibration–rotation interaction originates in the quadrupole–quadrupole interaction of both molecules.

If the temperature of the crystal is high enough sum- (Stokes) and difference bands (anti-Stokes) may occur. When the crystal symmetry allows no infrared or Raman active modes [i.e. the symmetry properties of the dipole moment or polarizability are such that the first- or second-order terms are zero in eqs. (1.5) and (1.6)], then the presence of defects or of a perturbation of translation symmetry may alter the symmetry classification causing spectra to be observable: for example, the fundamental modes of N_2 and O_2 are not IR active, but are measurable in sprayed-on films by modern FTIR (Jodl et al. 1987).

In the case of mixtures one can distinguish two cases:

(1) when the external modes of species A and species B exist simultaneously in the spectra but differ in their relative intensities with varying concentration, this is the so-called two-mode behaviour;

(2) when modes of type A change gradually towards the ones of type B as the fraction of B increase then we call this one-mode behaviour (for further details see section 2).

For a matrix-isolated atom or molecule, the phonon band of the matrix material contains additional features: if a defect with lighter mass (or heavier mass) is embedded, then a local mode (or in-band mode) will be observable outside the phonon density of states. Cohen and Klein (1984) have considered these effects theoretically as a function both of mass exchange and change in force constants and compared their model with the few known experimental data.

The reader should now be familiar with the necessary definitions, concepts, spectroscopic terms and recent examples. For more details the original literature should be consulted: Horton and Maradudin (1974), Wallace (1972), Fowler (1968), Califano et al. (1981), Barker and Sievers (1975), Briels et al. (1986), Klein and Venables (1976, 1977), Coufal et al. (1984) and Klein (1984).

1.3. Solid-state properties of matrix materials

In the three decades since the beginning of matrix isolation spectroscopy (MIS), various matrices have been successfully tested besides rare-gas solids (RGS); e.g. CO_2 is a useful high-temperature matrix ($T_{solid-liquid} \sim 220$ K), or CH_4, CCl_4, zeolite. For simplicity we will concentrate only on rare-gas solids to study specifically the solid-state effects in matrices, because these materials possess simple phases (fcc, hcp), no complicated external modes, etc.; when compared with solid CH_4 that has three different phases at $p = 0$, each with more than 10 Raman active lattice modes. Beyond the rare-gas solids only anisotropic media such as N_2 and O_2 will be considered, and even this extension is complicated enough. The discussion will focus on how the matrix influences the defect *and* how the defect influences the matrix, comparing isotropic and anisotropic guest/host particles.

The solid-state properties of these matrix materials are well known. The reader is referred to the following references for more details:

(1) Rare-gas solids; Klein and Venables (1976, 1977), Coufal et al. (1984), Klein (1984).

(2) Solid and liquid nitrogen; Scott (1976) and Jansen (1987).

(3) Solid oxygen; Krupskii et al. (1979), DeFotis (1981) and Meier (1984).

Relevant and interesting data are summarized for these materials in the following tables; they are organized as molecular properties including intra-molecular potentials (see table 3) and as solid-state properties (see table 4). The data are taken from the above mentioned references and, to a minor extent, from original publications. The following tables (table 4) contain structural and thermodynamic data, optical and lattice dynamical data and intermolecular potentials.

Table 2a
Correlation diagram for N_2 at the center of the Brillouin zone.

Mode	Molecular symmetry	Site symmetry	Factor group symmetry	Activity
	$D_{\infty h}$	S_6	T_h	

α-N_2

Mode	Molecular symmetry	Site symmetry	Factor group symmetry	Activity
ν_0	Σ_g^+	A_g	A_g	Raman
			E_g	Raman
R_x, R_y	Π_g	E_g	T_g	Raman
T_z	Σ_u^+	A_u	A_u	–
			E_u	–
T_x, T_y	Π_u	E_u	T_u	Infrared

	$D_{\infty h}$	D_{3h}	D_{6h}	

β-N_2

Mode	Molecular symmetry	Site symmetry	Factor group symmetry	Activity
ν_0	Σ_g^+	A_1'	A_{1g}	Raman
			B_{1u}	–
R_x, R_y	Π_g	E''	E_{1g}	Raman
			E_{2u}	Acoustic
T_z	Σ_u^+	A_2''	B_{2g}	–
			A_{2u}	Acoustic
T_x, T_y	Π_u	E'	E_{2g}	Raman
			E_{1u}	–

Table 2b
Correlation diagram for O_2 at the center of the Brillouin zone.

Mode	Molecular symmetry	Site symmetry	Factor group symmetry	Activity
	$D_{\infty h}$	C_{2h}	C_{2h}	

α-O_2

Mode	Molecular symmetry	Site symmetry	Factor group symmetry	Activity
ν_0	Σ_g^+	A_g	A_g	Raman
R_x, R_y	Π_g	B_g	B_g	Raman
T_z	Σ_u^+	B_u	B_u	Infrared
T_x, T_y	Π_u	A_u	A_u	Infrared

	$D_{\infty h}$	D_{3d}	D_{3d}	

β-O_2

Mode	Molecular symmetry	Site symmetry	Factor group symmetry	Activity
ν_0	Σ_g^+	A_{1g}	A_{1g}	Raman
R_x, R_y	Π_g	E_g	E_g	Raman
T_z	Σ_u^+	A_{2u}	A_{2u}	Infrared
T_x, T_y	Π_u	E_u	E_u	Infrared

Table 3
Molecular properties of matrix materials.

	Matrix material					
	Ne	Ar	Kr	Xe	N_2	O_2
Diameter (Å)	3.0	3.8	4.2	4.6	4.34	4.18
					3.39	3.18
Bondlength (Å)					1.09	1.21
Polarizability (Å^3)	0.39	1.63	2.46	4.02	1.76	1.60
Quadrupole moment ($\times 10^{26}$ esu cm^2)					-1.4	-0.4
Lennard-Jones parameter ε (cm^{-1})	24.33	83.98	118.18	155.03	66.3	81.8
σ (Å)	2.76	3.41	3.62	4.03	3.75	3.43
Ionization energy (eV)	21.56	15.8	14.0	12.1	15.6	12.2
Harmonic frequency (cm^{-1})					2359.61	1580.36
Anharmonicity (cm^{-1})					14.46	12.07
Rotational constant (cm^{-1})					2.01	1.45

On the basis of structural data (see fig. 4a) it is straight forward to determine the IR and Raman activities of the internal modes v_0, the rotational modes R_x, R_y, and the translational modes T_x, T_y, T_z (see table 2).

Single crystals of γ-O_2 have been established to have an eight-molecule cubic unit cell with orientationally ordered structure. The electron density distribution in the lattice suggests that the molecules in the crystal are found in two non-equivalent states: O_h^3 with $Z = 2$ molecules on site T_h and $Z = 6$ on site D_{2d}. But either the molecules on site T_h possess arbitrary orientation with respect to one C_3-symmetry axis or the molecules on site D_{2d} possess arbitrary orientation with respect to one C_2-symmetry axis; as a consequence we cannot apply group theory to determine spectroscopic activity.

Rare gases crystallize in general in fcc (O_h^5; $Z = 1$). Neither IR nor first-order Raman scattering are allowed by symmetry. (Second-order Raman scattering will be discussed later.) The hcp phase can be stabilized by small quantities of impurities: D_{6h}^4 with two molecules per primitive cell resulting in two allowed sites D_{3d} and D_{3h}. At the Γ-point of the hcp Brillouin zone the irreducible representation of the phonons is B_{1g}, E_{2g}, A_{2u}, E_{1u}. Only the doubly degenerate E_{2g} – quasi-optic TO phonon – is active in first-order Raman scattering. Using standard texts (Fately et al. 1972), the polarization tensor elements or dipole moment components for each point group for polarized spectroscopic measurements can be identified.

Figure 3 illustrates the crystal structure of solid N_2 and solid O_2 at zero pressure. Whereas the P–T diagram for O_2 is documented by Meier (1984), for N_2 by Scott (1976) and the P–V diagram for rare gases by Bobetic and Barker (1983), for N_2 by Chandrasekharan et al. (1983) and for O_2 by Etters et al.

Table 4a
Solid-state properties of matrix materials (structural and thermodynamic data).

	Matrix material					
	Ne	Ar	Kr	Xe	N_2	O_2
Space group (Z molecules per unit cell)	fcc $O_h^5(1)$ hcp $D_{6h}^4(2)$	fcc $O_h^5(1)$ hcp $D_{6h}^4(2)$	fcc $O_h^5(1)$ hcp $D_{6h}^4(2)$	fcc $O_h^5(1)$ hcp $D_{6h}^4(2)$	α-N_2 $T_h^6(4)$ β-N_2 $D_{6h}^4(2)$ γ-N_2 $D_{4h}^{14}(2)$	α-O_2 $C_{2h}^3(1)$ β-O_2 $D_{3d}^5(1)$ γ-O_2 $O_h^3(8)$
Phase transitions $T_{\alpha-\beta}$ (K)					35.61	23.8
$T_{\beta-\gamma}$ (K)						43.8
Lattice constant (Å) at $T = 15$ K	4.47	5.31	5.65	6.13	5.66	$a = 5.403$, $b = 3.429$, $c = 5.086$
Substitutional hole (Å)	3.16	3.75	3.99	4.34	3.99	3.64
Octahedral hole (Å)	1.31	1.56	1.65	1.80		
Tetrahedral hole (Å)	0.71	0.85	0.90	0.97		
Binding energy (eV/atom)	0.02	0.08	0.116	0.17		
Enthalpies of monovacancy formation (cal/mol)	478	1750	1780	2480		
Activation energy of self-diffusion (cal/mol)	947 (20 K)	4150 (80 K)	4800 (100 K)	7400 (150 K)		
Critical point T_c (K)	44.4	150.7	209.5	289.7	126.20	154.36
P_c (bar)	26.5	48.6	55.2	58.4	34	50.7
Triple point T_t (K)	24.6	83.8	115.8	161.4	63.15	54.4
P_t (bar)	0.43	0.69	0.73	0.81	0.13	0.002
Sublimation temperature (K) at $P \sim 10^{-6}$ Torr	9	31	42	58		

(1985). We will need these data when discussing anharmonicity (section 3) and high-pressure results (section 4).

In most matrix-isolation experiments the matrix material is commonly condensed directly from the gas phase. Figure 4 shows a section of the pressure–temperature diagram especially in the low-pressure, low-temperature region. As already pointed out samples with good physical quality – optically transparent, and small inhomogeneous linewidth – should, however, be grown from the liquid phase at elevated pressure; for RGS, N_2 and O_2 the gas loading system should be outfitted for $p < 50$ bar (see fig. 4). Bier et al. (1987) applied this technique successfully to matrix-isolated hydrogen.

On warming a crystal, the lattice expands and gives rise to a frequency shift to smaller energies for a given transition. This is the traditional explanation of the temperature shift. For comparison the zero-pressure dilation and the lattice constant as a function of temperature are shown in fig. 5. The fundamental

Table 4b
Solid-state properties of matrix material (optical data).

	Matrix material					
	Ne	Ar	Kr	Xe	N_2	O_2
Gap energy (eV)	21.69	14.15	11.60	9.28		
Ionization energy of solid (eV)					14.0	10.5
Refractive index	1.28	1.29	1.28	1.49	1.22	1.25
(at $T = 60$ K and $\lambda = 4880$ Å)						

Lattice vibrational frequencies (cm^{-1}) for N_2 at the center of the Brillouin zone.

Mode	Exp.[a]	Pure LJ pot.[b]	atom–atom EQQ[b]	Lattice dynamics[c]	Molecular dynamics[c]
α-N_2					
E_g	32.3	34.5	34.5	38	38
T_g	36.3	38.5	41.0	46	48
T_g	59.7	47.5	57.5	69	70
A_u	46.8	43.7	46.7	46	
E_u	48.4	51.8	55.0	49	(51)
T_u	54.0	47.0	51.1	45	
T_u	69.4	69.5	75.6	66	

	Exp.[d]	Time-dependent[e] Hartree
β-N_2		
E_{2g}	25	34–70
E_{1g}	50	

[a]Kjems and Dolling (1975).   [d]Medina and Daniels (1976).
[b]Scott (1976).   [e]van der Avoird et al. (1984).
[c]Cardini and O'Shea (1985).

Lattice vibrational frequencies (cm^{-1}) for O_2 at the center of the Brillouin zone.

Mode	Exp.[a]	Pure LJ pot.[b]	atom–atom EQQ[c]	Ab initio calc. includ. magn. interaction[d]
α-O_2				
B_g	43	42.7	40	39.9
A_g	74	43.6	46	72.2
β-O_2				
E_g	48	40.7		53.6

[a]Bier and Jodl (1984).   [c]Etters et al., (1983).
[b]Kobashi and Klein (1979).   [d]Jansen and van der Avoird (1987).

Table 4c
Solid-state properties of matrix materials (lattice dynamical data).

		Matrix material				
	Ne	Ar	Kr	Xe	α-N$_2$	α-O$_2$
Debye frequency (cm^{-1}) at $T=0$ K	75	93	72	64	69	104
Grüneisen parameter at $T=8$ K	2.9	2.6	2.8	2.8	1.7	5.1
Elastic constants (in kbar) C_{11}	16.49	42.4	51.4	52.7	28.0	
(at $T=10$ K) C_{12}	9.03	23.9	28.4	28.7	20.0	
C_{44}	9.28	22.5	26.8	29.5	13.5	
Compressibility at $T=$ const. (10^{-11} cm^2/dyne)	8.92	3.74	2.90	2.80	4.64	2.78
Thermal conductivity (Wm^{-1} K^{-1}) at 20 K	0.4	1.3	1.2	2	0.4	
Lennard-Jones parameter ε (cm^{-1})	29.3	99.46	138.8	195.8	55	161.2
(from solid-state data) σ (Å)	3.09	3.76	4.01	4.36	3.7	2.99

frequency is proportional to the second derivative of the molecular potential, which is inversely proportional to the lattice constant. Apart from this volume effect, i.e. $V = V(T)$, one has phonon–phonon interactions which are normally larger and physically more important in anharmonic systems (see sections 3 and 4).

For our purpose only some optical data (see table 4b) describing the matrix material are necessary. While the external modes of solid N$_2$ and O$_2$ are of interest; special emphasis will be placed on comparing recent experimental and theoretical values and documenting the large discrepancies that are encountered in spite of multi-parameter potential approaches or anharmonic calculations (see table 4b).

Some lattice dynamical values are gathered in table 4c. All calculations concerning the solid state of molecular crystals are based upon potentials often containing from 2 to 10 parameters, which are, of course, different from the ones of the free state, and which are expressed by different mathematical functions. These trial potentials are modelled in such a way that as many bulk properties as possible are reproduced as well as possible. For example, Aziz (1984) collected very carefully various potentials for pure and mixed rare gases. In addition, he reports the different combining rules, i.e. how the potential parameters for the mixed state AB are formed from the pure states A and B, and reviews the different interatomic potential functions, known and used in the literature, from a mathematical point of view.

The crucial test for all solid-state theories is the reproduction of details of the phonon dispersion curves and of the phonon density of states. Experimentally, they are found by inelastic neutron-scattering for the whole Brillouin zone and

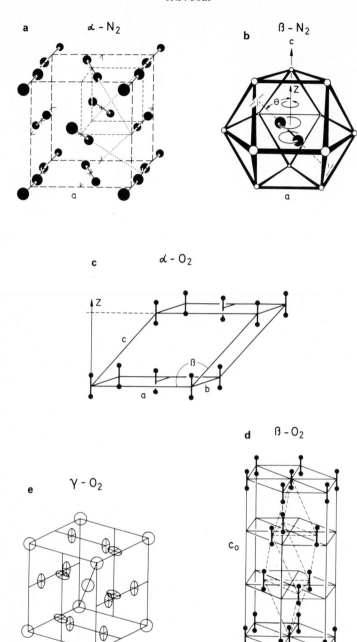

Fig. 3. Crystal structure of solid N_2 and O_2 at $p = 0$: α-N_2 and β-N_2 (Scott 1976); α-O_2, β-O_2 and γ-O_2 (DeFotis 1981). In β-N_2 and γ-O_2 the molecules are partially rotating and statistically disordered.

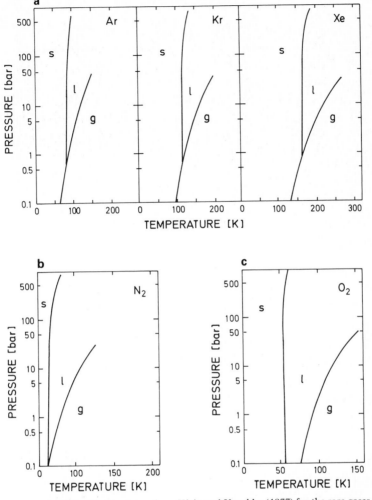

Fig. 4. Pressure–temperature diagram taken from Klein and Venables (1977) for the rare gases and from the gas encyclopaedia (1976) for nitrogen and oxygen.

by spectroscopic data at the zone center of the Brillouin zone. Theoretically they are calculated by harmonic lattice dynamics with anharmonic corrections and/or by classical molecular dynamics as outlined before. Figure 6 shows the phonon dispersion curves of argon [those of Ne, Kr, Xe were given by Klein and Venables (1976)] and solid nitrogen, whereas fig. 7 shows the respective one-phonon density of states. For oxygen either the dispersion curves or the density of states are estimated only by theory (Etters et al. 1985).

To conclude this section, two spectroscopic examples will be discussed:

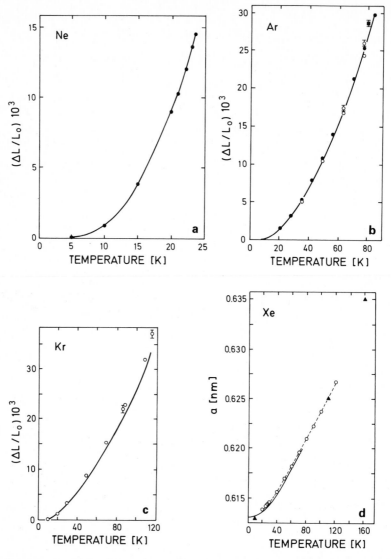

Fig. 5. Zero-pressure dilation for rare-gas solids by X-ray studies (line) and by theory (circles) according to Klein and Venables (1976).

(1) Pure rare-gas solids should possess only a Raman spectrum of second order due to their cubic symmetry. The spectral structure should be related to maxima in the phonon density of states. Comparing fig. 8 with fig. 7, the very weak feature in the Raman spectrum of Xe at 30 K extends up to about 90 cm^{-1} which is twice the frequency at the edge of the one-phonon density of states.

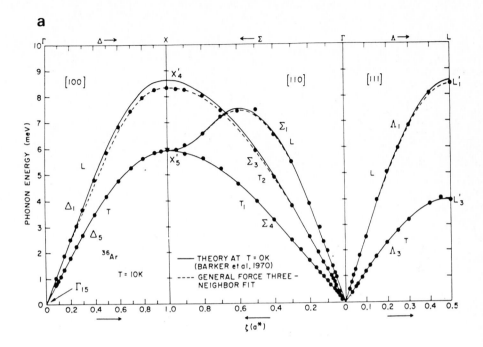

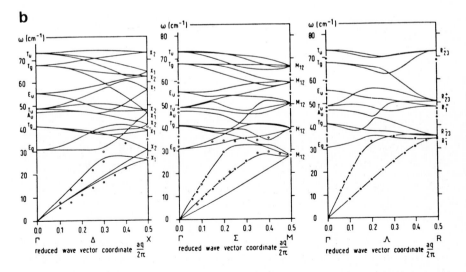

Fig. 6. Phonon dispersion curve for : (a) Ar at 10 K (Klein and Venables 1976), and (b) for N_2 at 15 K (Kjems and Dolling 1975, Briels et al. 1984). The dots are the experimental data, the solid lines are due to theory.

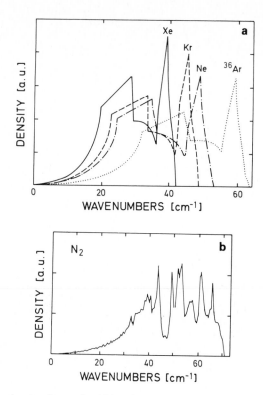

Fig. 7. One-phonon density of states for: (a) Ne, Ar, Kr and Xe (after Klein and Venables 1976), and (b) for N_2 (Cardini and O'Shea 1985).

(2) Nitrogen possesses a phase transition at $T_{\alpha-\beta} = 35.61$ K (see table 4a). According to the correlation diagram (table 2a), the fundamental mode ν_0 in α-N_2 should possess a crystal-field splitting into A_g/T_g in the Raman spectrum, whereas in β-N_2 only a single line A_{1g} should be seen. This is well documented by fig. 9a. Oxygen has a phase transition at $T_{\beta-\gamma} = 43.8$ K (see table 4a). Again, due to symmetry, the fundamental mode ν_0 should be one A_{1g} line in β-O_2 (table 2b) and should show a site splitting in γ-O_2. Figure 9b proves this expectation; even the intensity ratio of the doublet corresponds to the number of molecules on site T_h and D_{2d}: namely, 2:6.

When using O_2 and N_2 as matrix material, the existence of the splitting in Raman spectra is a proof of the good physical quality of the matrix, i.e. the matrix is not only perfectly optically transparent but possesses also long-range crystal symmetry. Moreover, both phase transitions, which are easily measureable spectroscopically, are excellently suited to calibrate thermocouples. In particular, this in situ temperature determination helps solving a puzzle which has been known since the early days of matrix-isolation technique; whether the

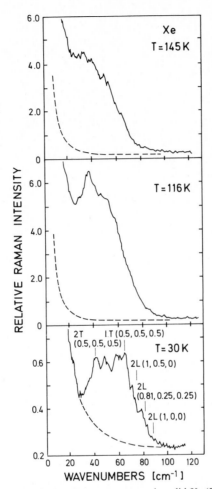

Fig. 8. Temperature-dependent Raman spectrum in solid Xe (Fleury et al. 1973).

measured temperature of the thin film is same as the local temperature – especially during laser irradiation.

These concluding remarks should form a transition to the second section where some critical remarks concerning a further classical problem in matrix-isolation spectroscopy namely the matrix shift will be considered.

2. Matrix-isolated case as the low-concentration limit of mixtures

Many concentration-dependent spectroscopic and non-spectroscopic measurements are known in the matrix-isolation literature. Preferentially in IR and

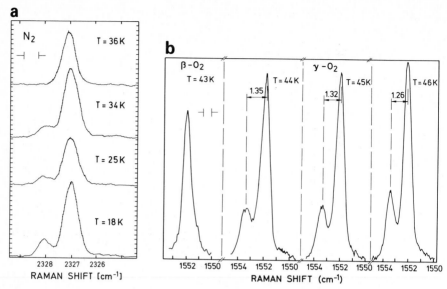

Fig. 9. Temperature-dependent Raman spectrum in solid N_2 (Bier 1986) and in solid O_2 (Bier and Jodl 1984).

FTIR the sensitivity is sufficient and one is able to vary the matrix/impurity ratio between 10^4 and 10^2. In Raman scattering, due to the very small cross-section, orders of magnitude smaller than in IR or optical studies, one must dope the matrix with several percent impurity concentration to gain enough signal. For this reason, we started to investigate mixtures of various concentrations which, in the limit of high dilution, become matrix-isolated systems. In general, these studies were aimed at determining the behaviour of the defect. The goal was to assign and separate monomer bands from polymer features in the spectra. Our focus here has shifted towards determining how the defect influences the solid-state behaviour of the matrix. Unfortunately, the structures of only very few mixed molecular systems containing atoms or small molecules have been carefully studied by X-rays or electron diffraction up to now.

2.1. Phase diagrams and known spectroscopic measurements

The article on mixed crystals by Kitaigorodsky (1984) gives an excellent introduction on the properties of mixtures, such as stability, solubility and phase diagrams of mixtures. Aziz (1984) compiled relevant parameters for interatomic potentials for pure and mixed rare gases.

Phase diagrams for matrix materials were studied by X-ray diffraction by Barrett and Meyer (1965); only the $Ar-N_2$ system is shown as an example in fig. 10. For many years it was unclear whether rare gases condense in fcc or hcp

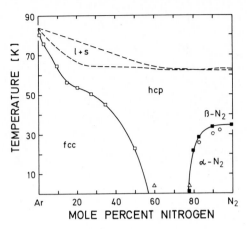

Fig. 10. Phase diagram of Ar–N$_2$ (Barrett and Meyer 1965); for completeness: Ar–O$_2$ (Barrett et al. 1966) and N$_2$–CO (Angwin and Wasserman 1966).

structures, whether the structure is dependent on the sample manufacturing conditions, or whether small amounts (ppm) of impurities stabilize hcp. Ahmad et al. (1981) reinvestigated the phase diagram of Ar–O$_2$ and Ar–N$_2$ on large single crystals, and found that a minimum solute concentration of about 5% was required to produce the hcp phase. Lower concentrations – the ordinary matrix-isolated case – resulted in fcc phase. The systems Ne–N$_2$, Ar–N$_2$, Xe–N$_2$ were also investigated by Curzon and Eastell (1971) by electron diffraction on thin films; these authors report the phase diagrams and the nearest neighbour distance as a function of concentration.

We use different techniques to produce mixed matrices. Each method has both advantages and disadvantages:

(1) thin-film condensation – very practical, but different melting points cause problems;

(2) crystal growing – time and gas consuming, concentration gradient but excellent single crystals;

(3) high-pressure technique (see section 4).

Only a few publications discuss the physical question and the quality of the results of the above-mentioned techniques. Girardet and Maillard (1982) studied HCl in RGS between 0.5 and 2%. The interaction between dopant monomers located in a wide range of positions are responsible for this absorption which appears as a distribution of Q branches corresponding to the distribution of dopants inside the crystal. The shape of the sum band and its evolution with concentration and temperature are thus explained in terms of an inhomogeneous broadening process (fig. 11a). Kiefte et al. (1982) reported highly resolved Raman spectra of N$_2$, O$_2$ and CO as dilute solutions in single

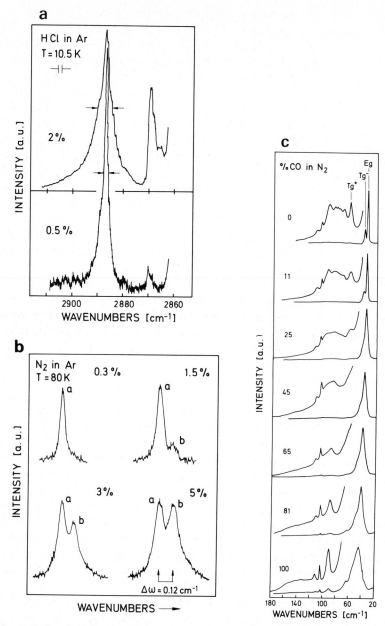

Fig. 11. Concentration-dependent measurements: (a) $R(0)$ and Q branches for HCl in Ar; note the statistical broadening on the high-frequency side at 2% (Girardet and Maillard 1982); (b) pure vibrational Raman spectra of N_2 in crystalline Ar for four concentrations; peak labelled a is due to monomer, peak b due to dimer band (Kiefte et al. 1982); (c) one- and two-phonon Raman spectra of N_2/CO mixed crystals (Witters and Cahill 1977).

crystals of Ar, in combination with X-ray structural analysis. Figure 11b shows the fundamental mode of isolated N_2 and shifted to higher frequencies, the same mode of an N_2-pair on substitutional site. The small frequency shift (0.12 cm^{-1}) of the dimer is surprising. Normally, the frequencies of aggregates are several cm^{-1} away from the fundamental of the single molecule. Witters and Cahill (1977) measured Raman spectra of N_2/CO and CO_2/N_2O mixed crystals in the lattice mode region: the respective molecules are isoelectronic and the structures are isomorphous. They found that the frequency either varies continuously with concentration between those of the pure components (one-mode behaviour) or that they form groups with each group being associated primarily with one of the components (two-mode behaviour). The spectrum is shown in fig. 11c, and has been analyzed by Molter (1984): The E_g and T_g^- mode show one-mode behaviour while the T_g^+ shows two-mode behaviour. We have reported concentration studies of CS_2 in N_2, Ar, Kr and Xe up to 20% by Raman and FTIR discussing the results in terms of sites, defect/defect interaction and matrix-material effects (Givan et al. 1986). To make reliable concentration-dependent statements on intensities, we suggested as an intensity standard the well-known Raman scattering cross-section of the N_2-fundamental mode.

Besides the delicacy with which results of measurements of intensities should be regarded (comparing one experiment with another) even the determination of, and conclusions drawn from, the so-called matrix shift are dubious as will be outlined in the next subsection.

2.2. Matrix shift

It is obvious that each energy level of a free molecule will be changed if the molecule is embedded in a lattice resulting in a matrix shift, $\Delta\omega_M$ (Jodl 1984 and other reviews quoted therein, or Knözinger and Schrems 1987). Theoretically, to calculate the change in energy levels and therefore $\Delta\omega_M$, one has only to modify the potential parameters: i.e. ε^{AB}, σ^{AB}, R^{AB} in an 6–12 Lennard-Jones pair potential on the basis of the pure parameters ε^A, ε^B, σ^A, σ^B, R^A and R^B.

As different transitions are different in energies, such as pure electronic, vibrational or rotational, only the relative matrix shift $\Delta\omega/\omega$ is of interest and is typically $< 1\%$. (The error in frequency determination is about 0.5 cm^{-1}.) Considering the series of matrix shifts for Ne, Ar, Kr, Xe and using experimental data it is too optimistic to consider this a crucial test of pure and combined potentials.

Interest in the matrix shift has been a long standing one: McCarty and Robinson (1959), Buckingham (1960), Friedmann and Kimel (1965), Barnes (1973) and Jodl et al. (1979). The idea is the following: the interaction between defect A and the matrix-particle B is reduced to two-body interactions and the influence by the nearest neighbour shells is represented by simple geometric factors. This interaction pair potential AB is separated into four contributions each with a different mathematical form:

(1) V_1^{AB} is the electrostatic interaction between the charge distribution of particle A and B.

(2) V_2^{AB} is the induction term; these forces are caused by the interaction

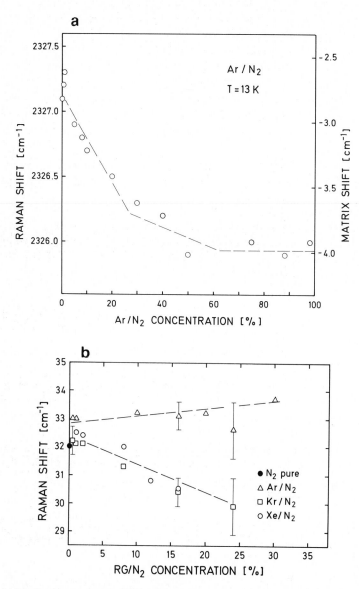

Fig. 12. (a) The fundamental mode of N_2 in Ar; (b) libron mode of N_2 in Ar, Kr, Xe at different host concentrations (Löwen et al. 1988a).

Table 5
Frequencies (cm^{-1}) of fundamental modes (± 0.5 cm^{-1}).

	Gas	Solid (~10 K)	Matrix					References
			Ar	Kr	Xe	N$_2$	O$_2$	
o-H$_2$	4155.3	4143.4	4136.9	4118.5	4110	4139		Bier et al. (1988)
p-H$_2$	4161.2	4151.8	4143.2	4124.4	4116	4147	4147	Bier et al. (1988)
N$_2$	2329.9	2327.1	2325.9	2324.4	2322.0			Löwen et al. (1988a)
O$_2$	1556.2	1552	1551			1554.5		Bier and Jodl (1984)
			(1536.6)	(1557.5)	(1587.1)	(1554.3)		Schoen and Broida (1960)
CO	2143.3	2138.1	2140.1	2138.2	2134.8	2139.8		Jodl (1984)
NO	1875.9		1873	1872	1867.4	1874.7		Bier (1986)
ν_1-CC	1388.2	1384	1385.0	1383.3	1380.2	1386.0		Jodl (1984)
ν_1-CS$_2$	658.0	653.8	657.7	654.5	653.9	656.9		Givan et al. (1986)
ν_1-H$_2$O	3651.7	3156	3640	3632	3623	3635		Däufer (1985)
ν_1-NO$_2$	1320.2	1320.9	1320.5	1321.6	1315.8			Bolduan et al. (1984)
ν_1-SO$_2$	1151.3	1143.8	1151.3	1150.0	1146.8			Conrad (1983)

between the permanent charge distribution of one molecule and the charge distribution induced in the other molecule.

(3) V_3^{AB} is the dispersive term, which is due to the long-range attraction between the instantaneous charge distribution of two molecules.

(4) V_4^{AB} is the repulsive term which represents the overlap of the electronic distribution resulting in a large short-range distortion.

Each of the four terms is described by parameters such as potential data, polarizability, bondlength, etc.; as a first step, the set of data for A and for B are the data from the free molecule or sometimes slightly modified; as a second step the parameters AB are combined in a specific way from the pure ones. Finally, the model required to estimate the *one* value for a matrix shift contains up to 10 parameters, and it is very sensitive to changes in those parameters which enter theory in higher orders.

Experimentalists have tried to present the series of shifts in Ne, Ar, Kr and Xe in a suitable manner, such as plotting the shifts versus matrix lattice constant d^{-n}, or versus polarizabilities of the different hosts, or versus combinations of the above-mentioned potential parameters. On the basis of four scattering values, one often found a linear relationship; the Ne-value was off the line which is understandable due to its greater quantum mechanical behaviour.

There is no principal difference between a guest molecule A embedded either in host A or in host B; therefore, the so-called solid shift $\Delta\omega_S = \omega_S - \omega_{gas}$ should be incorporated into the notion of a matrix shift. From this it is obvious that one has also to consider solid-state data: e.g., Zumofen and Dressler (1976) calculated the four terms of the total shift to be -2.9 cm^{-1} in comparison to the experimental value -3.0 cm^{-1} for N_2 in solid α-N_2. Table 5 lists the frequencies of fundamental modes of some small molecules.

Finally, the matrix shift as a function of concentration between the matrix-isolated case and the pure solid will be discussed: Maillard et al. (1977) studied HCl and HBr in nitrogen–rare-gas mixed crystals by IR measurements. The matrix shift of the HX fundamental as a function of N_2-concentration is non-linear; not unexpectedly as one has to consider a complicated three-component system ABC. We have concentrated on a very simple system, N_2 in RGS, and studied mixtures, clusters as well as the matrix-isolated case. The frequency shift of the fundamental mode and of the E_g libron of nitrogen as a function of Ar, Kr, Xe concentration is shown in fig. 12, which was discussed in detail by Löwen et al. (1988a). The saturation effect on the rare-gas rich side, the classical matrix-isolated case, is obvious in each matrix material; or in other words: the matrix shift is not sensitive to concentration up to a value of $\sim 20\%$, at least for this specific system. On the other hand, the measured matrix shift is sensitive to solid-state effects because the kink in the linearity (for example at 25 mol.% Ar in N_2, see fig. 12a) is clearly related to a phase transition: e.g. the α-N_2 to hcp-Ar transition at 78 mol.% N_2 in Ar (see fig. 10).

This concentration dependence of $\Delta\omega_S$ supplies more values than the $\Delta\omega_M$ of

3 to 4 rare gases. Therefore, we consider the reproduction of the form of the variation of the solid shift with RG impurity concentration as a test criterion for the quality of intermolecular potentials.

2.3. Destruction of local and global symmetry

Crystal-field splitting of the $v(N_2)$ in the Raman spectrum of solid α-N_2 is observed as a splitting of the pure vibrational band into two components (fig. 9a): qualitatively, adjacent molecules in the unit cell can vibrate in phase and out of phase.

In the present study, we have considered the question of the persistence of this band structure with increasing concentration of impurities in the solid or, in other words, how severely can the local symmetry of the oscillating molecules be disturbed before the crystal-field splitting is not observed any more. We refer to the two components of the $v(N_2)$ band by their A_g and T_g notations, although this is not strictly correct for crystal regimes where RG impurity atoms replace N_2 molecules. In fig. 13a we present scans of the $v(N_2)$ band in pure solid nitrogen and in mixed solids with 5, 10 and 30% Ar. The two components centered at 2327.1 and 2328.1 cm^{-1} are clearly resolved for the pure solid and are assigned to the A_g and T_g components, respectively. As expected, the splitting becomes less distinct with increasing argon concentrations with a shoulder still recognizable in the 10% Ar solid, and no resolved structure at higher argon concentrations. In Kr and Xe matrices the same result is obtained at similar RG concentrations.

The lack of structure in the band at higher concentrations is not due to inhomogeneous broadening. No significant increase in line width, compared to the value of 0.6 cm^{-1} for the pure solid, was observed and annealing by temperature cycling had no beneficial effect on band widths or structure. When not resolved, the lower intensity T_g component is also not covered by a higher background intensity or not observed because of poorer sample (and spectral) quality. The $v(^{14}N-^{15}N)$ isotopic band which has only 0.7% of the main band intensity could still be resolved even in the 30% Ar solid. Our results, therefore, indicate that at impurity concentrations of about 10%, the crystal structure is disrupted to an extent which inhibits the A_g-T_g splitting.

At concentrations of about 10% RG impurities, for a cubic crystal structure, approximately half of the N_2 molecules are in an environment which has one or more RG atoms as nearest neighbour. That this kind of statistical consideration (Behringer 1958, Harvey et al. 1964) is dominant as the cause of the described effect is also indicated by the observation that the concentration at which the splitting collapses is, essentially, independent of the identity of the RG impurity atom.

The two rotational degrees of freedom in the free molecule are transformed under the site- and crystal symmetry into eight librational degrees of freedom:

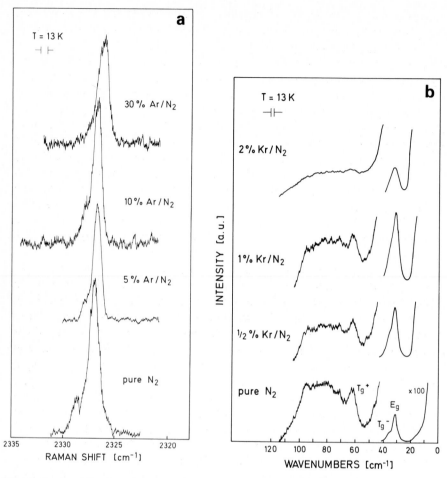

Fig. 13. Concentration-dependent Raman studies of: (a) the N_2 fundamental mode, and (b) the two-phonon sideband structure of N_2 (Löwen et al. 1988a).

E_g at 32 cm^{-1} and two T_g modes at 36.5 and 60 cm^{-1}. Additionally, a broad and weak feature in the region of 50 to 110 cm^{-1} was assigned to phonon combination bands. A typical scan of the low-frequency region of the Raman spectrum of solid N_2 is shown in Fig. 13b. Only the E_g band is of high enough intensity to be the subject of concentration dependence studies. The position of this band (peak intensity about four times that of the $v(N_2)$ vibrational band) as a function of RG concentration at 13 K is shown in fig. 12b for N_2/Ar, N_2/Kr and N_2/Xe mixed solids. The concentration range over which the measurements could be done is limited both by the reduction of intensity with decreasing fraction of nitrogen in the mixed solids and by the loss of structure in the band with increasing impurity concentration. The broad, weak Raman feature extending from about 50 to about 115 cm^{-1} was first reported by Anderson et al. (1970) and assigned by them to two-phonon combination transitions. Zumofen and Dressler (1976) could satisfactorily reproduce in their calculation both the position of the maximum and the total intensity distribution and ratio to the one-phonon transitions by considering the librational contributions only.

We have observed the two-phonon region as a wide, but clear band in the above region with a broad maximum at about 90 cm^{-1}. Our concentration dependence studies showed (fig. 13b) that for Ar, Kr and Xe impurities, at concentration of 2–4% and above, this feature smears out to be observed only as a structureless high-frequency tail of the E_g peak obscuring also the T_g^- and T_g^+ peaks. Thus, the two-phonon structure is the first solid-state feature to disappear as a result of RG doping. This is related to its being the feature most sensitive to perturbations in crystal order. As phonons are involved, long-range order plays a role and we can, therefore, state that in this case global symmetry is destroyed.

The aim of our experiments has been to observe the concentration dependence of the shape and intensity of the phonon sideband. The experimental difficulties should be pointed out. While the integrated intensity of the sideband in the pure solid is about three times that of the $^{15}N-^{14}N$ (natural abundance of 0.74%) vibrational transitions, its peak intensity is only half of that of the isotopic band. With increasing rare-gas impurity concentrations the sideband intensity decreases, on the one hand, while, on the other, the scattered-light background intensity increases due to the lower spectral quality of the layers produced. Stating that the phonon sideband is not observed for a given sample is, therefore, meaningful only if the $^{15}N-^{14}N$ band at 2289 cm^{-1} can still be resolved. As already mentioned we generally use the v-N_2 fundamental as a Raman scattering standard, in the case of very weak features we prefer the v-$^{15}N^{14}N$ fundamental mode. The determination from relative Raman intensities as a function of concentration, cross-sections and instrumental corrections is straightforward.

2.4. Defects in isotropic and anisotropic matrices

If nitrogen is employed as matrix material, one may suppose that N_2 has the same lattice constant as Kr (see table 4a), and, therefore, one expects the same behaviour; this consideration is correct for rather insensitive quantities, such as the matrix shift (see table 5). However, as already outlined, α-N_2 or α-O_2 condenses in a more specific structure than RGS, with other site symmetries. One of the reasons for this difference is the quadrupole moment and, in addition, the magnetic moment of O_2. Therefore, we expect spectral differences if the same defect is embedded in molecular crystals consisting of spheres or ordered ellipsoids. Some publications deal with N_2 as matrix material, but very few have considered O_2.

In the case of matrix-isolated CH_4 the rotational energy levels of the v_3 fundamental are split by the octahedral fields of Xe, Kr and Ar matrices (Regitz et al. 1984). The correct group theoretical analysis is quite complicated. Experimentally, it was shown by Jones et al. (1986) by highly resolved FTIR that the complicated, rich spectra for very diluted CH_4 in Ar, Kr or Xe are the same: the fundamental mode possesses a site splitting with rotational fine structure.

In the case of matrix-isolated CS_2 we showed (Givan et al. 1986) that the v_3 band structure of CS_2 is the same in Ar and in N_2, but much more complicated than in Kr. On the basis of very careful systematic investigations varying concentration, temperature and annealing procedures we conclude that, contrary to the accepted view, during deposition of samples, the most occupied site is not necessarily also the most stable site. Since more impurity molecules could occupy disturbed sites, such as substitutional or interstitial sites next to a vacancy, than stable sites. The situation is analogous to the situation in polymers or glasses.

In the case of matrix-isolated NH_3, Abouaf-Marguin et al. (1977) pointed out the substantial difference of N_2 as matrix material. For example: the v_2-NH_3 line is split into a doublet when the molecule is embedded into Kr and into a quadruplet when NH_3 is matrix-isolated in N_2. Girardet et al. (1984b) developed in a series of papers a very sophisticated model to explain completely the IR spectrum, reproducing frequencies and band intensities: the coupling of the proper vibration rotation eigenstates with nuclear spin states of the NH_3 molecule positioned at site O_h in an octahedral field (O_h^5) or positioned at site C_{3i} in a primitive cubic field (T_h^6). However, the difference in both spectra is caused mainly by the different inversion doubling that is the same in all the RGS and in the gas phase, but which is much smaller in N_2. The reason for that is modelled by Girardet et al. via inversion–translation tunneling on the basis of an equilibrium eccentricity which is different for NH_3 in Kr or in N_2.

In the case of matrix-isolated H_2, Bier et al. (1988) and Bier (1986) thoroughly studied the rotational, vibrational and rotational/vibrational excitations of H_2 and D_2 in Ar, Kr, Xe as well as in N_2 and O_2. The strong, sharp Raman line is due to $S_0(0)$ in comparison to the well-known gas-phase spectrum. This assignment confirms free rotation of H_2 molecules in the RGS. In α-N_2, for comparison, the spectrum shows a triplet instead of a single line, whose higher energetically component lost intensity during annealing cycles, whereas the remaining doublet increased by the same amount; this doublet transforms reversibly into a single strong line when warming up similar to the one in RGS. These authors suggest different interpretations to explain this behaviour of the H_2 rotation in RGS and in N_2, and favour one: the first-mentioned small component is attributed to be a site effect; the stronger component of the doublet is interpreted as a rotational translational coupled mode and the weaker component as a pure rotational mode. These type of studies on matrix-isolated H_2 ($T_{s-f} = 20.4$ K) must be executed in a closed cooled cell (Bier et al. 1988), otherwise the hydrogen diffuses out of the matrix or forms $(H_2)_x$-complexes, resulting in irreproducible spectra.

The systematic investigation of the different influences of isotropic and anisotropic matrices, especially with respect to solid-state effects – although in its infancy – will deliver rich information as already shown. Theoretical models are expected to be very complicated.

2.5. Phase-coexistence during condensation and segregation after annealing

Annealing matrices to improve spectra is a common and well-established technique. The reduction in linewidth and the number of matrix sites populated is often discussed in terms of perfect (homogeneous linewidth) and imperfect crystals (inhomogeneous linewidth). This qualitative picture often is insufficient to account for the observations as will be demonstrated by two examples.

Bier (1986) studied the crystal-field splitting of the fundamental mode of α-N_2 (A_g–T_g), by high-resolution Raman spectroscopy. By using a tandem inter-ferometer triple-monochromator arrangement, a resolution of < 0.001 cm^{-1} was achieved without losing signal. Thin films (80 μm) were very slowly sprayed on and irradiated by a single mode laser (450 mW; 5145 Å). The spectrum recorded with a free spectral range of 1.56 cm^{-1} and therefore a resolution of 0.05 cm^{-1} (fig. 14a) at $T = 15$ K showed three lines. The assignment is straightforward from the correlation diagram for α and β-N_2. The ordinary Raman spectrum is presented for comparison in fig. 9a. The temperature was raised progressively to 33 K, recooled to 15 K, warmed up to 33 and to 36 K. Careful analysis of the frequency shifts with respect to the fixed position of $\omega(A_g)$ yields a crystal-field splitting of 1.02 ± 0.02 cm^{-1}, a linewidth of 0.11 cm^{-1} and reversible temperature shifts of the usual size and sign. The total integrated intensity of all three components was constant ($\pm 10\%$) during temperature cycles, but the relative intensities varied irreversibly: directly after condensation at $T = 15$ K, the intensity of the fundamental line of β-N_2 (A_{1g}) was 20% as compared to its counterpart in the doublet of α-N_2 (A_g–T_g). This result characterises the sample thermodynamically. Although the α-N_2 phase was expected at 15 K, in fact a phase mixture was produced with the amount of β-N_2 in α-N_2 after condensation and the relative ratio of both phases during annealing procedures varying from one experiment to the other, a typical effect for mixtures. On the other hand, the samples possess good polycrystal quality as shown by the crystal-field splitting and small linewidth. As a consequence, slow spray-on samples in general do not possess the thermodynamically expected structure.

The second example deals with annealing procedures, i.e. condensation at 15 K, warming up to about 30 K (depending on the matrix material) for minutes

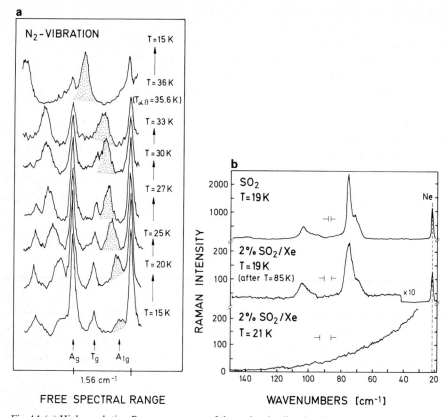

Fig. 14. (a) High-resolution Raman spectrum of the molecule vibration in a solid α–β-N$_2$ mixture of different temperatures (after Bier 1986). (b) Low-energy Raman spectrum of SO$_2$ in Xe compared to the Raman spectrum of pure solid SO$_2$ (after Conrad 1983).

or hours, and recooling again. While the spectra indeed improve, as is commonly reported, it is not clear that the guest molecules may still be regarded as matrix-isolated. Smaller linewidth and measurable $\Delta\omega_M$ is not enough evidence. Givan et al. (1986) clearly showed that the Raman spectrum of the fundamental region of an annealed 1% CS$_2$ doped matrix looks like that of pure solid CS$_2$. Even more convincing proof lies in the low-energy spectrum of 2% SO$_2$ in Xe (see fig. 14b): after condensation at 21 K no excitations below 100 cm^{-1} are measurable, as expected; whereas after annealing the spectrum contained exactly the same librational modes as in pure SO$_2$; aggregation clearly took place. To check whether an impurity is matrix-isolated, it is not sufficient to measure just the matrix shift of the fundamental but also the solid-state character of the matrix material and the "pure impurity" for comparison.

Conclusion and outlook

Using examples and results this section has outlined the idea, behind our approach to study the matrix-isolated case as the low concentration limit of mixed molecular crystals. The matrix shift $\Delta\omega_M$ is not a very sensitive parameter for describing the matrix-isolated case. Qualitative models to estimate the sign and size of $\Delta\omega_M$ are currently sufficient, because quantitative, elaborate models contain many parameters, and unproved assumptions, allowing almost any result to be reproduced. Systematic studies of mode coupling in spectra reveal information concerning short- and long-range order of the matrix lattice. Sample production and preparation should be more than simply producing glassy transparent films. The mode of preparation strongly influences the solid-state properties of the matrix. Sophisticated theories demand "good" matrix-isolated systems with sufficient solid-state behaviour; simple systems should, however, be studied first.

3. Anharmonic phenomena

Typical theoretical models, describing matrix-isolated systems, are based, among others, on the following assumptions: "... The nuclear motion of the guest molecule is harmonic/...Intramolecular vibrations are not coupled to each other via their interaction with the phonon bath/... The molecule–medium interaction is linear in the intramolecular displacements...". As we have already seen there are many spectroscopic manifestations that must be attributed to anharmonicity.

We now consider mechanical [eq. (1.3)] or electrical [eqs. (1.5) and (1.6)] anharmonicity in matrix-isolated molecules and discuss one example for each category: anharmonicity of the impurity molecule potential tuned by the matrix material, anharmonic contribution to the lattice dynamics of the pure matrix material with an emphasis on phonon–phonon interaction and anharmonic guest–host interaction. But first we have to deduce some conclusions from the basic anharmonic theory (section 1.3) with respect to measurable values in the spectra. For example, if we warm up a crystal then we expect volume expansion and increasing phonon–phonon interaction; how is this related to spectroscopic values?

The qualitative change of a spectrum, which is generated, e.g. by a librational excitation in molecular crystals, measured by Raman scattering at two different temperatures is illustrated in fig. 15a. Systematic studies reveal a frequency shift with temperature (fig. 15b) and a line broadening (fig. 15c) at a constant total integrated intensity (in fig. 15a). Other variables, such as pressure, could also be manipulated besides temperature. A series of phenomenological, empirical models exist to fit the shape of $\Delta\omega(T)$ and $\Gamma(T)$; mostly on the basis of the phonon occupation number $n_i(T) = [\exp(\hbar\omega_i/kT) - 1]^{-1}$ which can be expanded in a high-temperature limit ($\hbar\omega_i \ll kT$) and a low-temperature limit

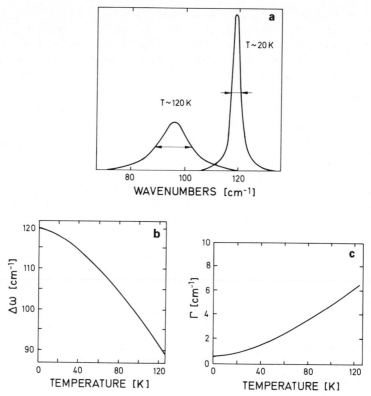

Fig. 15. (a) Qualitative low-energy Raman spectrum at two different temperatures; (b) total frequency shift, and (c) line broadening as a function of temperature.

$(\hbar\omega_i \gg kT)$. My aim here is now to trace back $\Delta\omega(T)$ and $\Gamma(T)$, or similar relations, to microscopic expressions describing the anharmonicity of a lattice.

3.1. Underlying principles

Anharmonicity in molecules is well covered in ordinary textbooks.

In the *harmonic* approximation the lattice potential is expanded and truncated after the third term [eq. (1.3)]. Under the conditions of zero-temperature, vanishing external forces, etc., the lattice frequencies $\omega_i = $ const. are obviously independent of volume V and temperature T. The *quasi-harmonic* approximation can explain effects which are connected with a change in crystal volume: $\omega_i = \omega_i(V)$; temperature causes an increase and pressure a decrease of crystal volume. The instantaneous locations of the atoms produce a static deviation, and the motion of the atoms a dynamical deviation. The *anharmonic* approximation describes an increasing interaction between the phonons as well as a rise in the mean vibration amplitude, therefore, $\omega_i = \omega_i(T, V)$. The so-called explicit dependence of the phonon energy on temperature carries with it the assumption of constant volume, i.e. $\omega_i = \omega_i(T)$. The thermal expansion is offset by hydrostatic pressure. This anharmonic contribution to phonon frequencies and linewidths arises from the third-, fourth- or higher-order terms in the expansion of the crystal potential.

The volume dependence in the quasi-harmonic approximation defines the Grüneisen parameters γ_i via the relation $\omega_i \sim V^{-\gamma_i}$. The mode Grüneisen parameter is then

$$\gamma_i = -\mathrm{d}(\ln \omega_i)/\mathrm{d}(\ln V). \tag{3.1}$$

The Grüneisen parameter for the bulk crystal $\bar{\gamma}$ is the mean square weighted value of the mode Grüneisen parameters and is related to thermodynamic values and, therefore, also measurable by other techniques,

$$\bar{\gamma} = \frac{\alpha}{\beta c}, \tag{3.2}$$

where

$$\alpha = \frac{1}{V}\left(\frac{\partial V}{\partial T}\right)_p, \qquad \beta = -\frac{1}{V}\left(\frac{\partial V}{\partial p}\right)_T,$$

and c is the heat capacity per unit volume. Theoretically, γ_i should be temperature independent, whereas experimental values naturally show a small temperature dependence, showing the limits of the quasi-harmonic approximation.

Sherman (1980, 1982) extended the mode Grüneisen concept, originally developed for phonon frequencies (< 100 cm^{-1}), and applied the definition also to internal modes (~ 1000 cm^{-1}) and including pressure dependence $\gamma = \gamma(p)$.

For matrix-isolated molecules, it was common to explain the frequency shift with temperature via lattice expansion or trapping cage expansion (see, e.g., Jones and Swanson 1981),

$$\Delta\omega_T \sim (R_0/R_T)^6, \tag{3.3}$$

where R_T and R_0 represent the matrix-atom–molecule contact distance (see, e.g., fig. 5). In these terms: the total temperature shift is explained by the quasi-harmonic approximation only.

Starting with $\omega = \omega(T, V(T))$ and using ordinary thermodynamic methods,

$$\mathrm{d}\omega = \left(\frac{\partial \omega}{\partial T}\right)_V \mathrm{d}T + \left(\frac{\partial \omega}{\partial V}\right)_T \mathrm{d}V, \tag{3.4}$$

we can write for the experimentally determined frequency shift

$$\left(\frac{\mathrm{d}\omega}{\mathrm{d}T}\right)_P = \left(\frac{\partial \omega}{\partial T}\right)_V + \left(\frac{\partial \omega}{\partial V}\right)_T\left(\frac{\mathrm{d}V}{\mathrm{d}T}\right)_P. \tag{3.5}$$

Incorporating the thermal expansion coefficient $\alpha_P = V^{-1}(\partial V/\partial T)_P$ and iso-thermal compressibility $\beta_T = -V^{-1}(\partial V/\partial P)_T$, which are both measurable, we find

$$\left(\frac{d\omega}{dT}\right)_P = \left(\frac{\partial \omega}{\partial T}\right)_V - \frac{\alpha_P}{\beta_T}\left(\frac{\partial \omega}{\partial P}\right)_T. \tag{3.6}$$

The meaning of the terms is: $d\omega/T$ is, as usual, the total frequency shift which is found by experiment to be negative with increasing temperature; $(\partial \omega/\partial P)_T$ is the measurable frequency shift at constant temperature, which is in general positive with increasing pressure; α_P and β_T are in general positive, and can be found in the literature; the product $(\alpha_P/\beta_T)(\partial \omega/\partial P)$ can also be expressed by the mode Grüneisen parameter and $V(T)$; the last term, the explicit temperature shift $(\partial \omega/\partial T)_V$, which is of interest in anharmonic considerations, is now determinable and is, in general, negative.

Only few systems have already been systematically investigated. For reason of classification, Zallen and Slade (1978) defined a useful value,

$$\theta := \frac{(\partial \omega/\partial T)_V}{d\omega/dT}. \tag{3.7}$$

In the case of ionic crystals, consisting of hard incompressible spheres, $\theta \sim 0$ because the explicit temperature shift, i.e. the phonon–phonon interaction, is almost zero and the volume term $(\alpha/\beta)(\partial \omega/\partial P)$ is dominant. For covalent crystals $\theta \sim 0.5$, and both terms contribute to the anharmonicity. Molecular crystals are very compressible, $\theta \sim 1$, i.e. the explicit temperature shift has about the same magnitude as the total temperature shift, and anharmonicity is described mainly by phonon–phonon interaction.

Anharmonicity depends, therefore, on two terms: the implicit term driven by volume dilation and the explicit term driven by phonon–phonon interaction. The last one is of greater interest. How is its size, which is determined indirectly from experiments [see eq. (3.6)] related to the anharmonicity in the potential? Expanding the potential (see later) according to eq. (1.3) we can write:

$$V = V_0 + V_1 + V_2 + V_3 + V_4 + \cdots, \tag{3.8}$$

and the Hamiltonian is divided into its harmonic and anharmonic part

$$H = H_0 + H_{anh.} \tag{3.9}$$

with

$$H_0 = T + V_2 \quad \text{and} \quad H_{anh.} = V_3 + V_4, \tag{3.10}$$

where T is the kinetic energy.

As described in the literature (Wallace 1972, Califano et al. 1981, Della Valle et al. 1983, Righini 1985) based on second-order perturbation theory, the anharmonic frequency is obtained for each mode i and wave vector k,

$$\omega_i(k) = \omega_i^0(k) + \omega_i'(k), \tag{3.11}$$

where ω_i^0 is the harmonic frequency and ω_i' is the complex anharmonic correction given by

$$\omega_i'(k) = \Delta_i(k) - i\Gamma_i(k), \tag{3.12}$$

where $\Delta_i(k)$ is the frequency shift and $2\Gamma_i(k)$ is the linewidth of the i-phonon at $k = 0$, the zone center. The real expressions for Δ_i and Γ_i contain, in addition to the phonon occupation numbers n and frequencies ω_i, the cubic (V_3) and quartic (V_4) terms of the anharmonic potential,

$$\Delta_i(0) = \text{fct}(V_3, V_4, n, \omega_i), \qquad \Gamma_i(0) = \text{fct}(V_3, n, \omega_i). \tag{3.13}$$

The cubic and quartic contributions to the anharmonic correction in the explicit temperature shift $\partial\omega/\partial T$ are different in size and sign. The temperature dependence of Δ_i and Γ_i is buried in $n(T)$.

The intermolecular potential V^{AB} is usually described by a two-body atom–atom Lennard-Jones type potential and is sometimes extended by an electric quadrupole–quadrupole term which considers anisotropy,

$$V = \sum_A \sum_B V^{AB}, \tag{3.14}$$

with

$$V^{AB} = V^{AB}_{\text{atom–atom}} + V^{AB}_{\text{EQQ}}, \quad \text{and} \tag{3.15}$$

$$V^{AB}_{\text{EQQ}} = \text{const.} \frac{Q_A Q_B}{R_{AB}^5} \quad \text{(geometric expression)}. \tag{3.16}$$

Recommended literature dealing with applications: Zallen and Slade (1978) studied molecular chalcogenides; Kobashi (1978) calculated α-N_2; Della Valle et al. (1983) investigated naphthalene; Antsygina et al. (1984) compared N_2, CO, CO_2 and N_2O; Righini (1985) re-calculated α-N_2; Häfner and Kiefer (1987) measured and calculated naphthalene again.

To calculate the explicit temperature shift experimentally we must first measure the total temperature shift (like in fig. 15b), the implicit term $(\alpha/\beta)(\partial\omega/\partial P)_T$ must then be determined; i.e. we have to measure the frequency shift as a function of pressure at the lowest accessible temperature (see section 4); the experimental coefficients α_P and β_T are based more or less on a knowledge of $V(T)$ and $P(V)$ (Klein and Venables 1976, 1977) that must often be extrapolated from the known data. So, for example, $P(V)$ is known for $T = 300$ K, but we need $P(V)$ at $T = 0$ K, or $V(T)$ is known for $P = 0$, but we need $V(T)$ at $P > 0$. As a result only very few systems had been carefully studied (see also section 4). Finally, $\Gamma(T)$ is, in principle, known experimentally (like fig. 15c) and a slightly modified curve, fig. 15b, showing $\partial\omega/\partial T$.

As outlined before the equations defining Δ_i and Γ_i [eq. (3.13)] are rather complicated, and the V_3 and V_4-terms are responsible for phonon decays to two or three new phonons, respectively, in such a way that conservation of energy, momentum and symmetry is satisfied.

Nevertheless, further information from the experimentally determined $\omega(T)$ and $\Gamma(T)$ can only be obtained if a specific model is chosen allowing the decay of a phonon with harmonic frequency ω_0 to two, three or four phonons, and the experimental data $\omega(T)$ and $\Gamma(T)$ to be fitted to a function of temperature. From these coefficients, one can finally conclude whether V_3 or V_4, or both, contribute to the anharmonicity. In this way, Jodl and Bolduan (1982) studied N_2 in Ar (see section 3.4), Balkanski et al. (1983) investigated silicon and Liarokapis et al. (1985) dealt with LaF_3.

We now discuss examples of anharmonicities in molecular crystals and we will begin with the question: does the matrix material influence typical anharmonic constants of the impurity molecule?

3.2. Within the impurity molecule

Anharmonicity of the intramolecular potential causes different phenomena, e.g. mode coupling or anharmonic rovibrator, and can be described by spectroscopic constants. These type of anharmonic corrections are small, $<1\%$ (see table 3: $\omega_e x_e$ of ω_e). Besides measuring the matrix frequency shift $\Delta\omega_M$, it is pertinent to ask if the anharmonic constants are also shifted. κ_{ij} was obtained from the overtone Raman spectra of all fundamental modes of NO_2, in Ar, Kr and Xe, and found to be altered up to 50% with respect to the gas-phase values (Jodl 1984). This experimental fact confirms again that we cannot simply adopt the free-molecule parameters in models describing matrix-isolated molecules.

As a second example we studied how mode coupling is influenced by the matrix material. This was accomplished by tuning the Fermi resonance of CO_2, CS_2 and N_2O by temperature, pressure and matrix material (Bier and Jodl 1987). The Fermi resonance between v_1 and $2v_2$ modes of linear three-atomic molecules is described by

$$(v_+ - v_-)^2 = (v_1 - 2v_2)^2 + 4W^2, \tag{3.17}$$

with the Fermi constant

$$W = -k_{122}/\sqrt{2}. \tag{3.18}$$

The anharmonic constant k_{122} is defined by the Taylor series expansion of the intramolecular potential

$$V \sim k_1 q_1^2 + k_2 q_2^2 + k_3 q_3^2 + \cdots + k_{111} q_1^3 + k_{122} q_1 q_2^2 + \cdots. \tag{3.19}$$

We measured the frequency splitting δ of the perturbed levels,

$$\delta = v_+ - v_-, \tag{3.20}$$

and their intensity ratio R,

$$R = I_+/I_-,\tag{3.21}$$

to calculate the two parameters which regulate the size of the Fermi resonance,

$$\Delta = |v_1 - 2v_2|,\tag{3.22}$$

the splitting of unperturbed levels and k_{122}. Comparison in different matrices is facilitated by defining the strength of the Fermi resonance S,

$$S = (W/\Delta)^2.\tag{3.23}$$

Data showing this effect in matrix-isolated CO_2 and CS_2 (1% CO_2 or CS_2 in N_2, Ar, Kr, Xe) compared with the pure molecular crystals are presented in table 6 (Bier and Jodl 1987), together with data for pure solid N_2O (Naßhahn 1987). The strength S of the Fermi resonance is changed enormously: implying that either k_{122} is constant and Δ changes, or k_{122} and Δ are simultaneously influenced. The authors interpret the variation of the Fermi resonance on the basis of the isotropy (RGS) or anisotropy (N_2) of the matrices, of lattice symmetry arguments, etc.

The Fermi resonance of CS_2 in mixed Ar–Kr matrices was also studied and δ was plotted as a function of concentration (Givan et al. 1986). The departures from a linear relation suggest that the larger and more polarizable Kr is predominant in determining the extent of resonance.

3.3. Of the matrix lattice

The concept of Fermi resonance can also be applied successfully to describe the "vibron–vibron resonant transfer coupling" in solid γ-O_2. As already mentioned, γ-O_2 contains eight molecules per primitive cell: $Z = 2$ molecules (index a) on site T_h, and $Z = 6$ molecules (index b) on site D_{2h}. In the harmonic model, in which molecule a and molecule b – neighbouring on non-equivalent sites – do not interact, the levels for the fundamental modes are equally spaced around $\frac{1}{2}(V_{aa} + V_{bb})$ and the intensity ratio is proportional to the number of molecules $R = I_a/I_b = N_a/N_b = 2:6$, where V_{aa} and V_{bb} denote the eigenvalue of V of the unperturbed eigenfunctions of the molecules on site a and b, respectively. In the anharmonic model, in which both molecules interact via a coupling V_{ab}, the energies of the two coupled oscillators are shifted [similar to eq. (3.17)] and

$$E = \tfrac{1}{2}(V_{aa} + V_{bb}) \pm \tfrac{1}{2}[(V_{aa} - V_{bb})^2 + 4V_{ab}^2]^{1/2},\tag{3.24}$$

with $\Delta = V_{bb} - V_{aa}$ being the splitting of unperturbed levels and $\delta = [(V_{aa} - V_{bb})^2 + 4V_{ab}^2]^{1/2}$ the total splitting of the perturbed levels. And the intensity exchanged between a and b is described by a mixing parameter c,

$$R = I_a/I_b = \frac{N_a}{N_b} \frac{[1 - c(N_b/N_a)^{1/2}]^2}{[1 + c(N_a/N_b)^{1/2}]^2}.\tag{3.25}$$

Table 6
Influence of matrix material on Fermi-resonance quantities of CO_2 and CS_2 (for notation used see the text).

Resonance	CO_2					CS_2					N_2O
	Pure	Ar	Kr	Xe	N_2	Pure	Ar	Kr	Xe	N_2	Pure
ν_+ (cm^{-1})	1384.6	1385.3	1383.7	1383.7	1384.7	653.8	657.0	654.5	653.5	656.0	1293
ν_- (cm^{-1})	1273.3	1281.8	1280.1	1277.7	1278.7	802.4	807	805	800	810	1164
ν_2 (cm^{-1})	659.3	662	660	659.4	658	396	397	396	390	397	588
$|\delta|$ (cm^{-1})	111.3	103.5	103.6	106	106	148.6	150	150.5	146.5	154	129
R	2.5	1.7	1.9	2.1	4.2	15	11	6	4	14	20
$|\Delta|$ (cm^{-1})	47.7	27	32	38	65	130	125	108	88	134	114
$|k_{122}|$ (cm^{-1})	71.1	71	70	70	59	51	59	75	83	54	40
S	1.1	3.5	2.4	1.7	0.4	0.08	0.11	0.24	0.44	0.08	0.1

The coupling V_{ab} [Fermi constant, see eq. (3.18)] is determined by the respective force constant, i.e. the second derivative of the interaction potential V with respect to the normal coordinates,

$$V_{ab} \sim k_{ab} \sim \frac{\partial^2 V}{\partial q_a \, \partial q_b}. \tag{3.26}$$

If the temperature is increased, the site splitting in the Raman spectra (see fig. 9b) decreases (Bier and Jodl 1984); and it increases with increasing pressure (Jodl et al. 1985). By measuring δ and R as a function of T and P, Δ and V_{ab} can be determined. At $P = 0$ we find that $\Delta = 1.3$ cm^{-1} and $I_a/I_b = 1:3$ when $44 < T < 55$ K; consequently, c and V_{ab} are equal to zero, which describes the uncoupled case. If we increase the pressure (up to 50 kbar) the coupling constant V_{ab} increases from 0 to 2 cm^{-1} demonstrating how the anharmonicity of the potential V is altered. Δ increases in the same manner as a function of P; i.e. the usual pressure shift of unperturbed levels (see section 4). Therefore, the strength of the resonant *intermolecular* vibron–vibron coupling, which is defined by $(V_{ab}/\Delta)^2$ in analogy to eq. (3.23), increases from 0 to 0.15 in the range of 0–5 GPa. This implies a very weak coupling as compared to the value of the *intramolecular* Fermi resonance in CO_2, CS_2 and N_2O in table 6.

In addition to vibron–vibron mode coupling, vibron–phonon mode coupling may occur as outlined. According to eq. (1.6), $(\partial^2 \alpha/\partial Q \, \partial q)$, this interaction is described by the electric anharmonicity. We investigated and discussed the behaviour already in pure N_2, O_2, CO molecular crystals by FTIR studies (Jodl et al. 1987) and by Raman studies (Bier and Jodl 1984, Bier 1986). Figure 16, trace a, shows the FTIR spectrum of CO_2 (Löwen et al. 1988b); note the appearance of the normally IR inactive mode v_+ (or v_1), induced by disorder, and which is in Fermi resonance with v_- (or $2v_2$) – see section 3.2. Also visible are the so-called zero-phonon line and to higher frequencies the phonon sideband on the Stokes side. The assignment of the maxima in the structured sideband is straightforward, using the inelastic neutron scattering data of Powell et al. (1972). Since the sideband is proportional to the weighted phonon density of states (see section 1.3) it is pertinent to compare our results with recent anharmonic calculations on CO_2 by Cardini et al. (1987). Figure 16, trace b, shows the calculated phonon density of states by lattice dynamics, whereas fig. 16, trace c, is the one obtained by molecular dynamics. The first reflects our experimental data much better with respect to frequency of phonons and to relative intensities. From these data, we were able to define a coupling strength from the temperature-dependent Huang–Rhys factor S equivalent to the Debye–Waller factor in Mössbauer spectroscopy. This factor is derived from the intensity ratio of the whole band I (zero-phonon line and sideband) to the pure vibrational line I_0 (zero-phonon line),

$$S = \ln(I/I_0). \tag{3.27}$$

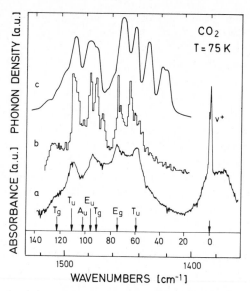

Fig. 16. Curve a shows the v^+-CO_2 mode and phonon sideband by FTIR at 75 K (Löwen et al. 1988b); curve b the calculated phonon density of states by lattice dynamics, and curve c the phonon density by molecular dynamics (Cardini et al. 1987).

The reader is referred to the literature developed, in which this theory has been applied to semiconductors (Sapozhnikov 1976), to colour centers (Fitchen 1968) and to molecular crystals (Califano et al. 1981). Lin (1976) deduced this Huang–Rhys factor from a detailed model of vibration relaxation. We determined this vibron–phonon coupling strength for several matrix materials: $S \sim 0.03$ for α-O_2, $S \sim 0.06$ for β-O_2 (Bier and Jodl 1984); and $S \sim 0.02$ for α-N_2 (Bier 1986); according to an arbitrary classification, $S < 0.1$ defines a weak coupling. These measurements were obtained from Raman scattering in which the zero-phonon line is allowed and Raman active.

The final portion of this subsection will deal with anharmonic effects in pure phonon modes in matrix materials. As outlined in the preceding subsection 3.1, the quantity $d\omega/dT$ depends on two anharmonic terms: the implicit term $(\alpha/\beta)(\partial\omega/\partial P)$ due to volume expansion and the explicit term $(\partial\omega/\partial T)_V$ produced by phonon–phonon interaction [see eq. (3.6) and definition and meaning of θ, eq. (3.7)]. The line width $\Gamma(T)$ is also related to cubic and quartic anharmonic terms of the potential [eq. (3.13)]. Table 7 contains relevant data for α-O_2, α-N_2, N_2 in Ar and, for comparison, CO_2 and CS_2; θ is in the range 0.7 to 1 for all systems, implying that phonon–phonon interaction is the main contribution to anharmonicity in these systems. But the data must be considered with care because while $d\omega/dT$ and $\partial\omega/\partial P$ can be measured quite accurately, α_P and β_T are extrapolated data or, as in the case of CS_2, simply unavailable. In those cases, only the trends in $\partial\omega/\partial T$ and θ are meaningful. The linewidths $\Gamma(T)$ (similar to fig. 15c) for the following systems are known: B_g mode in α-O_2 (Bier and Jodl 1984); E_g mode in α-N_2 (Bier 1986); N_2 in Ar (Jodl and Bolduan 1982); B_{1g} mode in CS_2 (Bier et al. 1985); and E_g mode in CO_2 (Olijnyk et al. 1987a). However, it is not feasible to determine 6–10 fitting parameters from $\partial\omega/\partial T$, which is uncertain due to the uncertainty in α/β, and from $\Gamma(T)$, which is dependent on crystal quality, and hence difficult to determine the contributions of V_3 and V_4 separately to the anharmonicity.

Concluding remarks: Experimental values (like the ones in table 7) are, on the whole, not complete and precise enough to date; moreover, theoretical models currently do not address the fact that translational modes and librational modes require different anharmonic contributions calculated either via lattice dynamics or via molecular dynamics on the basis of different intermolecular potentials. In other words, more systematic work is needed in this very interesting area. The reader may wish to regard α-N_2 as a case study: experiments by Medina and Daniels (1976); lattice dynamics with an atom–atom Lennard-Jones potential by Kobashi (1978); lattice dynamics with an EQQ potential by Righini (1985); molecular dynamics by Cardini and O'Shea (1985).

3.4. The guest–host interaction

The guest–host interactions should be reflected in the following spectroscopic features:

(1) Features in the spectra of pure matrices, which arise from anharmonicity, are, in general, very weak; e.g., the ratio of the phonon sideband to the allowed zero-phonon line is about 10^{-2}.

Table 7

Total temperature shift and adapted pressure shift of phonons in several molecular crystals. The pure temperature shift and θ are calculated from eqs. (3.6) and (3.7).

Substance (mode)	$d\omega/dT$ (cm^{-1}/K)	$(\alpha/\beta)(\partial\omega/\partial P)$ (cm^{-1}/K)	$\partial\omega/\partial T$ (cm^{-1}/K)	θ	References
α-O_2 (A_g), $T = 10$ K	-0.1	$+7 \times 10^{-4}$	-0.1	1	Jodl et al. (1985)
α-N_2 (ν_0), $T = 15$ K	-0.01	$+10^{-4}$	-0.01	1	Jodl and Bolduan (1982)
N_2 in Ar (ν_0), $T = 5$ K	-0.02	$+10^{-3}$	-0.02	1	Jodl and Bolduan (1982)
CO_2 (E_g), $T = 20$ K	-0.06	$+0.03$	-0.03	0.5	Olijnyk et al. (1987a)
CS_2 (B_{1g}), $T = 10$ K	-0.05	$+0.85$ (cm^{-1}/kbar)			Bier et al. (1985) Bolduan et al. (1986)

(2) In typical matrices the guest:host ratio is about 1:100, and the zero-phonon line is decreased by that amount.

(3) Features in the spectra, which arise from the anharmonicity of the guest/host interaction, will be even weaker. Only in cases in which the guest/host coupling is resonantly enhanced is it observed.

Of course, both electric and mechanical anharmonic terms of the vibron–vibron, vibron–phonon, and phonon–phonon combination bands in the guest/host interaction can play a role. Three representative examples follow:

(1) Free rotations of a matrix-isolated molecule, in general, are hindered by the matrix. Spectroscopically, many such examples are known, e.g. CO (Dubost and Charneau 1976) or HCl (Friedmann and Kimel 1965) and H_2 (Bier et al. 1987). A number of well-established models for hindered rotation also exist: rotation–translation–coupling (Friedmann and Kimel 1965), libration–phonon–coupling (Blaisten-Barojas and Allavena 1976), hindered rotator (Beyeler 1974) or rotating cage (Manz 1980). Common to each model is the basic idea that for a heteronuclear molecule in the octahedral potential of the matrix the center of mass and the center of interaction do not coincide, causing a coupling of modes. In general, these models are described in terms of ordinary molecular physics such as oblate or prolate rotators instead of solid-state physics (e.g. dynamic Jahn–Teller effect).

(2) In molecules such as CO or NO vibrational energy transfer tends to accumulate the vibrational energy in the high-lying levels (up to $v \sim 30$). This is the so-called anharmonic vibron–vibron pumping; the VV-transfer is phonon assisted, or can be a resonant pumping (zero phonon). Phonons from the matrix material must be emitted or absorbed to compensate for the energy mismatch between transitions in each of the interacting molecules due to isotopic effects and/or vibrational anharmonicity. These effects are thoroughly studied experimentally by Dubost and Charneau (1976, 1979), by Galaup et al. (1985) and theoretically explained by Blumen et al. (1978) and later on reviewed by Dubost (1984).

(3) In the last example we discuss how the phonon density of states or directly observable values like the IR absorption band or Raman scattering band are modified, if pure systems are doped. The simplest but most effective model is the linear chain consisting of matrix particles with mass M connected by springs (force constant f), containing only one impurity (mass M') occupying a substitutional lattice site. The dynamics of the pure chain is calculated conventionally by solving the equation of motion. In the case of the impure chain, the changes in mass and in force constants are separated out and we introduce them as perturbations. We define two defect parameters,

$$\varepsilon = 1 - M'/M \quad \text{and} \quad \lambda = 1 - f'/f, \tag{3.28}$$

which relate to the mass change and force constant change in an impure chain.

In this very simple model, all possible additional modes can be determined ($\omega^2 = f/M$): ω_{local} is a mode larger than the cut-off frequency which can be modeled by $M' < M$; $\omega_{resonant}$ and $\omega_{in\text{-}band}$ lie within the one-phonon regime and can be described by a suitable ε and λ. Figure 17 shows the results obtained with a 48 atom chain and 24 unit cells, so that adding one impurity atom corresponds to ~ 4 at.% defects; this is a very heavy doping compared to most matrix-isolation experiments. The procedure involves determining by experiment the modified phonon spectrum of the guest/host system, which fixes ε. The theoretical spectrum is then fitted to the experimental spectrum (position of band maximum) and λ is determined. For example, Keeler and Batchelder (1972) found a defect-induced IR inband mode ω_i at 28 cm^{-1} for 0.15% Ne in Ar; Cohen and Klein (1984) were able to calculate $\omega_i = 20$ cm^{-1} with $\lambda = 0.924$; then $\omega_i = 35$ cm^{-1} with $\lambda = 0.755$; and finally $\omega_i = 28$ cm^{-1} with $\lambda = 0.83$.

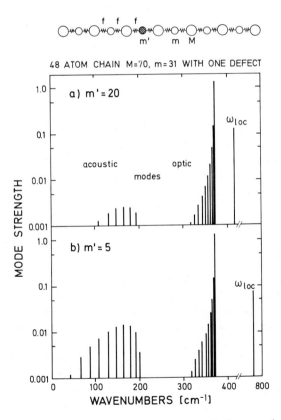

Fig. 17. Linear chain consisting out of heavy and light atoms producing acoustic and optic modes; the light atom is replaced by a lighter atom causing a high-frequency local mode as well as absorption throughout the bands (according to Barker and Sievers 1975).

Some basic literature in this field are the articles by Maradudin (1965), Mannheim (1972, and references therein); and the excellent review article by Barker and Sievers (1975), who developed in a clear and systematic way the concept of lattice vibrations associated with defects, applied to ionic compounds; the review by Cohen and Klein (1984) who considered the dynamics of impure rare-gas crystals. Several examples from the matrix-isolation literature are:

– H_2 in Kr by mid-IR absorption (De Remigis and Welsh 1970);
– Ne or Kr in Ar by far-IR absorption (Keeler and Batchelder 1972);
– optical emission of Eu in Ar (Jakob et al. 1977);
– CO in N_2 by high resolution IR absorption (Dubost et al. 1982);
– Ag_2 in Xe by Raman scattering (Bechtold et al. 1984);
– UV emission of CO in RGS (Barhdt et al. 1987).

As an illustration (fig. 18) we show the optical emission of Eu isolated in Ar, a specific electronic transition (the zero-phonon line) at about 3655 Å and a vibrational sideband is plotted as a function of temperature; two in-band modes in Stokes and anti-Stokes region are discernible. Table 8 contains several examples, and data for RG in RGS were collected by Cohen and Klein (1984).

The agreement obtained with the linear chain model is convincing, i.e. the parameter range of λ is plausible. One should consult the original spectra, however, since the maximum of a broad, weak structure in the spectra is often condensed to a single frequency in table 8.

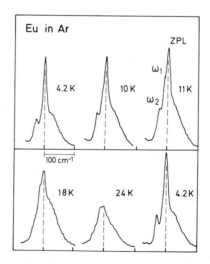

Fig. 18. Optical emission spectrum of Eu in Ar as a function of temperature is plotted (after Jakob et al. 1977); zero-phonon line (ZPL) and in-band modes ω_1 and ω_2 are marked.

Table 8
Impurity-induced lattice modes ω (in cm^{-1}), with ω_i the in-band mode; ω_r the resonant mode and ω_ℓ the local mode.

Impurity	Matrix						References
	Ne	Ar	Kr	Xe	N_2	O_2	
H_2		$\omega_r=22$	$\omega_r=16$		$\omega_r=27$	$\omega_r=18$	De Remigis and Welsh (1970)
		$\omega_\ell=122$	$\omega_\ell=106$				Bier et al. (1988)
N_2		$\omega_r=39$					De Remigis et al. (1971)
		$\omega_\ell=70$					
HCl		$\omega_\ell=73$	$\omega_\ell=63$	$\omega_\ell=45$			Verstegen et al. (1966)
CO	$\omega_\ell=83$	$\omega_\ell=80$	$\omega_\ell=68$	$\omega_\ell=53$			Dubost (1976)
NO	$\omega_i=40$	$\omega_i=38$					Bourcey and Roncin (1975)
Eu		$\omega_i=12$					Jakob et al. (1977)
		$\omega_i=26$					
Ag_2		$\omega_r=46$	$\omega_r=40$				Bechtold et al. (1984)
			$\omega_\ell=57$	$\omega_\ell=47$			

Conclusion and outlook

Anharmonic phenomena in molecules, in the matrix material, and in the guest/host interaction are well understood theoretically, in principle. Spectroscopic features, which are related to anharmonicity, are in general very weak and one must apply several cross-checks to be sure in the assignment of these features. On the one hand, there is a lack of complete solid-state data, describing the pure matrix, which are necessary for theoretical modes; however, on the other hand, there are only a few matrix-isolated systems studied carefully thus far.

Using numerous examples, I tried to point out how anharmonic phenomena show up in the spectra; how they can be manipulated by parameters like temperature, pressure, matrix material; how measured $\omega(T)$ and $\Gamma(T)$ data are related to the anharmonic parts of the potential; how a suitable potential (intramolecular, intermolecular) must be properly chosen; and how complicated this matter already is for simple matrix-isolated systems.

In the future, either the known classical experiments on simple systems, such as RG in RGS or H_2 in RGS, should be repeated by modern techniques (FTIR or HRRS), or the commonly used thin-film preparation techniques should be replaced by those leading to high-quality single crystals. More low-temperature experiments, in conjunction with high-pressure experiments, should be performed to get direct information about anharmonicity in molecular crystals.

4. Matrices at high pressure

4.1. Present state of high-pressure spectroscopy and how high pressure is related to MIS

High-pressure has been used for studying solids since the pioneering work of Bridgeman (1910–1950). Hydrostatic pressures greater than a few kbar (1 kbar $= 0.1$ Gigapascal $= 10^9$ dyne/cm^2) produce stronger and, in principle, cleaner perturbations than a variation in temperature. For example, the lattice constant of Ar is reduced by only about 3% on cooling from 80 to 5 K (see fig. 5); by contrast the lattice constant of α-O_2 is reduced about 30% on pressurized to 70 kbar at $T < 10$ K. Furthermore, changing the temperature is complicated by the simultaneous action of thermal expansion and phonon population, whereas hydrostatic pressure produces solely a volume change.

The rationale for high-pressure studies on molecular crystals are:
– *technically:* The molecular gas can act as a pressure transmitting medium. Traditionally, the sample must be immersed into a special liquid mixture, Ar, or, very recently, Xe and He as a pressure transmitting medium; hydrostaticity is expected.
– pressure-induced *chemistry* can occur to form new compounds such as black CS_2 and yellow CO.
– pressure-induced *physics:* (1) To form new phases in molecular crystals; (2) If pressure of the order of Mbar is applied to rare-gas solids pressure-induced metallization should take place. At these pressures, the mechanical energy applied to the system is of the same order as the electronic band gaps. As a result the electronic band structure will be enormously deformed and Xe, for example, should become electrically conducting, O_2 should turn black and normal hydrogen shows a soft vibron due to the change in chemical bonding. (3) All spectroscopic values in combination with electronic transitions, vibrons, librons, phonons, magnons, etc., show, of course, pressure effects: there are frequency shifts, changes in lineshape and in linewidth, selection-rule changes as well as line splittings, and pressure-tuned resonant processes, such as phonon transitions or Fermi resonance.

For an overview of high-pressure research we recommend: Klement and Jayaraman (1966), Swenson (1977), Sherman and Wilkinson (1980), Hazen and Finger (1982), Jayaraman (1983), Weinstein and Zallen (1984) and Jayaraman (1986).

How high pressure is related to MIS

From the *chemical* point of view: isomerization of matrix-isolated species as a function of pressure is conceivable; investigation of reaction products of matrix-isolated systems driven by high pressure at room temperature, etc. (Swanson 1987). From the *physical* point of view: the direct test of combined potentials by comparing experimental and theoretical frequency shifts as a function of pressure; site effects (splitting), local distortions (inhomogeneous linewidth) dephasing mechanisms as a function of high pressure at low temperatures, anharmonicities of guest/host interaction (described in terms of section 3).

High-pressure studies in pure and doped condensed gases have been reviewed by Vu (1980) who investigated pure RG and RG doped with small molecules (H_2, HCl and CH_4) and molecular crystals consisting of small molecules like H_2, N_2, HBr. The pressure range is limited either to a few bar (compressed gas) or to a few kbar only. The data are presented phenomenologically and in terms of molecular physics. Work done by my group (Jodl 1984) on matrix-isolated N_2 at high pressure is discussed in section 3.

My aim in this section is to present a few specific examples of pure and doped molecular crystals at elevated pressure to illustrate the result of high pressure. Some remarks concerning the technique, sample preparation of matrix-isolated systems and the combination of diamond anvil cell (DAC) with Raman scattering and FTIR apparatus will be made. This specific research field – matrices at high pressure – is in its infancy and many problems are still unsolved.

4.2. Technical aspects

We have used two types of high-pressure cells: gas cells ($p < 10$ kbar), whose application in matrix isolation is described by Jodl and Holzapfel (1979), and diamond anvil cells ($p < 500$ kbar). In the latter, two high-quality diamonds press on a sample that is located between them. This is shown in fig. 19 together with the mechanical device used to hold and direct the diamonds. This holding device differs from one DAC-type to the other. By lowering the effective area and indirectly the sample volume this cell can reach pressures up to 1–2 Mbar today: the sample volume in gas cells is about 1 cm^3, in DAC it is about 200 μm in diameter times 50 μm in height ($\ll 1$ mm^3).

Different methods are available to measure the local pressure: in the DAC the ruby technique is most often used (Piermarini and Block 1975). A tiny chip of ruby, 5–10 μm in diameter, is placed in the pressure medium along with the sample (in our case the medium and the sample are one). Ruby fluorescence is excited by a low-power laser so as to avoid local heating. An intense doublet with $\lambda = 6927$ and 6942 Å at $P = 0$ is observed, which shifts to higher wavelength as pressure is increased. The shift is temperature independent and linear with pressure (0.365 Å/kbar), allowing pressure to be determined to better than ± 0.5 kbar up to Mbar region. The low sample temperatures are also difficult to determine with accuracy, both because of laser heating, and the effects of thermal contact between sample and DAC and DAC and cryostat. Careful indirect measurements (phase transition temperature, Stokes to anti-Stokes Raman intensities, etc.) must be used to confirm that the temperature of the DAC is the same as the temperature of the sample inside. A temperature range

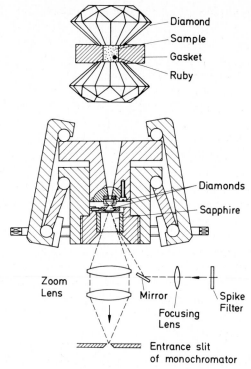

Fig. 19. Opposed diamond-anvil configuration, the basic part of the DAC; a sectional view of Syassen–Holzapfel cell; optical geometry at backward scattering geometry (after Hirsch and Holzapfel 1981).

between 5 and 300 K is available to us. Only few groups are capable of combining high pressures with low temperatures. In most DAC systems, change in pressure is only possible at room temperature.

The DAC may be loaded in one of two ways. In the first technique, the DAC is placed in the cryostat and cooled to the desired temperature, at which the gas sample is liquid (for Xe 165 K). The cell area is sealed off by a wall of indium and a capillary inserted in the wall brings in the liquid gas, which is trapped afterwards in the gasket hole by pressurizing the DAC (Jodl et al. 1985).

In the second technique the DAC is inserted into an additional high pressure bomb outside the cryostat and pressurized with the sample gas $P < 5$ kbar. This gas fills the space outside and inside the cell and also between the diamonds. The DAC is mechanically closed from outside the bomb, the sample is trapped, and bomb pressure is released. Then the cell is withdrawn from the bomb, located under a microscope for single-crystal growing, and afterwards clamped to the helium-cooled copper flange inside the cryostat.

Comparing both techniques, each method – cryogenic and gas loading – has its advantages and disadvantages. However, for mixed molecular crystals and for matrix-isolated cases gas loading is the only suitable technique; for example, consider species A and B with different melting points [1% H_2 ($T_{s-\ell} = 20$ K) in Ar ($T_{s-\ell} = 83$ K)]. A second advantage of our gas loading technique is that we pressurize the pure and doped gas at first well above the critical pressure at $T = 300$ K (see fig. 4 and table 4a) and then cool the DAC down along a certain trajectory in the pressure–temperature diagram. This high-density gas/liquid state contains no defects allowing one to produce molecular crystals of high quality, whose spectral lines would have a minimum of inhomogeneous broadening. We have been successful in loading Ar, N_2, Xe and He (Däufer 1987).

Figure 19 shows a schematic view of our optical geometry to Raman scattering. Laser alignment, Rayleigh scattering versus Raman scattering, the weakness of the Raman signal which becomes still weaker due to sample size (laser spot < 50 μm) are some of the difficulties encountered. Recently, also IR studies at high pressure have become possible as a result of the IR microscope attachment to FTIR spectrometers (IR beam < 50 μm). Another problem is the limited spectral range of sapphire and diamond. Nevertheless, satisfactory results, e.g. for CO_2 (Johannsen et al. 1988), are possible.

4.3. Application to pure systems

Before investigating matrix-isolated systems, it is natural to study the pure molecular crystals at elevated pressure first. Mixtures and matrix-isolated species are much more complicated systems from the solid-state physics point of view. Their spectra will be much weaker and sample preparation will be more complicated. The rare-gas solids have been studied thoroughly over the years because they were used as pressure transmitting medium. Swenson (1977) reviewed the medium pressure range, < 100 kbar. Because IR and Raman transitions are forbidden, no spectroscopic information is available on these systems.

Solid H_2 and D_2 have been favourite molecules in high-pressure studies because they revealed new and interesting physics (see section 4.4). They are the probe species in establishing high-pressure records (1–2 Mbar, today). Specifically, the melting curve P (< 200 kbar) and T (< 500 K) was measured by Diatschenko et al. (1985); and the theory of the properties of solid H_2, HD and D_2 has been developed by Van Kranendonk (1983), on the basis of experimental work by Silvera (1980).

Solid N_2 has already been extensively described in this chapter: low-pressure data in section 1.4 and anharmonicities in section 3.3. The basic data for lattice dynamics and molecular dynamics theory (by Cardini and O'Shea (1985) and references therein) are the Raman data of Medina and Daniels (1976) and IR data by Obriot et al. (1978), at $p < 20$ kbar and low temperatures. Recent higher pressure data up to 1.3 Mbar (Reichlin et al. 1985), show a rich splitting of the fundamental modes of this solid, related to phase transitions at 200, at 660 and 1000 kbar. While the pressure behaviour of nitrogen appears to be complicated, the Raman results indicate that neither a pressure-induced metallization takes place, as in xenon, nor that the molecular binding is altered appreciably, as in hydrogen.

The Raman spectra obtained at high-pressure change dramatically in comparison to the ones obtained at low pressure: the lines broaden and the band maxima decrease in intensity. The combination of DAC with Raman scattering technique is now so well developed that spectra obtained at moderate pressures are as good as those obtained under ambient conditions.

Although phase transitions and the evolution of new structures are usually investigated by X-ray diffraction, the Raman spectra often contain clues indicating a phase transition. This will be illustrated by solid CO_2. Figure 20 shows the frequency shift as a function of pressure at different temperatures. Additional external modes show up at about 60–70 kbar. Line splittings, kinks in the frequency shift of internal modes versus pressure and changes in intensities and linewidth are further indications of a phase transition. Some of these changes result from modifications in the selection rules and Raman activity brought about by changes in symmetry (see section 1.3).

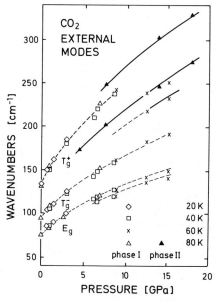

Fig. 20. Frequency shift as a function of pressure taken from Raman spectra of solid CO_2 at ambient temperatures (Olijnik et al. 1987a).

Variable pressure is also an excellent tool for testing anharmonic models of the solid state. Sherman (1980, 1982) predicted the pressure dependence of mode Grüneisen parameters theoretically: Figure 21a shows this trend in which the internal modes and external ones should reach $\gamma_i = 1$ in the high-pressure limit. Several molecular crystals were investigated: O_2 (Jodl et al. 1985), CS_2 (Bolduan et al. 1986) and CO_2 (fig. 21b) and in each case, the predicted trend was observed. The mode Grüneisen parameters [eq. (3.1)] contain the respective mode frequency in the denominator, causing it to be very small for internal modes at low pressures. Its value of unity at high pressures means that the intramolecular and intermolecular interactions, well separated at low pressure, tend to be equal at higher pressure. Determining $\gamma(p)$ is even sensitive to phase transitions, as in the case of O_2 (Jodl et al. 1985) or to isotopic effects as in the case of CS_2 (Bolduan et al. 1986).

High-pressure data can serve as a very sensitive and direct test for the intra- and intermolecular potentials modelled at $T = 0$ K as long as the pressure dependence of the volume or of the internal distances of the material are known: By considering $\omega \sim \partial^2 V/\partial x^2$ with $x = q$ or Q, and $x = x(P)$, one can plot and determine the pressure dependence of the frequencies and compare them with

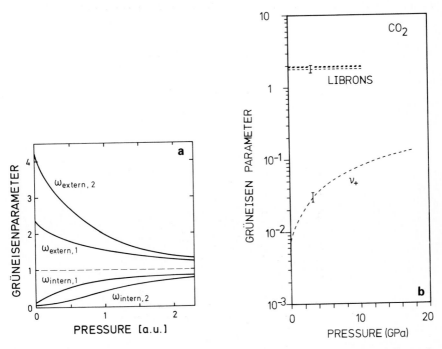

Fig. 21. (a) Mode Grüneisen parameter for external and internal modes as a function of pressure by theory (after Sherman 1982); (b) by experiment for CO_2 (Olijnyk et al. 1987a).

those determined from Raman studies at low temperatures. Figure 22 demonstrates this for the internal modes (upper part) and the external modes (lower part) of solid O_2. Without going into details (Jodl et al. 1985) the theoretical curves (dashed, dashed-dotted lines) for the pressure dependence of vibrons and librons in α-O_2 are obtained from atom–atom Lennard-Jones type potentials, or atom–atom r^{-6}/exp-type potentials, in addition to electric quadrupole/quadrupole terms and/or in addition to magnetic interaction terms.

For pure systems we discussed high-pressure induced effects, such as frequency shifts of elementary excitations, line shape changes, selection-rule changes accompanying phase transitions, and pressure-tuned resonant processes. Now we will consider high-pressure effects in matrix isolated species.

4.4. Application to matrix-isolated systems

Raman signals in these systems are, in general, weak and high-pressure measurements of Raman scattering (or by FTIR) present a challenge. In

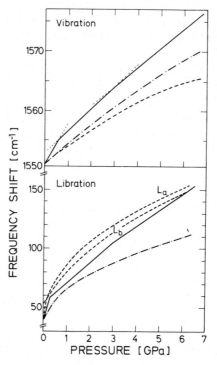

Fig. 22. Comparison of theory with experiment for the pressure dependence of the vibron and librons in α-O_2. Solid line, experimental data at 10 K; dashed line, theory of Helmy; dashed-dotted line English–Venables potential (after Jodl et al. 1985).

addition, loading the doped sample into the gas pressure cell or DAC is rather difficult, as previously discussed.

Matrix-isolated hydrogen has been experimentally investigated by IR in the range $2 < T < 80$ K and $P < 10$ kbar (Vu 1980). Of interest had been pressure-induced IR absorption in the region of the vibration/rotation fundamental $(4000-5000 \text{ cm}^{-1})$; local modes due to the lighter mass of H_2 in RGS showed up and were carefully studied. Very recently, matrix-isolated H_2 under pressure was theoretically considered by Silvi et al. (1986). The reason for that project was the fact that solid H_2 and D_2 showed a strange pressure behaviour (Sharma et al. 1980, Wijngaarden et al. 1982, Mao et al. 1985). Figure 23a shows the pressure dependence of the vibron of H_2 up to 700 kbar; it has since been measured up to 1.5 Mbar. The frequency increases with pressure to 300 kbar as expected, then it decreases. Different models have been proposed to explain this turn over or softening, among them the dissociation of the molecular bond leading to an atomic (metallic) phase, or a gradual increase in the intramolecular bond length at high pressure, or charge transfer and electron correlation effects. Silvi et al. (1986) have calculated the frequency shift as a function of pressure of the H_2 fundamental mode on the basis of new elaborate H–Ar combined potentials. They found a continuous increase (fig. 23b) and made further speculations from their matrix H_2 high-pressure data to explain the pure H_2 high-pressure data (see fig. 23a).

Besides changes in chemical bonding, lattice distortions take place both in doped or in compressed systems. For Kr embedded in Ar, the next-nearest shells expand $+1\%$ (i.e. change in lattice constant), and for Xe in Ar $+6\%$ as measured by EXAFS (Malzfeldt 1985). Therefore, doping causes local distortion of several percent whereas application of pressure produces a decrease in volume of about 30% at 70 kbar and $T < 10$ K in α-O_2. Obviously, it is important to

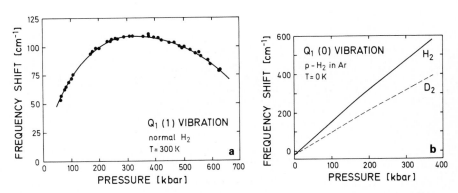

Fig. 23. (a) Experimentally determined frequency shift as a function of pressure in pure H_2 (Sharma et al. 1980); (b) theoretically estimated frequency shift as a function of pressure in p-H_2 in Ar (Silvi et al. 1986).

consider matrix-isolated species at high pressure in the context of lattice distortions. Maillard et al. (1981) and Obriot et al. (1986) studied HCl in Kr at 4 K up to 5 kbar using a piston-cylinder device. IR spectra revealed conventional frequency shift as a function of pressure. Their determination allows statements to be made regarding the effect of pressure on structure: for example (fig. 24) the lattice distortion increases monotonically by about 0.01 Å for HCl as pressure increases from 0 to 5.4 kbar due to a decrease of the Ar-lattice constant from 3.75 to 3.60 Å.

Very recently, Swanson (1987) succeeded in executing matrix-isolation FTIR spectroscopy at high pressure. His group had chosen the well-understood matrix-isolated SF_6, which was the aim of their own investigations over years. By cryogenic loading, they filled $SF_6:Xe = 1:1000$ in a special type of DAC and registered the v_3 mode as graphed in fig. 25, in which spectra at room and low temperature for two pressures are compared. Preliminary analysis reveals change in linewidth, in lineshape and large frequency shifts with pressure; the last effect indicating the increasing importance of $Xe–SF_6$ repulsive potential on the internal modes of SF_6.

To conclude this subsection, results on matrix-isolated CO_2, NO_2 and N_2 in Ar will be presented. Besides the anharmonic considerations of section 3.1 and 3.4 we studied a further high-pressure relationship – the bond anharmonicities [see for details Jodl and Bolduan (1982)]. A scaling law, in which the frequency spectrum uniformly expands as the crystal contracts [see eq. (3.1)] $\Delta\omega/\omega \sim -\gamma\Delta V/V$, combines the different mode Grüneisen parameters γ_i (internal), γ_e (external) with bondscaling parameters γ' which are of order unity (Zallen and

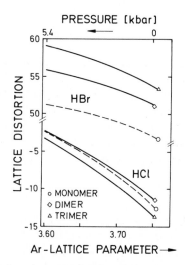

Fig. 24. Lattice distortion ($10^3\Delta a$ (Å) with a (Å) lattice parameter) around an impurity HCl or HBr in Ar (Maillard et al. 1981).

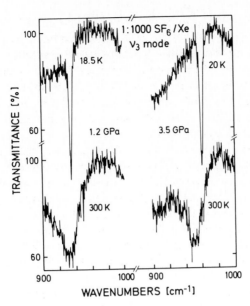

Fig. 25. High-pressure FTIR studies of v_3-SF$_6$ in Xe at 20 K and 300 K (Swanson 1987).

Slade 1978, Sherman 1980),

$$\gamma_i \sim \left(\frac{k_e}{k_i}\right)\frac{a}{r_i}\gamma_i',$$ (4.1)

$$\gamma_e \sim \frac{a}{r_e}\gamma_e',$$ (4.2)

where k_e/k_i is the intermolecular to intramolecular force constant ratio, while γ_i and γ_e are the respective mode Grüneisen parameter, a the lattice constant, r_i the bondlength of molecule and r_e the bondlength between defect and lattice particle. Data analysis for N_2 in N_2, and N_2 in Ar shows

(1) that eqs. (4.1) and (4.2) hold,

(2) that they can be interpreted to the extent that bondscaling parameters γ' are measures of the bond anharmonicity,

$$\gamma_i' = -\frac{r_i}{k_i}\left(\frac{\partial k_i}{\partial r_i}\right),$$ (4.3)

and (3) that bondscaling parameters are of the order of unity, whereas the observed Grüneisen parameters for external to internal modes differ by a factor of 10^2 to 10^4. This last statement is interpretable as a natural consequence of the disparate forces co-existing in molecular crystals or matrix-isolated species, i.e. the intra- and intermolecular forces.

If pressure is applied to a three-dimensional crystal, both the lattice constant and bondlength are compressed. In a linear-chain model with equilibrium requirements, the following relationship (Sherman 1980) exists,

$$k_i \, \delta r_i = k_e \, \delta r_e. \tag{4.4}$$

Adapting this to our systems, we deduce

$$\frac{\Delta r}{r} = \frac{k_e}{k_i} \frac{a}{r} \frac{\Delta a}{a}. \tag{4.5}$$

Now if the lattice constant of N_2 or Ar is changed by pressure by about $\Delta a/a = 10^{-2}$ per kbar (Swenson 1977) we find a relative change in molecular bondlength of $\Delta r/r \sim 10^{-4}$ per kbar; this result is of the same size as the lattice distortion shown in fig. 24.

Conclusion and outlook

By introducing the novel technique of DAC combined with spectroscopic apparatus for pure and doped molecular crystals, high-pressure research of importance to solid-state physics has been done:
– by varying pressure and temperature independently, one can distinguish between different anharmonic phenomena;
– frequency shifts of internal and external modes as a function of pressure are a direct test of respective intra- and intermolecular potentials;
– lattice distortions induced by doping and by pressurizing. Unfortunately, only few experiments of this sort currently exist for molecular crystals, consisting of small molecules, and high-pressure data are rare; therefore, only preliminary general statements can be made at present, and more molecular systems should be studied.

We are now engaged in carefully studying the gas loading technique of DAC, especially with He, and we are trying to prove its feasibility; next the system H_2 in Ar will be measured to verify the theoretically predicted dependence of fig. 23b; in the long term we intend to study systematically small mixed molecular crystals at high pressure and low temperature by Raman scattering and by FTIR.

References

Abouaf-Marguin, L., M.E. Jacox and D.E. Milligan, 1977, J. Mol. Spectrosc. **67**, 34.
Ahmad, S.F., H. Kiefte and M.J. Clouter, 1981, J. Chem. Phys. **75**(2), 5848.
Ahmad, S.F., H. Kiefte, M.J. Clouter and M.D. Whitmore, 1982, Phys. Rev. B **26**(8), 4239.
Allavena, M., and K. Seiti, 1986, J. Chem. Phys. **85**(8), 4614.
Allavena, M., H. Chakroun and D. White, 1982, J. Chem. Phys. **77**(4), 1757.
Anderson, A., T.S. Sun and C.A. Dankersloot, 1970, Can. J. Phys. **48**, 2265.
Angwin, M.J., and J. Wassermann, 1966, J. Chem. Phys. **44**, 417.

Antsygina, T.N., V.A. Slusarev, Y.A. Freiman and A.I. Erenburg, 1984, J. Low Temp. Phys. **56**(3/4), 331.

Aziz, R.A., 1984, Interatomic Potentials for Rare Gases: Pure and Mixed Interactions, in: Inert Gases, ed. M.W. Klein (Springer, Berlin) ch. 2.

Bahrdt, J., P. Gurtler and N. Schwenter, 1987, J. Chem. Phys. **86**(11), 6108.

Balkanski, M., R.F. Wallis and E. Haro, 1983, Phys. Rev. B **28**(4), 1928.

Ball, D.W., Z.H. Kafafi, L. Fredin, R.H. Hauge and J.L. Margrave, 1987, A Bibliography of Matrix Isolation Spectroscopy (Rice University Press, Houston, TX).

Barker, A.S., and A.J. Sievers, 1975, Optical Studies of the Vibrational Properties of Disordered Solids, in: Rev. Mod. Phys. **47**(2), S1.

Barnes, A.J., 1973, Theoretical Treatment of Matrix Effects, in: Vibrational Spectroscopy of Trapped Species, ed. H.E. Hallam (Wiley, London) ch. 4.

Barrett, C.S., and L. Meyer, 1965, J. Chem. Phys. **42**(1), 107.

Barrett, C.S., L. Meyer and J. Wassermann, 1966, J. Chem. Phys. **44**(3), 998.

Bechtold, P.S., U. Kettler and W. Krasser, 1984, Solid State Commun. **52**(3), 347.

Behringer, R.E., 1958, J. Chem. Phys. **29**, 537.

Beyeler, H.U., 1974, J. Chem. Phys. **60**, 4123.

Bier, K.D., 1986, Solid State Aspects and Anharmonic Phenomena in Matrix Isolation Raman Spectroscopy (Thesis, University of Kaiserslautern, Kaiserslautern, FRG).

Bier, K.D., and H.J. Jodl, 1984, J. Chem. Phys. **81**, 1192.

Bier, K.D., and H.J. Jodl, 1987, J. Chem. Phys. **86**(8), 4406.

Bier, K.D., H.J. Jodl and A. Loewenschuss, 1985, Chem. Phys. Lett. **115**(5), 34.

Bier, K.D., H.J. Jodl and H. Däufer, 1988, Can. J. Phys. **66**, 708.

Blaisten-Barojas, E., and M. Allavena, 1976, J. Phys. C **9**, 3121.

Blumen, A., S.H. Lin and J. Manz, 1978, J. Chem. Phys. **69**(2), 881.

Bobetic, M.V., and J.A. Barker, 1970, Phys. Rev. B **2**(10), 4169.

Bobetic, M.V., and J.A. Barker, 1983, Phys. Rev. B **28**(12), 7317.

Bobetic, M.V., J.A. Barker and M.L. Klein, 1972, Phys. Rev. B **5**(8), 3185.

Bogani, F., and P.R. Salvi, 1984, J. Chem. Phys. **81**(11), 4991.

Bolduan, F., H.J. Jodl and A. Loewenschuss, 1984, J. Chem. Phys. **80**(5), 1739.

Bolduan, F., H.D. Hochheimer and H.J. Jodl, 1986, J. Chem. Phys. **84**(12), 6997.

Bourcey, E., and J.Y. Roncin, 1975, J. Mol. Spectrosc. **55**, 31.

Briels, W.J., A.P.J. Jansen and A. van der Avoird, 1984, J. Chem. Phys. **81**(9), 4118.

Briels, W.J., A.P.J. Jansen and A. van der Avoird, 1986, Dynamics of Molecular Crystals, in: Advances in Quantum Chemistry, Vol. 18, ed. P.-O. Löwdin (Academic Press, New York) p. 131.

Buckingham, A.D., 1960, Trans Faraday Soc. **56**, 753.

Cabana, A., G.B. Savitsky and D.F. Hornig, 1963, J. Chem. Phys. **39**(11), 2942.

Califano, S., V. Schettino and N. Neto, 1981, Lattice Dynamics of Molecular Crystals, Lecture Notes in Chemistry, Vol. 26 (Springer, Berlin).

Cardini, G., and S.F. O'Shea, 1985, Phys. Rev. B **32**(4), 2489.

Cardini, G., P. Procacci and R. Righini, 1987, Chem. Phys. **117**, 355.

Chandrasekharan, V., R.D. Etters and K. Kobashi, 1983, Phys. Rev. B **28**(2), 1095.

Cohen, S.S., and M.L. Klein, 1974, J. Chem. Phys. **61**(8), 3210.

Cohen, S.S., and M.L. Klein, 1984, Dynamics of Impure Rare-gas Crystals, in: Inert Gases, Springer Series in Chemical Physics, Vol. 34, ed. M.L. Klein (Springer, Berlin) ch. 3.

Conrad, H.M., 1983, Matrix Isolation Raman Spectroscopy of SO_2 (Thesis, University of Kaiserslautern, Kaiserslautern, FRG).

Coufal, H., E. Lüscher, H. Micklitz and R.E. Norberg, 1984, in: Rare Gas Solids, Springer Tracts in Modern Physics, Vol. 103, ed. G. Höhler (Springer, Berlin).

Curzon, A.E., and M.J. Eastell, 1971, J. Phys. C **4**, 689.

Däufer, H., 1985, Raman Spectroscopy of Matrix Isolated Water (Thesis, University Kaiserslautern, Kaiserslautern, FRG).

Däufer, H., 1987, private communication.

De Fotis, G.D., 1981, Phys. Rev. B **23**, 4714.

De Remigis, J., and H.L. Welsh, 1970, Can. J. Phys. **48**, 1622.

De Remigis, J., H.L. Welsh, R. Bruno and D.W. Taylor, 1971, Can. J. Phys. **49**, 3201.

Della Valle, R.G., P.F. Fracassi, R. Righini and S. Califano, 1983, Chem. Phys. **74**, 179.

Diatschenko, V., C.W. Chu, D.H. Liebenberg, D.A. Young, M. Ross and R.L. Mills, 1985, Phys. Rev. B **32**(1), 381.

Dubost, H., 1976, Chem. Phys. **12**, 139.

Dubost, H., 1984, Spectroscopy of Vibrational and Rotational Levels of Diatomic Molecules in Rare Gas Crystals, in: Inert Gases, Springer Series in Chemical Physics, Vol. 34, ed. M.L. Klein (Springer, Berlin) ch. 4.

Dubost, H., and R. Charneau, 1976, Chem. Phys. **12**, 407.

Dubost, H., and R. Charneau, 1979, Chem. Phys. **41**, 329.

Dubost, H., R. Charneau and M. Harig, 1982, Chem. Phys. **69**, 389.

Etters, R.D., A.A. Helmy and K. Kobashi, 1983, Phys. Rev. B **28**, 216.

Etters, R.D., K. Kobashi and J. Belak, 1985, Phys. Rev. B **32**(6), 4097.

Fateley, W.G., F.R. Dollish, N.T. McDevitt and F.F. Bentley, 1972, Infrared and Raman Selection Rules for Molecular and Lattice Vibrations (Wiley, New York).

Fitchen, D.B., 1968, Zero Phonon Transitions, in: Physics of Color Centers, ed. W.B. Fowler (Academic Press, New York) ch. 5.

Fleury, P.A., J.M. Worlock and H.L. Carter, 1973, Phys. Rev. Lett. **30**(13), 591.

Forstmann, F., D.M. Kolb, D. Leutloff and W. Schulze, 1977, J. Chem. Phys. **66**(7), 2806.

Fowler, W.B., ed., 1968, Physics of Color Centers (Academic Press, New York).

Friedmann, H., and S. Kimel, 1965, J. Chem. Phys. **43**(11), 3925.

Galaup, J.P., J.Y. Harbec, R. Charneau and H. Dubost, 1985, Chem. Phys. Lett. **120**(2), 188.

Gas encyclopedia, 1976, Air Liquid (Elsevier, Amsterdam).

Girardet, C., and A. Lakhlifi, 1985, J. Chem. Phys. **83**(11), 5506.

Girardet, C., and A. Lakhlifi, 1986, Chem. Phys. **110**, 447.

Girardet, C., and D. Maillard, 1982, J. Chem. Phys. **77**, 5941; J. Chem. Phys. **77**, 5923.

Girardet, C., L. Abouaf-Marguin, B. Gauthier-Roy and D. Maillard, 1984a, Chem. Phys. **89**, 415.

Girardet, C., L. Abouaf-Marguin, B. Gauthier-Roy and D. Maillard, 1984b, Chem. Phys. **89**, 431.

Givan, A., A. Loewenschuss, K.D. Bier and H.J. Jodl, 1986, Chem. Phys. **106**, 151.

Häfner, W., and W. Kiefer, 1987, J. Chem. Phys. **86**(8), 4582.

Harvey, K.B., H.F. Shurvell and J.R. Henderson, 1964, Can. J. Phys. **42**, 911.

Hazen, R.M., and L.W. Finger, 1982, Comparative Crystal Chemistry (Wiley, New York).

Henry, C.H., and C.P. Slichter, 1968, Moments and Degeneracy in Optical Spectra, in: Physics of Color Centers, ed. W.B. Fowler (Academic Press, New York) ch. 6.

Hirsch, K.R., and W.B. Holzapfel, 1981, Rev. Sci. Instrum. **52**(1), 52.

Horton, G.K., and A.A. Maradudin, 1974, Dynamical Properties of Solids, Vol. 1 (North-Holland, Amsterdam).

Horton, G.K., and A.A. Maradudin, 1975, Dynamical Properties of Solids, Vol. 2 (North-Holland, Amsterdam).

Jakob, M., H. Micklitz and K. Luchner, 1977, Phys. Lett. A **61**(4), 265.

Jansen, A.P.J., 1987, Theoretical Approach to the Optical, Thermodynamic and Magnetic Properties of Solid Nitrogen and Solid Oxygen (Thesis, University of Nijmegen, Nijmegen, The Netherlands).

Jansen, A.P.J., and A. van der Avoird, 1987, J. Chem. Phys. **86**(6), 3583.

Jayaraman, A., 1983, Rev. Mod. Phys. **55**(1), 65.

Jayaraman, A., 1986, Rev. Sci. Instrum. **57**(6), 1013.

Jodl, H.J., 1984, Raman Spectroscopy on Matrix Isolated Species, in: Vibrational Spectra and Structure, Vol. 13, ed. J.R. Durig (Elsevier, Amsterdam) ch. 6.

Jodl, H.J., and F. Bolduan, 1982, J. Chem. Phys. **76**(7), 3352.

Jodl, H.J., and W.B. Holzapfel, 1979, Rev. Sci. Instrum. **50**(3), 44.

Jodl, H.J., G. Theyson and R. Bruno, 1979, Phys. Status Solidi b **94**, 161.

Jodl, H.J., F. Bolduan and H.D. Hochheimer, 1985, Phys. Rev. B **31**(11), 7376.

Jodl, H.J., H.W. Löwen and D. Griffith, 1987, Solid State Commun. **61**(8), 503.

Johannsen, P.G., H.J. Jodl and H.W. Löwen, 1988, FTIR Studies on CO_2 at High Pressure., to be published.

Jones, L.H., and B.I. Swanson, 1981, J. Chem. Phys. **74**(6), 3216.

Jones, L.H., S.A. Ekberg and B.I. Swanson, 1986, J. Chem. Phys. **85**(6), 3203.

Jortner, J., 1974, Electronic Excitations in Molecular Crystals, in: Proc. Int. Conf. on Vacuum Ultraviolet Radiation Physics, eds E. Koch, R. Haensel and C. Kunz (Vieweg, Hamburg, FRG) p. 263.

Keeler, G.J., and D.N. Batchelder, 1972, J. Phys. C **5**, 3264.

Kiefte, H., M.J. Clouter, N.N. Rich and S.F. Ahmad, 1982, Can. J. Phys. **60**, 1204.

Kikas, J., and M. Rätsep, 1982, Phys. Status Solidi b **112**, 409.

Kitaigorodsky, A.I., 1984, Mixed Crystals, Series in Solid-State Sciences, Vol. 33, ed. M. Cardona (Springer, Berlin).

Kjems, J.K., and G. Dolling, 1975, Phys. Rev. B **11**, 1639.

Klein, M.L., ed., 1984, Inert Gases – Potentials, Dynamics and Energy Transfer in Doped Crystals, in: Springer Series in Chemical Physics, Vol. 34 (Springer, Berlin).

Klein, M.L., and J.A. Venables, eds, 1976, Rare Gas Solids, Vol. 1 (Academic Press, London).

Klein, M.L., and J.A. Venables, eds, 1977, Rare Gas Solids, Vol. 2 (Academic Press, London).

Klement, W., and A. Jayaraman, 1966, Prog. Solid State Chem. **3**, 289.

Knözinger, E., and O. Schrems, 1987, Low Frequency Vibrational Spectroscopy of Molecular Complexes, in: Vibrational Spectra and Structure, Vol. 16, ed. J.R. Durig (Elsevier, Amsterdam) ch. 3.

Kobashi, K., 1978, Mol. Phys. **36**(1), 225.

Kobashi, K., and M.L. Klein, 1979, J. Chem. Phys. **71**, 843.

Kobashi, K., Y. Kataoka and T. Yamamoto, 1975, Can. J. Chem. **54**, 2154.

Krupskii, I.N., A.I. Prokhvatilov, Y.A. Freiman and A.I. Erenburg, 1979, Sov. J. Low Temp. Phys. **5**(3), 130.

Kühle, H., J. Bahrdt, R. Fröhlich and N. Schwentner, 1985, Phys. Rev. B **31**(8), 4854.

Langen, J., K.P. Lodemann and U. Schurath, 1987, Chem. Phys. **112**, 393.

Lascombe, J., ed., 1987, Dynamics of Molecular Crystals, Proc. 41st Int. Meeting of the Société Francaise de Chimie, Grenoble 1986 (Elsevier, Amsterdam).

Liarokapis, E., E. Anastassakis and G.A. Kourouklis, 1985, Phys. Rev. B **32**(12), 8346.

Lin, S.H., 1976, J. Chem. Phys. **65**(3), 1053.

Löwen, H.W., H.J. Jodl, A. Loewenschuss and H. Däufer, 1988a, Can. J. Phys. **66**, 808.

Löwen, H.W., K.D. Bier, H. Jodl, A. Löwenschuss and A. Givan, 1988b, Vibron–phonon sidebands in the FTIR spectra of the molecular crystal CO_2 (submitted to J. Chem. Phys.).

Maillard, D., A. Schriver and J.P. Perchard, 1977, J. Chem. Phys. **67**(9), 3917.

Maillard, D., A. Schriver, F. Fondere, J. Obriot and C. Girardet, 1981, J. Chem. Phys. **75**(3), 1091.

Malzfeldt, W., 1985, Structure Analysis of RG Atoms in RGS by EXAFS Measurements (Thesis, University Kiel, Kiel, FRG).

Mannheim, P.D., 1968, Phys. Rev. **165**(3), 1011.

Mannheim, P.D., 1972, Phys. Rev. B **5**(2), 745.

Manz, J., 1980, J. Am. Chem. Soc. **102**, 1801.

Mao, H.K., P.M. Bell and R.J. Hemley, 1985, Phys. Rev. **55**(1), 99.

Maradudin, A.A., 1965, Rep. Prog. Phys. **28**, 331.

McCarty, M., and G.W. Robinson, 1959, Mol. Phys. **2**, 415.

Medina, F.D., and W.B. Daniels, 1976, J. Chem. Phys. **64**, 150.

Meier, R.J., 1984, On the Magnetic and Optical Properties of Condensed Oxygen (Thesis, University of Amsterdam, Amsterdam, The Netherlands).

Miller, R.E., and J.C. Decius, 1973, J. Chem. Phys. **59**(9), 4871.

Molter, M., 1984, Raman Spectroscopy on Molecules in Nitrogen Matrices (Thesis, University Kaiserslautern, Kaiserslautern, FRG).

Montano, P.A., W. Schulze, B. Tesche, G.K. Shenoy and T.I. Morrison, 1984, Phys. Rev. B **30**(2), 672.

Naßhahn, K., 1987, Raman Matrix Isolation Spectroscopy on NH_3 and N_2O (Thesis, University of Kaiserslautern, Kaiserslautern, FRG).

Nanba, T., M. Sagara and M. Ikezawa, 1980, J. Phys. Soc. Jpn. **48**(1), 228.

Niemann, W., W. Malzfeldt, P. Rabe and R. Haensel, 1987, Phys. Rev. B **35**(3), 1099.

Obriot, J., F. Fondere and P. Marteau, 1978, Infrared Phys. **18**, 607.

Obriot, J., F. Fondere and P. Marteau, 1986, J. Chem. Phys. **85**(9), 4925.

Olijnyk, H., H. Däufer, H.J. Jodl and H.D. Hochheimer, 1987a, J. Chem. Phys. **88**(7), 4204.

Olijnyk, H., H. Däufer, H.J. Jodl and H.D. Hochheimer, 1987b, High Pressure Raman Studies of Solid CO_2 and N_2O, in: High Pressure Science and Technology, Abstracts XIth AIRAPT Int. High Pressure Conf., July 12–17, Kiev (Institute for Superhard Materials of the UkrSSR Academy of Sciences, Kiev, USSR) p. 223.

Piermarini, G.J., and S. Block, 1975, Rev. Sci. Instrum. **46**, 973.

Powell, B.M., G. Dolling, L. Piseri and P. Martel, 1972, in: Neutron Inelastic Scattering 1972, Proc. Symp. on Neutron Inelastic Scattering; held by the International Atomic Energy Agency in Grenoble, France, 6–10 March 1972 (International Atomic Agency, Vienna) p. 207.

Regitz, E., A. Loewenschuss, K.D. Bier and H.J. Jodl, 1984, J. Chem. Phys. **80**(12), 5930.

Reichlin, R., D. Schiferl, S. Martin, C. Vanderborgh and R.W. Mills, 1985, Phys. Rev. Lett. **55**, 1464.

Righini, R., 1985, Physica B **131**, 234.

Rössler, U., 1976, Band Structure and Excitations, in: Rare Gas Solids, Vol. 1, eds M.L. Klein and J.A. Venables (Academic Press, London) ch. 8.

Sapozhnikov, M.N., 1976, Phys. Status Solidi b **75**, 11.

Schoen, L.J., and H.P. Broida, 1960, J. Chem. Phys. **32**(4), 1184.

Scott, T.A., 1976, Phys. Rep. **27**(3), 89.

Sharma, S.K., H.K. Mao and P.M. Bell, 1980, Phys. Rev. Lett. **44**, 886.

Sherman, W.F., 1980, J. Phys. C **13**, 4601.

Sherman, W.F., 1982, J. Phys. C **15**, 9.

Sherman, W.F., and G.R. Wilkinson, 1980, Raman and Infrared Studies of Crystals at Variable Temperature and Pressure, in: Advances in Infrared and Raman Spectroscopy, Vol. 6, eds R.J.H. Clark and R.E. Hester (Heyden, London) ch. 4.

Silvera, I.F., 1980, Rev. Mod. Phys. **52**, 393.

Silvi, B., V. Chandrasekharan, M. Chergui and R.D. Etters, 1986, Phys. Rev. B **33**(4), 2749.

Steinmetz, A., A. Schrimpf, H.J. Stöckmann, E. Görlach, R. Dersch, G. Sulzer and H. Ackermann, 1987, Z. Phys. D **4**, 373.

Stöckmann, H.J., 1984, Z. Phys. B **54**, 229.

Swanson, B.I., 1987, private communication.

Swanson, B.I., and L.H. Jones, 1981, J. Chem. Phys. **74**(6), 3205.

Swanson, B.I., and L.H. Jones, 1983, High Resolution Infrared Studies of Site Structure and Dynamics for Matrix Isolated Molecules, in: Vibrational Spectra and Structures, Vol. 12, ed. J. Durig (Elsevier, Amsterdam) ch. 1.

Swenson, C.A., 1977, High Pressure and Thermodynamics of Rare Gas Solids, in: Rare Gas Solids, Vol. 2, eds M.L. Klein and J.A. Venables (Academic Press, London) ch. 13.

van der Avoird, A., W.J. Briels and A.P.J. Jansen, 1984, J. Chem. Phys. **81**(8), 3658.

van Kranendonk, J., 1983, Solid Hydrogen (Plenum Press, New York).

van Kranendonk, J., and G. Karl, 1968, Rev. Mod. Phys. **40**(3), 531.

Verstegen, J., M.P.H. Goldring, S. Kimel and B. Katz, 1966, J. Chem. Phys. **44**, 3216.

Vu, H., 1980, High Pressure Studies, in: Matrix Isolation Spectroscopy, ed. A.J. Barnes (Reidel, Dordrecht) ch. 11.

Wallace, D.C., 1972, Thermodynamics of Crystals (Wiley, New York).

Weinert, C.M., F. Forstmann, H. Abe, R. Grinter and D.M. Kolb, 1982, J. Chem. Phys. **77**(7), 3392.

Weinert, C.M., F. Forstmann, R. Grinter and D.M. Kolb, 1983, Chem. Phys. **80**, 95.

Weinstein, B.A., and R. Zallen, 1984, Pressure-Raman Effects in Covalent and Molecular Solids, in: Light Scattering in Solids IV, Topics in Applied Physics, Vol. 54, eds M. Cardona and G. Güntherodt (Springer, Berlin) ch. 8.

Werthamer, N.R., 1970, Phys. Rev. B **1**, 572.

Wijngaarden, R.J., Ad. Lagendijk and I.F. Silvera, 1982, Phys. Rev. B **26**(9), 4957.

Witters, K., and J.E. Cahill, 1977, J. Chem. Phys. **67**(6), 2405.

Zallen, R., and M.L. Slade, 1978, Phys. Rev. B **18**(10), 5775.

Zumofen, G., and K. Dressler, 1976, J. Chem. Phys. **64**(12), 5198.

Subject Index

Materials Index

423